EDUCATION FOR ALL

CONTEXTS OF LEARNING
Classrooms, Schools and Society

Managing Editors:

Bert Creemers, *GION, Groningen, the Netherlands*
David Reynolds, *School of Education, University of Newcastle upon Tyne, England*
Sam Stringfield, *Center for the Social Organization of Schools, Johns Hopkins University, U.S.A.*

EDUCATION FOR ALL

ROBERT E. SLAVIN

(Co-director of the Center for
Research on the Education of
Students Placed at Risk,
Johns Hopkins University,
Baltimore, U.S.A.)

SWETS & ZEITLINGER PUBLISHERS

LISSE   ABINGDON   EXTON (PA)   TOKYO

Library of Congress Cataloging-in-Publication Data

Slavin, Robert E.
Education for all / Robert E. Slavin.
    p.   cm.
Includes bibliographical references (p.   ) and index.
ISBN 9026514727. – ISBN 9026514735 (pbk.)
1. Educational change – United States. 2. Teaching – United States.
3. School management and organization – United States. 4. Group work
in education – United States. 5. Education – Research – United States.
I. Title.
LA210.S47 1996
370 .973 – dc20
                                                 96-12890
                                                   CIP

Cip-gegevens Koninklijke Bibliotheek, Den Haag

Slavin, Robert E.

Education for all / Robert E. Slavin. - Lisse : Swets & Zeitlinger
Met index, lit. opg.
ISBN 90-265-1472-7 geb.
ISBN 90-265-1473-5 pbk.
NUGI 724
Trefw.: onderwijs ; essays.

Cover design: Ivar Hamelink
Lay-out and graphics: G.V.K.
Printed in the Netherlands by Grafisch Produktiebedrijf Gorter, b.v.,
Steenwijk

ISBN 90 265 1472 7
ISBN 90 265 1473 5
NUGI 724

Acknowledgments

Portions of the following chapters were adapted from previously published articles, listed below, and are reprinted with permission.

CHAPTER 1
Slavin, R.E. (1995). A Model of Effective Instruction. *The Educational Forum, 59*(2), 166-176.

CHAPTER 2
Slavin, R.E. (1995). *Cooperative Learning: Theory, Research, and Practice* (2nd Ed.). Boston: Allyn & Bacon.

CHAPTER 3
Stevens, R.J., & Slavin, R.E. (1995). The Cooperative Elementary School: Effects on Student Achievement, Attitudes, and Social Relations. *American Educational Research Journal, 32*, 321-351.

CHAPTER 4
Slavin, R.E., Madden, N.A., Dolan, L.J., Wasik, B.A., Ross, S., Smith, L., & Dianda, M. (1996). Succes for All: A Summary of Research. *Journal of Education for Students Placed at Risk, 1.*

CHAPTER 5
Slavin, R.E., Madden, N.A., Dolan, L.J., & Wasik, B.A. (1994). Roots and Wings: Inspiring Academic Excellence. *Educational Leadership, 52*(3), 10-13.

CHAPTER 6
Slavin, R.E. (1986). Best-evidence synthesis: An Alternative to Meta-analytic and Traditional Reviews. *Educational Researcher, 15*(9), 5-11.

CHAPTER 7
Slavin, R.E. (1900). Ability Grouping and Student Achievement in Elementary Schools: A Best-evidence Synthesis. *Review of Educational Research, 57*, 347-350.

CHAPTER 8
Slavin, R.E. (1990). Achievement Effects of Ability Grouping in Secondary Schools: A Best-evidence Synthesis. *Review of Educational Research, 60*(3), 471-499.

CHAPTER 9
Gutiérrez, R., & Slavin, R.E. (1992). Achievement Effects of the Nongraded Elementary School: A Best-evidence Synthesis. *Review of Educational Research, 62*, 333-376.

CHAPTER 10
Slavin, R.E. (1987). Mastery Learning Reconsidered. *Review of Educational Research, 57*, 175-213.

Preface

This book came about as a result of a conversation between my colleague, Sam Stringfield, and Martin Scrivener of Swets & Zeitlinger Publishers. They thought it might be a good idea to ask me to assemble what I thought to be the most important "milestones" in my research, with introductions and commentary to set these articles in context. When they presented this idea to me I wasn't very interested at first, but finally began to see the potential of such a volume. The conventions of scholarly publishing don't allow authors to say why they wrote what they did, to set the stage for their article and explain the personal history and philosophy behind it. In addition, there is no forum for discussing the aftermath of a publication, what happened after it appeared, sometimes as a result of its appearance.

This book is intended to give educators and educational researchers a single source for my most important work and to explain how they came about and what I think they mean. I had fun writing it, if for no other reason than that it let me write in the first person about ideas I usually don't get to discuss. I hope it will prove to be a useful resource for educators interested in my work.

The work I've done over the years has benefited from financial and intellectual support from many sources, particularly from the Office of Educational Research and Improvement (OERI) of the U.S. Department of Education, and from the Carnegie Corporation of New York, the Pew Charitable Trusts, and the New American Schools Development Corporation. This book was written under funding from OERI (Grant no. R117D-40005). However, any opinions I have expressed are my own and do not represent the policies or positions of any of my funders.

Table of Contents

X

CHAPTER 1

Education for All: Research and Reform Toward Success for All Children

There is a quiet revolution underway in educational thought and practice. From early on, education has played a role in sorting children into different categories – high, middle, and low ability groups, gifted, special education, and so on. This sorting paradigm is still dominant today, but is beginning to be challenged on a broad front in research, policy, and practice. The sorting paradigm depends in large part on a belief that children have relatively unchangeable intellectual capabilities and that the best that schools can do is provide instruction to each child's innate talents. This assumption is itself being challenged, as are schooling practices based on it (Oakes, 1992).

An alternative to the sorting paradigm is practices based on the belief that all children can learn to high levels. From this viewpoint, the responsibility of educational institutions is to hold all students to high standards and then provide top quality instruction designed to meet the needs of diverse groups of students, backed up by supportive services to help students who would struggle with a high-expectations curriculum to succeed in it. A name for this alternative paradigm is *talent development* (Boykin, 1996). This term is intended to communicate the idea that all children have talents capable of being developed to meet high standards. Another way to express the same idea is *success for all*, which is the name of a school restructuring program my colleagues and I have developed and researched (see Chapter 4) but is also a statement of the philosophical basis of our work: the idea that schools must be organized not just to advance all children from their starting point, but also to place a high floor under the achievement of all, including those who are placed at risk by societal, institutional, family, or personal factors. The Center for Research on the Education of Students Placed at Risk, which I co-direct from Johns Hopkins University with my colleague Wade Boykin at Howard University, bases its work on these talent development/success for all paradigms, seeking to identify programs and practices capable of ensuring success as students face obstacles to healthy development and negotiate key hurdles imposed by the schools themselves. Dr. Boykin likes to speak of CRESPAR's mission as finding ways to place all children at promise for success rather than at risk for failure.

Sorting mechanisms used in schools might be dispassionately discussed as a technical issue of efficient school organization if it were not the case that the students who are sorted into the lowest categories are disproportionately those from poor and minority families. It is this reality that gives the debate about sorting versus talent development political, social, and moral urgency. Every advanced democracy states the belief that all children, regardless of family background, should have an equal opportunity to rise to the highest levels of wealth and power, and all recognize the education system as the primary mechanism by which children of the poor can transcend their origins. Americans in particular admire individuals who have overcome impoverished family origins, from Abraham Lincoln to Colin Powell. Yet the United States is one of the only advanced democracies in the world in which children who happen to live in high-poverty areas are likely to receive substantially less funding than children a mile away in wealthy neighborhoods. In all societies, the persistent correlation between social class and educational performance indicates a chasm between national ideals and daily reality.

Support for the sorting paradigm continues largely because of a perceived lack of practical alternatives to it. Few if any educators or policy makers believe that it is desirable to have large numbers of students failing to achieve at an adequate level. If there were practical, replicable educational methods capable of placing a high floor under the achievement of all students while at the same time accelerating the achievement of all, including those at the highest levels of performance, educators and policy makers would embrace them, even if (perhaps especially if) they undermined a sorting ideology.

The most important task for educational research and reform, then, is to create, evaluate, disseminate, and institutionalize instructional methods, materials, and school organization practices capable of enabling educators to bring children to high levels of achievement. Fundamental changes in schooling and society are unlikely to take place until we have available tools and techniques capable of substantially improving our capacity to ensure that any child born can achieve his or her full potential.

The Role of Research in Education Reform

Fundamental change in the practice of education and the belief systems around it will require changes on many fronts, including changes in support for school-level change and professional development, assessment and accountability systems capable of telling educators and the public when reform is working, and other changes at the governmental and policy levels (see Chapter 11). However, education reformers often lose sight of the fact that outcomes for children cannot improve significantly unless there is widespread and lasting change in the daily interactions between teachers and students. Teachers as a rule work very hard and care deeply about their students. "Motivating" them to work harder, through various rewards and sanctions, is not going to reform educational practice. The only way to substantially increase the effectiveness of education practice, to ensure that the tireless efforts of dedicated teachers actually pay off in greater student learning, it is to put into teachers' hands methods and materials capable of ensuring success for every child.

What this conclusion implies is that research and development must be at the heart of education reform. We cannot move forward in educational practice without bringing into widespread use methods and materials known to be markedly more effective than methods in current use. This seems obvious, but the education reform movement currently under way has very little research demonstrating effects on students of particular policies and practices. Most educational innovations currently in widespread use are supported only by anecdotal evidence, from journalistic accounts of individual schools "turned around" by a given method to reports of outstanding gains on test scores in one or more schools in a given year. When there are dozens or hundreds of schools implementing a program and all are collecting routine test data and other indicators, it is inevitable that at any particular moment a few schools are making impressive progress. However, the same is true of any set of dozens or hundreds of schools *not* involved in any particular reform. Not until we have planned, rigorous, conscientiously carried out studies comparing innovative schools to control groups on fair and meaningful measures can we have any confidence that a program is effective. Not until that program is applied many times in a variety of locations and is found to be more effective than control groups in its replication sites can we have confidence that a program is replicable. Until we have programs that are known with a high degree of certainty to be effective and replicable, education reform is mere faddism, with the reform of the moment prevailing because of its attractiveness rather than its effectiveness (see Slavin, 1989, 1990). A lack of reliable evidence and a lack of confidence in the evidence that does exist is what produces the famous pendulum of educational innovation, a pattern of boom-to-bust enthusiasms that is characteristic of any field of endeavor in which fashion, not evidence, drives changes in practice.

The Purpose of This Book

This book was written to assemble in one place some of the key mileposts in my own work and that of my colleagues over the past quarter century to try to build a research base for reform in educational practice. The book is divided into two sections. The first, "Research on Effective Instructional Strategies," traces the program of development and research that led from cooperative learning strategies for elementary and secondary classrooms to the cooperative elementary school to Success for All to Roots and Wings. Each of these developments builds on the ones that came before it, incorporating and adding additional elements to the earlier work. In the fashion of Russian nesting dolls or the construction of Mayan temples (in which the current layer usually was laid on top of an earlier temple), each of the successful programs my colleagues and I have developed are incorporated in the next, more comprehensive programs. The difference is that unlike Russian dolls or Mayan temples, all previous structures in our case are easily visible in the later ones. It is easy to see Success for All in a Roots and Wings school, easy to see the cooperative elementary school in a Success for All school, and easy to see a range of cooperative learning programs in all of our school-level interventions.

The second section, "Best-Evidence Syntheses," presents quantitative reviews of research on several key issues of school and classroom organization. These reviews use a review technique called best-evidence synthesis, which is intended to combine the systematic literature search methods and quantification of study outcomes characteristic of meta-analysis with the extended discussion of key substantive and methodological issues and the expanded description of important individual studies characteristic of the best narrative reviews. These reviews touch on some of the most controversial issues of school and classroom organization: ability grouping, the nongraded primary school, and mastery learning. A theme that ties these topics together is that all deal with school responses to diversity. The ability grouping reviews directly confront the sorting paradigm and the sorting mechanisms that follow from it, asking whether there is an educational justification for various forms of ability grouping in elementary and secondary schools. The reviews of research on nongraded primary schools and mastery learning examine two potential alternatives to traditional between-class ability grouping strategies.

Finally, a concluding chapter discusses implications of my own work and that of other reformers for educational policy and practice. It also proposes areas of research and development most needed to achieve the goal of success for all students.

Pragmatic and Theoretical Background

There is an ideological base to my work which I will freely admit. I am a radical pragmatist. Following Dewey, I want for all children what the best and wisest parents want for their own children. I want schools to be able to ensure that virtually all students are confident, joyful, self-aware, and caring learners. I want to help create schools in which children are active, happy, and productive, in which every child has a high level of basic skills and knowledge as well as strategies and predispositions for problem solving, creativity, and self-expression. I want to help schools create conditions in which students will know a great deal, know how to do a great deal, and are able to learn whatever they don't know or can't do by themselves.

I care much more about these outcomes than I do about how they are achieved. Obviously, children should not be in dehumanizing or abusive environments, but within a broad range of humane possibilities I believe that evidence, tempered with judgment, should be a guide to practice. Given the primitive state of evidence for most educational decisions, using evidence today requires a particularly large dose of judgment, but I do believe that most dilemmas of educational practice can in principle be resolved by rigorous research, and that where convincing research does exist it should be taken seriously.

Being a radical pragmatist in an ideological and fad-driven field creates many seeming contradictions in my own conclusions. For example, I have written a great deal in opposition to typical between-class ability grouping plans because they have no important benefits for children, but I have at the same time supported use of the Joplin Plan (flexible cross-grade grouping for reading), of certain within-class grouping plans, and of acceleration programs for gifted students, because there is evidence that I find convincing in favor of each (see Chapters 7-9). I believe that first grade reading programs should have a strong phonetic base, but that upper-elementary reading instruction should emphasize literature, creative writing, and cooperative learning. My beliefs about what is effective practice for particular instructional purposes are not, of course, based on any a priori set of values and preference, but on my reading of the best available evidence. They can and do change as new and better evidence becomes available.

Of course, pragmatic decisions must be informed by a set of values, especially values concerning the most important goals of education. Yet I believe that the vast majority of educators, parents, and policy makers want pretty much the same thing for children. The question of how we achieve those outcomes most effectively for all children should be a question for research and experience, not for ideological debate.

Being a pragmatist does not, however, imply being atheoretical. While a pragmatist does demand a close fit of theory to evidence (rather than the other way around), theory is still important in making sense of evidence from diverse sources and in predicting fruitful areas for future inquiry.

A theoretical model that is both derived from my work over the years and has guided it in many ways is described in the following sections.

A Model of Effective Instruction[1])

In the past twenty years, research on teaching has made significant strides in identifying teaching behaviors associated with high student achievement (Brophy and Good,

---

1   This section is adapted from Slavin, 1987,1994,1995a.

1986; Rosenshine and Stevens, 1986). However, effective instruction is not just good teaching. If it were, we could probably find the best lecturers, make video tapes of their lessons, and show them to students (see Slavin, 1994). Consider why the video teacher would be ineffective. First, the video teacher would have no idea what students already knew. A particular lesson might be too advanced for a particular group of students, or it may be that some students already know the material being taught. Some students may be learning the lesson quite well, while others are missing key concepts and falling behind because they lack prerequisite skills for new learning. The video teacher would have no way to know who needed additional help, and would have no way to provide it in any case. There would be no way to question students to find out if they were getting the main points and then to reteach any concepts students were failing to grasp.

Second, the video teacher would have no way to motivate students to pay attention to the lesson or to really try to learn it. If students were failing to pay attention or were misbehaving, the video teacher would have no way to do anything about it. Finally, the video teacher would never know at the end of the lesson whether or not students actually learned the main concepts or skills.

The case of the video teacher illustrates the point that teachers must be concerned with many elements of instruction in addition to the lesson itself. Teachers must attend to ways of adapting instruction to students' levels of knowledge, motivating students to learn, managing student behavior, grouping students for instruction, and testing and evaluating students. These functions are carried out at two levels. At the school level, the principal and/or central administrators may establish policies concerning grouping of students (e.g., tracking), provision and allocation of special education and remedial resources, and grading, evaluation, and promotion practices. At the classroom level, teachers control the grouping of students within the class, teaching techniques, classroom management methods, informal incentives, frequency and form of quizzes and tests, and so on. These elements of school and classroom organization are at least as important for student achievement as the quality of teachers' lessons.

I have described a model of effective instruction which attempts to identify the critical elements of schools and classroom organization and their interrelationships. This model, based on the work of John Carroll (1963, 1989), focuses on the *alterable* elements of Carroll's model, those which teachers and schools can directly change (see Slavin 1984; 1987a; 1994). The components of this model are as follows:

1. *Quality of Instruction*. The degree to which information or skills are presented so that students can easily learn them. Quality of instruction is largely a product of the quality of the curriculum and of the lesson presentation itself.
2. *Appropriate Levels of Instruction*: The degree to which the teacher makes sure that students are ready to learn a new lesson (that is, they have the necessary skills and knowledge to learn it) but have not already learned the lesson. In other words, the level of instruction is appropriate when a lesson is nether too difficult nor too easy for students.
3. *Incentive*: The degree to which the teacher makes sure that students are motivated to work on instructional tasks and to learn the material being presented.
4. *Time*: The degree to which students are given enough time to learn the material being taught.

The four elements of this QAIT (Quality, Appropriateness, Incentive, Time) model have one important characteristic: *All four* must be adequate for instruction to be effective. Again, effective instruction is not just good teaching. No matter how high the quality of instruction, students will not learn a lesson if they lack the necessary prior skills or information, if they lack the motivation, or if they lack the time they need to learn the lesson. On the other hand, if the quality of instruction is low, then it may make little difference how much students know, how motivated they are, or how much time they have. Each of the elements of the QAIT model is like a link in a chain, and the chain is only as strong as its weakest link. In fact, it may be hypothesized that the four elements are *multiplicatively* related, in that improvements in multiple elements may produce substantially larger learning gains than improvements in any one.

Effective Classroom Organization

Given a relatively fixed set of resources, every innovation in classroom organization solves some problems but also creates new problems which must themselves be solved. Tradeoffs are always involved. Understanding the terms of these tradeoffs is critical for an understanding of how to build effective models of classroom organization.

The QAIT model is designed primarily to clarify the tradeoffs involved in alternative forms of classroom organization. This article presents a perspective on what is known now about each of the QAIT elements, and more importantly explores the theoretical and practical ramifications of the *interdependence* of these elements for effective school and classroom organization.

Quality of Instruction

Quality of instruction refers to the activities we think of first when we think of teaching: lecturing, discussing, calling on students, and so on. It also includes the curriculum and books, software, or other materials. When instruction is high in quality, the information being presented makes sense to students, is interesting to them, is easy to remember and apply.

The most important aspect of instructional quality is the degree to which the lesson makes sense to students. For example, teachers must present information in an organized orderly way (Kallison, 1986), note transitions to new topics (Smith and Cotton, 1980), use clear and simple language (Land, 1987), use many vivid images and examples (Hiebert et al., 1991; Mayer & Gallini, 1990), and frequently restate essential principles (Maddox and Hoole, 1975). Lessons should be related to students' background knowledge, using such devices as advance organizers (Pressley et al., 1992) or simply reminding students of previously learned material at relevant points in the lesson. The teacher's enthusiasm (Abrami, Leventhal, and Perry, 1982) and humor (Kaplan and Pascoe, 1977) can also contribute to quality of instruction, as can use of media and other visual representations of concepts (Hiebert, Wearne, and Taber, 1991; Kozma, 1991).

Clear specification of lesson objectives to students (Melton, 1978) and a substantial correlation between what is taught and what is assessed (Cooley and Leinhardt, 1980) contribute to instructional quality, as does frequent formal or informal assessment to see that students are mastering what is being taught (Crooks, 1988; Kulik and Kulik, 1988) and immediate feedback to students on the correctness of their performances (Barringer and Gholson, 1979).

Instructional pace is partly an issue of quality of instruction and partly of appropriate levels of instruction. In general, content coverage is strongly related to student achievement (Dunkin, 1978; Barr and Dreeben, 1983), so a rapid pace of instruction may contribute to instructional quality. However, there is obviously such a thing as too rapid an instructional pace (see Leighton and Slavin, 1988). Frequent assessment of student learning is critical for teachers to establish the most rapid instructional pace consistent with the preparedness and learning rate of all students.

Appropriate Levels of Instruction

Perhaps the most difficult problem of school and classroom organization is accommodating instruction to the needs of students with different levels of prior knowledge and different learning rates. If a teacher presents a lesson on long division to a heterogeneous class, some students may fail to learn it because they have not mastered such prerequisite skills as subtraction, multiplication, or simple division. At the same time, there may be some students who know how to divide before the lesson begins, or learn to do so very rapidly. If the teacher sets a pace of instruction appropriate to the needs of the students lacking prerequisite skills, then the rapid learners' time will be largely wasted. If the instructional pace is too rapid, the students lacking prerequisite skills will be left behind.

There are many common means of attempting to accommodate instruction to students' diverse needs, but each method has drawbacks that may make the method counterproductive. Various forms of ability grouping seek to reduce the heterogeneity of instructional groups (see Chapters 7 and 8 in this book). Special education and remedial programs are a special form of ability grouping designed to provide special resources to accelerate the achievement of students with learning problems. However, between-class ability grouping plans, such as tracking, can create low-ability classes for which teachers have low expectations and maintain a slow pace of instruction, and which many teachers dislike to teach (Good and Marshall, 1984; Oakes, 1985; 1987; Rowan and Miracle, 1983; Slavin, 1987b; 1990a). Similar problems make self-contained special education classes of questionable benefit to students with learning handicaps (see Leinhardt and Bickel, 1987; Leinhardt and Pallay, 1982; Madden and Slavin, 1983). Within-class ability grouping, such as the use of reading and mathematics groups, creates problems of managing multiple groups within the classroom, reduces direct instruction to each student, and forces teachers to assign large amounts of unsupervised seatwork to keep students engaged while the teacher is working with a reading or mathematics group (Anderson, Evertson, and Brophy, 1979; Barr, 1992).

As noted in Chapters 7 and 8 of this book, research on assignment of students to ability-grouped classes finds no achievement benefits for this practice at the elementary or secondary levels (see Slavin, 1987b; 1990a; Oakes, 1985; 1987). However, forms of ability grouping in which elementary students remain in heterogeneous classes most of the day but are regrouped into homogeneous reading or mathematics classes can be instructionally effective if teachers actually adapt their level and pace of instruction to meet the needs of the regrouped classes. In particular, the Joplin Plan and certain nongraded plans in which elementary students are regrouped for reading or mathematics across grade lines and instructional level is based on performance level rather than age (see Chapter 9 of this book) can be instructionally effective (Slavin, 1987b; Gutíerrez and Slavin, 1992). Also, research on within-class ability grouping finds this practice to increase student mathematics achievement, particularly when the number of groups used is small and management techniques designed to ensure smooth transitions and high time-on-task during seatwork are used (Slavin, 1987b; Slavin and Karweit, 1985).

Group-based mastery learning (Bloom, 1976; Block and Burns, 1976; Guskey and Gates, 1985) discussed in Chapter 10 of this book, is an approach to providing levels of instruction that does not use permanent ability groups but rather regroups students after each skill is taught on the basis of their mastery of that skill. Students who attain pre-set criteria (e.g., 80%) on a formative test work on enrichment studies while non-masters receive corrective instruction. In theory, mastery learning should provide appropriate levels of instruction by ensuring that students have mastered prerequisite skills before they receive instruction in subsequent skills. However, within the confines of traditional class periods, the time needed for corrective instruction may slow the pace of instruction for the class as a whole. Studies of group-based mastery learning conducted in elementary and secondary schools over periods of at least four weeks have found few benefits of this approach in comparison to control groups given the same objectives, materials, and time as the mastery learning groups (Slavin, 1987c).

The most extreme form of accommodation to individual differences short of one-to-one tutoring is individualized instruction, in which students work entirely at their own level and rate. Individualized instruction certainly solves the problem of providing appropriate levels of instruction, but it creates serious problems of classroom management, often depriving students of adequate direct instruction. Research on individualized instruction has not generally found positive effects on student achievement (Hartley, 1977; Horak, 1981). However, Team Assisted Individualization, a form of individualized instruction which also incorporates the use of cooperative learning groups, has been found to consistently increase student achievement in mathematics (Slavin, Leavey, and Madden, 1984; Slavin, Madden, and Leavey, 1984; Slavin and Karweit, 1985; Slavin, 1985).

Incentive

Thomas Edison once wrote that "genius is one percent inspiration and ninety-nine percent perspiration." The same could probably be said for learning. Learning is work. This is not to say that learning must be drudgery, but it is certainly the case that students must exert themselves to pay attention, to study, and to conscientiously perform the tasks assigned to them, and they must somehow be motivated to do these things. This motivation may come from the intrinsic interest value of the material being learned, or may be created through the use of extrinsic incentives, such as praise, grades, stars, and so on (see Stipek, 1993).

If students want to know something, they will be more likely to exert the necessary effort to learn it. This is why there are students who can rattle off the names and statistics relating to every player on their favorite sports team, but do not know their multiplication facts. Teachers can create intrinsic interest in material to be taught by arousing student curiosity, for example by using surprising demonstrations, by relating topics to students' personal lives, or by allowing students to discover information for themselves (Brophy, 1987; Malone and Lepper, 1988).

However, not every subject can be made intrinsically interesting to every student at all times. Most students need some sort of extrinsic incentive to exert an adequate level of effort on most school tasks. For example, studies of graded versus pass-fail college courses find substantially higher achievement in classes that give grades (Gold, Reilly, Silberman, and Lehr, 1971; Hales, Bain, and Rand, 1971). At the elementary level, informal incentives, such as praise and feedback, may be more important than the formal grading system (see Brophy, 1987). One critical principal of effective use of classroom incentives is that students should generally be held accountable for what they do. For example, homework that is checked has been found to contribute more to students achievement than homework that is assigned but not checked (Cooper, 1989). Also, questioning strategies that communicate high expectations for students, such as waiting for them to respond (Rowe, 1974) and following up with students who do not initially give full responses (Brophy and Evertson, 1974) have been found to be associated with high achievement (see Good, 1987).

Several methods of providing formal incentives for learning have been found to be instructionally effective. One practical and effective method of rewarding students for appropriate, learning-oriented behavior is home-based reinforcement (Barth, 1979), provision of daily or weekly reports to parents on student behavior. Another is group contingencies (Dolan et al., 1993; Hayes, 1976), in which the entire class or groups within the class are rewarded on the basis of the behavior of the entire group.

Cooperative learning methods, discussed in Chapters 2 and 3, involve students working in small learning groups to master academic material. Forms of cooperative learning that have consistently increased student achievement have provided rewards to heterogeneous groups based on the learning of their members (Slavin, 1995b). This incentive system motivates students to encourage and help one another to achieve. Rewarding students based on improvement over their own past performance has also been found to be an effective incentive system (Natriello, 1987; Slavin, 1980).

In addition to being a product of specific strategies designed to increase student motivation, incentive is also influenced by quality of instruction and appropriate levels of instruction. Students will be more motivated to learn about a topic that is presented in an interesting way, that makes sense to them, that they feel capable of learning. Further, a student's motivation to exert maximum effort will be influenced by their perception of the difference between their probability of success if they do exert themselves and their probability of success if they do not (Atkinson and Birch, 1978; Slavin, 1977; 1994). That is, if a student feels sure of success or, alternatively, of failure, regardless of his or her efforts, then incentive will be very low. This is likely to be the case if a lesson is presented at a level much too easy or too difficult for the student. Incentive is high when the level of instruction is appropriate for a student, so that the student perceives that with effort the material can be mastered, so that the payoff for effort is perceived to be great.

Time

Instruction takes time. More time spent teaching a subject does not always translate into additional learning, but if instructional quality, appropriateness of instruction, and incentives for learning are all high, then more time on instruction is likely to pay off in greater learning.

The amount of time available for learning depends largely on two factors: *Allocated time* and *engaged time*. Allocated time is the time scheduled by the teacher for a particular lesson or subject and then actually used for instructional activities. Allocated time is mostly under the direct control of the school and teacher. In contrast, engaged time, the time students actually engage in learning tasks, is not under the direct control of the school or the teacher. Engaged time, or time-on-task, is largely a product of quality of instruction, student motivation, and allocated time. Thus, allocated time is an alterable element of instruction (like quality, appropriateness, and incentive), but engaged time is a mediating variable linking alterable variables with student achievement.

While allocated time must be an essential element in any model of classroom organization, research on this variable has found few consistent effects on student achievement. For example, research on hours in the school day and days in the school year has found few relationships between these time variables and student achievement (Frederick and Walberg, 1980; Karweit, 1989). The Beginning Teacher Evaluation Study found no effect of allocated time in specific subjects on student achievement in those subjects when time was measured at the class level (Marliave, Fisher, and Dishaw, 1978).

On the other hand, research on engaged time generally finds positive relationships between of time students are on task and their achievement, but even with this variable results are inconsistent (see Karweit, 1989).

Studies of means of increasing student time on task generally go under the heading of classroom management research. Process-product studies (see, for example, Brophy and Good, 1986) have established that teachers' use of effective management strategies is associated with high student achievement. However, several experimental studies focusing on increasing time on-task have found that it is possible to increase engaged time and still have no significant effect on student achievement (Emmer and Aussiker, 1990; Slavin, 1986; Stallings and Krasavage, 1986).

A Model of Alterable Elements of Instruction and Student Achievement

The QAIT model can be conceptualized in terms of intermediate effects on time-related variables. Figure 1.1 depicts a model of how alterable elements of instruction might affect student achievement.

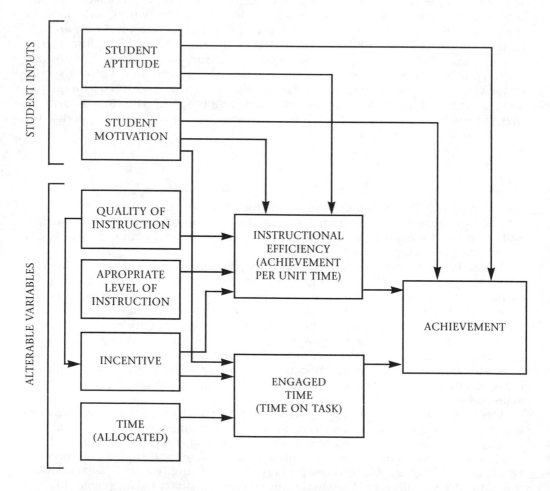

Figure 1.1 Model Relating Alterable Elements of Instruction to Student Achievement (from Slavin, 1987a)

In Figure 1.1, two types of independent variables are presented: *Student inputs* and *alterable variables*. Student inputs refer to factors over which the school has little control in the short run: Student aptitude (including their prior knowledge of a subject) and those aspects of motivation to learn that students bring from home (as distinct from the motivation created by classroom practices). The alterable variables are the QAIT elements discussed earlier. Of course, student inputs are not unchangeable, but can be affected by classroom practices. For example, student aptitude to learn a specific lesson may be strongly influenced by background knowledge resulting from earlier instruction, by specific training in thinking, problem solving, or study skills, or by general intellectual stimulation or learning skills provided by the school. Student motivation to learn is also largely a product of past experiences in school. However, in the context of any given lesson, the student inputs can be considered fixed, while the alterable variables can be directly altered by the school or teacher.

The effects of the alterable variables on student achievement are held to be mediated by two time-related variables: *Instructional efficiency* and *engaged time*, or time-on-task. Instructional efficiency can be conceptualized as the amount of learning per time. For example, students will learn more in a ten-minute lesson high in instructional efficiency than in a lesson of similar length low in instructional efficiency. Engaged time is the amount of time students are actually participating in relevant learning activities, such as paying attention to lectures and doing assignments. Instructional efficiency and engaged time are multiplicatively related to student achievement; obviously, if either is zero, then learning is zero.

The QAIT model can be easily related to instructional efficiency and engaged time. Instructional efficiency is a product of the quality of instruction (e.g., organization and presentation quality of the lesson), appropriate levels of instruction (students have prerequisite skills but have not already learned the lesson), and incentive (students are interested in learning the lesson). Of course, aptitude and motivation also contribute to instructional efficiency for any given student. Engaged time is primarily a product of allocated time and incentive.

The relationship between improvements in each of the four alterable elements and effects on student achievement is held to be multiplicative. If any of the elements is at zero, learning will be zero. Above zero, the argument that the effects of the four elements are multiplicative rests in part on an assumption that effects of increasing each element are greatest at low levels and ultimately reach maximum levels. For example, motivation to learn will reach a maximum in terms of affecting student achievement at some point. Effects of quality and appropriateness of instruction are similarly likely to reach a point of diminishing returns. Time on-task not only cannot be increased beyond 100% of time allocated, but it is doubtful whether increases beyond, say, 90% produce significant increases in learning. This may explain why several studies which produced substantial gains in time on-task have produced minimal effects on student achievement (see Emmer and Aussiker, 1990; Slavin, 1986).

The substantive implication of a multiplicative relationship among the QAIT elements is that it may be more effective to design instruction to produce moderate gains in two or more elements than maximize gains in only one. To increase a plant's growth, moderate increases in light, water, and fertilizer are likely to be more productive than large increases in only one of these elements. By the same token, substantial increases in any one element of the QAIT model, leaving all others unaffected, is likely to be less effective than more moderate, across-the-board improvements.

Another implication of the assumption that there is a point of diminishing returns in the achievement effect of each of the QAIT elements is that different types of programs or teaching emphases will work differently in different settings depending on pre-existing levels of each. For example, an emphasis on increasing time on-task is likely to be more effective in classrooms low on this variable than in those beginning at 80-90% levels. Highly motivated students may profit more from programs focusing on providing appropriate levels of instruction than from motivationally focused programs, and so on.

Conclusion

The most important implication of the QAIT model is that teachers need to focus on each of the four elements of effective instruction – quality, appropriateness, incentive, and time – if they expect to make a substantial difference in student achievement. In particular, teachers need to be sure that if they solve problems relating to one element they do not cause new problems relating to another. For example, Slavin (1987a) explained the failure of individualized instruction programs of the 1970's by noting that what they gained in appropriate levels of instruction they often lost in quality of instruction, motivation, and time on task.

Clearly, effective teaching is an art and must be sensitive to context and to the particular needs of a given group of students. Yet research on effective strategies is useful to teachers in helping them make informed and intelligent choices of teaching strategies designed to accelerate student achievement.

The remainder of this book presents reviews of research on instructional methods and school organization methods designed to ensure the success of all students, and concludes with a discussion of the implications of this research for policy and practice.

SECTION I:

Research on Effective Instructional Strategies

CHAPTER 2

Cooperative Learning and Student Achievement

Background and Commentary

Cooperative learning is where I began as a researcher and reformer in education. I came upon this interest in an unusual way. In the late 1960's I was a psychology major at Reed College in Portland, Oregon, and was very interested in applications of psychology to education. At that time, there was a widespread belief in education reform circles that children would never learn anything unless it was relevant in their own lives. I was skeptical that the entire curriculum could be made relevant to children, but I did have an idea of how to *simulate* relevance. In high school and college I had been involved in simulation gaming, and could see the enormous power of a complex simulation to engage participants' motivation and interest. I thought that if curriculum could be embedded in simulations, students would be motivated to master it. I teamed up with my future wife and colleague, Nancy Madden, to try out this idea. We got a small grant to spend a summer developing a simulation-based approach to lab science for junior high school. We developed a program called WorldLab (this name and idea has appeared again in our most recent work – see Chapter 5). In WorldLab, students worked in small groups to oversee the technological development of a newly colonized planet, Zupita. To do this, they had to learn the underlying science (and scientific method) and conduct experiments to ultimately design solutions to their countries' problems. For example, to make a steel mill they had to separate iron from iron ore in the laboratory; to control pollution they had to experiment with a variety of substances capable of adsorbing hydrogen sulfide

In developing WorldLab and implementing it in (ultimately) three schools, we saw the power of the simulation, but we also saw the power of groups. Group members helped each other, encouraged each other, and taught each other. Reed required an undergraduate thesis, and both Nancy and I did ours on our WorldLab experiments. Only later did we find out that other researchers elsewhere were heading in similar directions. The only work we knew about that had anything to do with cooperative learning was at Johns Hopkins University, led by James Coleman. Coleman, the author of The *Adolescent Society* (1961), had written about how the competitive nature of the classroom led students to discourage each other from academic excellence. This led him to propose and later develop group simulation games and academic tournaments intended to give academic excellence the same status enjoyed in secondary schools by intermural sports.

After a year of student teaching in high school social studies and a year of teaching special education, I went to Johns Hopkins, hoping to work with Coleman. He left the year I came, but I did work with others who had been doing research on cooperative learning, principally David DeVries, the original developer of Teams Games Tournament (TGT). In 1975, DeVries left Johns Hopkins and I took his job.

l   Throughout the 1970's, I was working primarily on the development and evaluation of Student Teams-Achievement Divisions, or STAD, a relatively simple and broadly applicable cooperative method. In 1978 my colleagues and I began to disseminate STAD, TGT, and a revision of Elliot Aronson's Jigsaw method. The research on these simple, "generic" methods was producing positive achievement outcomes (Slavin, 1995a) as well as positive effects on such outcomes as intergroup relations (see Slavin, 1995b) and acceptance of mainstreamed academically handicapped students (Slavin, 1995a).

In 1979, Nancy Madden was finishing her Ph.D. in clinical psychology (we had married in 1973), and she began working with me at Johns Hopkins. We were concerned that the simple cooperative methods, such as STAD, TGT, and Jigsaw, did too little to integrate learning with curriculum, and as such were usually used to supplement traditional practices rather than serve as the primary mode of instruction. We were also concerned about the difficulties inherent in using one pace of instruction when students were far apart in achievement levels. These concerns led first to the development of Team Assisted Individualization in Mathematics (TAI; Slavin, Madden, Leavey, 1984) and, somewhat later, to Cooperative Integrated Reading and Composition, or CIRC, in collaboration with our colleague Robert Stevens (Stevens, Madden, Slavin, & Farnish, 1987).

The review of research on the achievement outcomes of cooperative learning presented in this chapter is adapted from the second edition of *Cooperative Learning: Theory, Research, and Practice*. However, this review has a pedigree older than the book. I reviewed laboratory and classroom research on cooperation and competition in 1977 (Slavin, 1977), and then wrote a comprehensive review of research on practical applications of cooperative learning in 1983 (Slavin, 1983a, b). I've been revising and updating the 1983 reviews for several years, as the body of research and theory supporting cooperative learning has steadily expanded. I think it is fair to say that no classroom instructional method has ever been as extensively and rigorously evaluated as cooperative learning; in 1995 I located 99 studies that compared experimental and control groups over a period of at least four weeks. If I had dropped the methodological adequacy and duration requirements I established for this review, I'm sure I could have included more than 300 individual studies.

Despite this extraordinary number of high-quality studies, there is still a great deal more research worth doing on cooperative learning (see Slavin, in press). Cooperative learning has entered widespread use in many quite diverse forms, and there are always more questions to be answered about what conditions are necessary for cooperative learning to achieve its effects and how it can be made more effective and applied to broader categories of educational objectives.

"You'll all recall," started Mr. Dunbar, "how last week we figured out how to compute the area of a circle and the volume of a cube. Today you're going to have a chance to discover how to compute the volume of a *cylinder*. This time, you're really going to be on your own. At each of your lab stations you have five unmarked cylinders of different sizes. You also have a metric ruler and a calculator, and you may use water from your sink. The most important resources you'll have to use, however, are your minds and your partners. Remember, at the end of this activity everyone in every group must be able to explain not only the formula for volume of a cylinder, but also precisely how you derived it. Any questions? You may begin!" The students in Mr. Dunbar's middle school math/science class got right to work. They were seated around lab tables in groups of four. One of the groups, the Squares, started off by filling up all its cylinders with water.

"OK," said Miguel, "we've filled up all of our cylinders. What do we do next?"

"Let's measure them," suggested Margarite. She took the ruler and asked Dave to write down the measurements.

"The water in this little one is 36 millimeters high and ... just a sec ... 42 millimeters across the bottom."

"So what?" asked Yolanda. "We can't figure out the volume this way. Let's do a little thinking before we start measuring everything."

"Yolanda's right," said Dave. "We'd better work out a plan."

"I know," said Miguel, "let's make a hypo, hypotha, what's it called?"

"Hypothesis," said Yolanda. "Yeah! Let's guess what we think the solution is."

"Remember how Mr. Dunbar reminded us about the area of a circle and the volume of a cube? I'll bet that's an important clue."

"You're right, Miguel," said Mr. Dunbar, who happened to be passing by. "But what are you guys going to do with that information?"

The Squares were quiet for a few moments. "Let's try figuring out the area of the bottom of one of these cylinders," ventured Dave. "Remember that Margarite said the bottom of the little one was 42 millimeters? Give me the calculator ... now how do we get the area?"

Yolanda said, "I think it was pi times the radius squared."

"That sounds right. So 42 squared ..."

"Not 42, 21 squared," interrupted Margarite. "If the diameter is 42, the radius is 21."

"OK, OK, I would have remembered. Now, 21 squared is ... 441, and pi is about 3.14, so my handy dandy calculator says ... 1,384.7."

"So what?" said Yolanda.

"That doesn't tell us how to figure the volume!"

Margarite jumped in excitedly. "Just hang on for a minute, Yolanda. Now, I think we should multiply the area of the bottom by the height of the water."

"But why?" asked Miguel.

"Well," said Margarite, "when we did the volume of a cube we multiplied length times width times height. Length times width is the area of the bottom. I'll bet we could do the same with a cylinder!"

"The girl's brilliant!" said Miguel. "Sounds good to me. But how could we prove it?"

"I've got an idea," said Yolanda. She emptied the water out of all the cylinders and filled the smallest one to the top. "This is my idea. We don't know what the volume of this cylinder is, but we do know that it's always the same. If we pour the same amounts of water into all four cylinders and use our formula, it should always come out to the same amount!"

"Let's try it!" said Miguel. He poured the water from the small cylinder into a larger one, refilled it, and poured it into another of a different shape. The Squares measured the bases and the heights of the water in their cylinders, wrote down the measurements, and tried out their formula. Sure enough, their formula always gave them the same answer for the same volume of water. The Squares discussed the steps they had gone through to arrive at a solution and checked to be sure that any group member could explain what they had done and what they had found out. In great excitement they called Mr. Dunbar to come see what they were doing. Mr. Dunbar asked each of the students to explain what they did. "Terrific!" he said. "You not only figured out a solution, but everyone in the group participated and understood what you did. Now I'd like you to help me out. I've got a couple of groups that are really stumped. Do you suppose you could help them? Don't give them the answer, but help them get on track. How about Yolanda and Miguel helping with the Brainiacs, and Dave and Margarite help with the Dream Team. OK? Thanks!"[1]

Mr. Dunbar's class is involved in a revolution in classroom instruction. The name of this revolution is cooperative learning. Cooperative learning refers to a variety of teaching methods in which students work in small groups to help one another learn academic content. In cooperative classrooms, students are expected to help each other, to discuss and argue with each other, to assess each other's current knowledge and fill in gaps in each other's understanding. Cooperative work rarely replaces teacher instruction, but rather replaces individual seatwork, individual study, and individual drill. When properly organized, students in cooperative groups work with each other to make certain that everyone in the group has mastered the concepts being taught. In Mr. Dunbar's class, the Squares know they are not finished until every student in the group knows how to generate the formula. Their success as a group depends on their ability to ensure that everyone has grasped the key ideas.

Cooperative learning is not a new idea in education, but until recently it has only been used by a few teachers for limited purposes, such as occasional group projects or reports. However, research over the last twenty years has identified cooperative learning methods that can be used effectively at every grade level to teach every type of content, from math to reading to writing to science, from basic skills to complex problem solving. Increasingly, cooperative learning is being used as teachers' main way of organizing classrooms for instruction.

There are many reasons that cooperative learning is entering the mainstream of educational practice. One is the extraordinary research base (summarized in this book) supporting the use of cooperative learning to increase student achievement, as well as such other outcomes as improved intergroup relations, acceptance of academically handicapped classmates, and increased self-esteem. Another reason is the growing realization that students need to learn to think, to solve problems, and to integrate and apply knowledge and skills, and that cooperative learning is an excellent means to that end. The Squares in Mr. Dunbar's class illustrate how students working together can have the "AHA!" experiences that build deep understanding. While cooperative learning works well in classes that are homogeneous, including classes for the gifted, special education classes, and even classes for the severely and profoundly "average," it is especially needed in classes with a wide range of performance levels. Cooperative learning can help make diversity a resource rather than a problem. As schools are moving away from ability grouping toward more heterogeneous grouping, cooperative learning becomes increasingly important. Further, cooperative learning has wonderful benefits for relationships between students of different ethnic backgrounds and between mainstreamed special education students and their classmates, adding another critical reason to use cooperative learning in diverse classrooms.

What's Wrong with Competition?

One of the most important reasons that cooperative learning methods were developed is that educators and social scientists have long known about the detrimental effects of competition as it is usually used in the classroom. This is not to say that competition is always wrong; if properly structured, competition between well matched com-

---

1   Adapted from Slavin, 1994.

petitors can be an effective and harmless means of motivating people to do their best. Yet the forms of competition typically used in classrooms are rarely healthy or effective. To illustrate this, consider the following vignette.

"Class," said Ms. James, "who remembers what part of speech words such as 'it,' 'you,' and 'he' belong to?"

Twenty hands shoot up in Ms. James's fifth-grade class. Another ten students try to make themselves small in hopes that Ms. James won't call on them. She calls on Eddie.

"Proverb?"

The class laughs. Ms. James says, "No, that's not quite right." The students (other than Eddie, who is trying to sink into the floor in embarrassment) raise their hands again. Some of them are halfway out of their seats, calling "Me! Me!" in their eagerness. Finally, Ms. James calls on a student. "Elizabeth, can you help Eddie?"

Think about the scene being played out in Ms. James's class, a common sequence of events at every level of schooling, in every subject, in all sorts of schools. Whether or not she is conscious of it, Ms. James has set up a competition between the students. The students want to earn her approval, and they can do this only at the expense of their classmates. When Eddie fails, most of the class is glad: students who know the answer now have a second chance to show it, while others know that they are not alone in their ignorance. The most telling part of the vignette is when Ms. James asks if Elizabeth can "help" Eddie. Does Eddie perceive Elizabeth's correct answer as help? Does Elizabeth? Of course not.

Consider what is going on below the surface in Ms. James's class. Most of the class is hoping that Eddie (and also Elizabeth) will fail. Their failure makes their classmates look good in comparison, and makes it easier for others to succeed. Because of this, students will eventually begin to express norms or values opposed to doing too well academically. Students who try too hard are "teacher's pets," "nerds," "grinds," and so on. Students are put in a bind: their teachers reward high achievement, but their peer group rewards mediocrity. As students enter adolescence, the peer group becomes all-important, and most students accept their peers' belief that doing more than what is needed to get by is for suckers. Research clearly shows that academic success is not what gets students accepted by their peers, especially in middle and high school.

Typical classroom competition can be detrimental for another reason. Recall the ten students who tried to make themselves invisible when Ms. James asked her question. For most low achievers a competitive situation is a poor motivator; for some it is almost constant psychological torture. Students enter any class with widely divergent skills and knowledge. Low-achieving students may lack the prerequisites to learn new material. For example, students who never learned to subtract well may have difficulty learning long division. For this and other reasons, success is difficult for many students, but comes easily for others. Success is defined on a relative basis in the competitive classroom. Even if low achievers learn a great deal, they are still at the bottom of the class if their classmates learn even more. Day in, day out, low achievers get negative feedback on their academic efforts. After a while, they learn that academic success is not in their grasp, so they choose other avenues in which they may develop a positive self-image. Many of these avenues lead to antisocial, delinquent behavior.

How can teachers avoid the problems associated with classroom competition? How can students help one another learn and encourage one another to succeed academically?

Think back to Ms. James's class. What if Eddie and Elizabeth and two other students had been asked to work together as a team to learn parts of speech, and the teams were rewarded on the basis of the learning of all team members? Now the only way for Eddie and Elizabeth to succeed is if they make certain that they have learned the material and that their teammates have done so. Eddie and Elizabeth are now motivated to help and encourage each other to learn. Perhaps most important, they are rooting for each other to succeed, not to fail.

This is the essence of *cooperative learning* (Slavin, 1983a,b). In cooperative learning methods, students work together in four-member teams to master material initially presented by the teacher. For example, in a method called Student Teams-Achievement Divisions, or STAD (Slavin, 1986a), a teacher might present a lesson on map reading, then give students time to work with maps and to answer questions about them in their teams. The teams are heterogeneous – made up of high, average, and low achievers, boys and girls, and students of different ethnic groups. After having a chance to study in their teams, students take individual quizzes on map reading. The students' quiz scores are added up. All teams whose average scores meet a high criterion receive special recognition, such as fancy certificates or posting of their team picture in the classroom.

The idea behind this form of cooperative learning is that if students want to succeed as a team, they will encourage their teammates to excel and will help them to do so. Often, students can do an outstanding job of explaining difficult ideas to one another by translating the teacher's language into kid language.

Of course, cooperative learning methods are not new. Teachers have used them for many years in the form of laboratory groups, project groups, discussion groups, and so on. However, recent research in the United States and other countries has created systematic and practical cooperative learning methods intended for use as the main element of classroom organization, has documented the effects of these methods, and has applied them to the teaching of a broad range of curricula. These methods are now being used extensively in every conceivable subject, at grade levels from kindergarten through college, and in all kinds of schools throughout the world.

Cooperative Learning Methods

Social psychological research on cooperation dates back to the 1920s (see Slavin, 1977a), but research on specific applications of cooperative learning to the classroom did not begin until the early 1970s. At that time, four independent groups of researchers began to develop and research cooperative learning methods in classroom settings. At present, researchers all over the world are studying practical applications of cooperative learning principles, and many cooperative learning methods are available. Some of the most extensively researched and widely used cooperative learning methods are introduced in this chapter, and these and other methods are described in more detail later in this book.

Student Team Learning

Student Team Learning methods are cooperative learning techniques developed and researched at Johns Hopkins University. More than half of all studies of practical cooperative learning methods involve these methods.

All cooperative learning methods share the idea that students work together to learn and are responsible for their teammates' learning as well as their own. In addition to the idea of cooperative work, Student Team Learning methods emphasize the use of team goals and team success, which can be achieved only if all members of the team learn the objectives being taught. That is, in Student Team Learning the students' tasks are not to *do* something as a team but to *learn* something as a team.

Three concepts are central to all Student Team Learning methods – team rewards, individual accountability, and equal opportunities for success. Teams may earn certificates or other *team rewards* if they achieve above a designated criterion. Teams do not compete to earn scarce rewards; all (or none) of the teams may achieve the criterion in a given week. *Individual accountability* means that the team's success depends on the individual learning of all team members. Accountability focuses the activity of the team members on helping one another learn and making sure that everyone on the team is ready for a quiz or any other assessment that students take without teammate help. *Equal opportunities for success* means that students contribute to their teams by improving on their own past performance. This ensures that high, average, and low achievers are equally challenged to do their best, and that the contributions of all team members are valued.

Research on cooperative learning methods (summarized in Chapter 2) has indicated that team rewards and individual accountability are essential for basic skills achievement (Slavin, 1983a,b, 1989). It is not enough to simply tell students to work together; they must have a reason to take one another's achievement seriously. Further, research indicates that if students are rewarded for doing better than they have in the past, they will be more motivated to achieve than if they are rewarded for doing better than others, because rewards for improvement make success neither too difficult nor too easy for students to achieve (Slavin, 1980a).

Five principal Student Team Learning methods have been developed and extensively researched. Three are general cooperative learning methods adaptable to most subjects and grade levels: Student Teams-Achievement Divisions (STAD), Teams-Games-Tournaments (TGT), and Jigsaw II. The remaining two are comprehensive curricula designed for use in particular subjects at particular grade levels: Cooperative Integrated Reading and Composition (CIRC) for reading and writing instruction in grades 2-8, and Team Accelerated Instruction (TAI) for mathematics in grades 34. All five methods incorporate team rewards, individual accountability, and equal opportunities for success, but in different ways.

Student Teams-Achievement Divisions (STAD)

In STAD, students are assigned to four-member learning teams that are mixed in performance level, gender, and ethnicity. The teacher presents a lesson, and then students work within their teams to make sure all team members have mastered the lesson. Then, all students take individual quizzes on the material, at which time they may not help one another.

Students' quiz scores are compared to their own past averages, and points are awarded to each team based on the degree to which students meet or exceed their own earlier performances. These points are then added to form team scores, and teams that meet certain criteria may earn certificates or other rewards. The whole cycle of activities, including teacher presentation, team practice, and quiz, usually takes 3-5 class periods.

STAD has been used in every imaginable subject, from mathematics to language arts to social studies and science, and has been used from grade two through college It is most appropriate for teaching well-defined objectives, such as mathematical computations and applications, language usage and mechanics, geography and map skills, and science concepts.

The main idea behind Student Teams-Achievement Divisions is to motivate students to encourage and help each other master skills presented by the teacher. If students want their team to earn *team rewards*, they must help their teammates to learn the material. They must encourage their teammates to do their best, expressing norms that learning is important, valuable, and fun. Students work together after the teacher's lesson. They may work in pairs and compare answers, discuss any discrepancies, and help each other with any misunderstandings. They may discuss approaches to solving problems, or they may quiz each other on the content they are studying. They work with their teammates, assessing their strengths and weaknesses to help them succeed on the quizzes.

Although students study together, they may not help each other with quizzes. Every student must know the material. This individual accountability motivates students to do a good job explaining to each other, as the only way for the team to succeed is for all team members to master the information or skills being taught. Because team scores are based on students' improvement over their own past records (equal opportunities for success), all students have the chance to be a team "star" in a given week, either by scoring well above their past record or by getting a perfect paper, which always produces a maximum score regardless of students' past averages.

STAD is a general method of organizing the classroom rather than a comprehensive method of teaching any particular subject; teachers use their own lessons and other materials. Worksheets and quizzes are available for most school subjects for grades 3-9, but most teachers use their own materials to supplement or replace these materials.

Team-Games-Tournaments (TGT)

Teams-Games-Tournaments, originally developed by David DeVries and Keith Edwards, was the first of the Johns Hopkins cooperative learning methods. It uses the same teacher presentations and team work as in STAD, but replaces the quizzes with weekly tournaments, in which students play academic games with members of other teams to contribute points to their team scores. Students play the games at three-person "tournament tables" with others with similar past records in mathematics. A "bumping" procedure keeps the games fair. The top scorer at each tournament table brings sixty points to his or her team, regardless of which table it is; this means that low achievers (playing with other low achievers) and high achievers (playing with other high achievers) have equal opportunities for success. As in STAD, high-performing teams earn certificates or other forms of team rewards.

TGT has many of the same dynamics as STAD, but adds a dimension of excitement contributed by the use of games. Teammates help one another prepare for the games by studying worksheets and explaining problems to one another, but when students are playing the games their teammates cannot help them, ensuring **individual accountability**. The same materials used in STAD may also be used in TGT – the STAD quizzes are used as games in TGT. Some teachers prefer TGT because of its fun and activity, while others prefer the more purely cooperative STAD, and many combine the two.

Jigsaw II

Jigsaw II is an adaptation of Elliot Aronson's (1978) Jigsaw technique. In it, students work in the same four-member, heterogeneous teams as in STAD and TGT. The students are assigned chapters, short books, or other materials to read, usually social studies, biographies, or other expository material. Each team member is randomly assigned to become an "expert" on some aspect of the reading assignment. For example, in a unit on Mexico, one student on each team might become an expert on history, another on economics, a third on geography, and a fourth on culture. After reading the material, experts from different teams meet to discuss their common topics, and then they return to teach their topics to their teammates. Finally, there is a quiz or other assessment on all topics. Scoring and team recognition based on improvement are the same as in STAD.

Team Accelerated Instruction (TAI)

Team Accelerated Instruction (Slavin Leavey, & Madden, 1986) shares with STAD and TGT the use of four-member mixed ability learning teams and certificates for high-performing teams. However, STAD and TGT use a single pace of instruction for the class, where TAI combines cooperative learning with individualized instruction. Also, STAD and TGT apply to most subjects and grade levels, but TAI is specifically designed to teach mathematics to students in grades 3-6 (or older students not ready for a full algebra course).

In TAI, students enter an individualized sequence according to a placement test and then proceed at their own rates. In general, members of a team work on different units. Teammates check each other's work using answer sheets and help one another with any problems. Final unit tests are taken without teammate help and are scored by student monitors. Each week, teachers total the number of units completed by all team members and give certificates or other **team rewards** to teams that exceed a criterion score based on the number of final tests passed, with extra points for perfect papers and completed homework.

Because students are responsible for checking each other's work and managing the flow of materials, the teacher can spend most of the class time presenting lessons to small groups of students drawn from the various teams who are working at the same level in the mathematics sequence. For example, the teacher might call up a decimals group, present a lesson on decimals, and then send the students back to their teams to work on decimal problems. Then the teacher might call the fractions group, and so on. TAI has many of the motivational dynamics of STAD and TGT. Students encourage and help one another to work hard because they want their teams to succeed. **Individual accountability** is assured because the only score that counts is the final test score, and students take final tests without teammate help. Students have **equal opportunities for success** because all have been placed according to their level of prior knowledge; it is as easy (or difficult) for a low achiever to complete three subtraction units in a week as it is for a higher-achieving classmate to complete three long division units.

However, the individualization that is part of TAI makes it quite different from STAD and TGT. In mathematics, most concepts build on earlier ones. If the earlier concepts were not mastered, the later ones will be difficult or impossible to learn, a student who cannot subtract or multiply will be unable to master long division, a student who does not understand fractional concepts will be unable to understand what a decimal is, and so on. In TAI, students work at their own levels, so if they lack prerequisite skills they can build a strong foundation before going on. Also, if students can progress more rapidly, they need not wait for the rest of the class.

Cooperative Integrated Reading and Composition (CIRC)

CIRC is a comprehensive program for teaching reading and writing in the upper elementary and middle grades (Madden, Slavin, & Stevens, 1986). In CIRC, teachers use novels or basal readers. They may or may not use reading groups, as in traditional reading classes. Students are assigned to teams composed of pairs of students from two or more different reading levels. Students work in pairs within their teams on a series of cognitively engaging activities, including reading to one another, making predictions about how narrative stories will be resolved, summarizing stories to one another, writing responses to stories, and practicing spelling, decoding, and vocabulary. Students also work in their teams to master main idea and other comprehension skills. During language arts periods, students engage in a writer's workshop, writing drafts, revising and editing one another's work, and preparing for publication of team or class books. In most CIRC activities, students follow a sequence of teacher instruction, team practice, team pre-assessments, and quiz. Students do not take the quiz until their teammates have determined that they are ready. Team rewards and certificates are given to teams based on the average performance of all team members on all reading and writing activities. Because students work on materials appropriate to their reading levels, they have equal opportunities for success. Students' contributions to their teams are based on their quiz scores and independently written compositions, which ensures individual accountability.

Other Cooperative Learning Methods

Group Investigation
Group Investigation, developed by Shlomo and Yael Sharan at the University of Tel Aviv, is a general classroom-organization plan in which students work in small groups using cooperative inquiry, group discussion, and cooperative planning and projects (Sharan and Sharan, 1992). In this method, students form their own two-to-six-member groups. The groups choose topics from a unit being studied by the entire class, break these topics into individual tasks, and carry out the activities necessary to prepare group reports. Each group then presents or displays its findings to the entire class.

Learning Together
David and Roger Johnson at the University of Minnesota developed the Learning Together model of cooperative learning (Johnson and Johnson, 1987; Johnson, Johnson, & Smith, 1991). The methods they have researched involve students working in four- or five-member heterogeneous groups on assignment sheets. The groups hand in a single sheet, and receive praise and rewards based on the group product.

Complex Instruction
Elizabeth Cohen (1986) and her colleagues at Stanford University have developed and researched approaches to cooperative learning that emphasize use of discovery-oriented projects, particularly in science, math, and social studies. A major focus of Complex Instruction is on building respect for all of the abilities students have. Projects in Complex Instruction require a wide variety of roles and skills, and teachers point out how every student is good at something that helps the group succeed. Complex Instruction has particularly been used in bilingual education and in heterogeneous classes containing language minority students, where materials are often available in Spanish as well as English.

Structured Dyadic Methods
While most cooperative learning methods involve groups of about four members who have considerable freedom in deciding how they will work together, there is an increasing body of research on highly structured methods in which pairs of students teach each other. There is a long tradition of laboratory research showing how scripted pair learning, in which students take turns as teacher and learner to learn procedures or extract information from text, can be very effective in increasing student learning (Dansereau, 1988). Pair learning strategies have also been used over longer time periods in classrooms. One method, called Classwide Peer Tutoring (Greenwood, Delquadri, & Hall, 1989), has peer tutors follow a simple study procedure. Tutors present prob-

lems to their tutees. If they respond correctly the tutees earn points. If not, tutors provide the answer and the tutee must write the answer three times, reread a sentence correctly, or otherwise correct their error. Every ten minutes the tutors and tutees switch roles. Dyads earning the most points are recognized in class each day. A similar method, Reciprocal Peer Tutoring (Fantuzzo, King, & Heller, 1992), also alternates tutor and tutee roles within dyads, but gives tutors specific prompts and alternative problems to use if tutees make errors.

Ms. Logan's physical science class is a happy mess. Students are working in small groups at lab stations filling all sorts of bottles with water and then tapping them to see how various factors affect the sound. One group has set up a line of identical bottles and put different amounts of water in each so that tapping the bottles in sequence makes a crude musical scale. "The amount of water in the bottle is all that matters," one group member tells Ms. Logan, and her groupmates nod in agreement. Another group has an odd assortment of bottles and has carefully measured the same amount of water into each. "It's the shape and thickness of the bottles that makes the difference," says one group member. Other groups are working more chaotically, filling and tapping large and small, narrow and wide, and thick and thin bottles with different amounts of water. Their theories are wild and varied. After a half hour of experimentation, Ms. Logan calls the class together and asks group members to describe what they did and what they concluded. Students loudly uphold their group's point of view. "It's the amount of water!" "No, it's their shape!" "It's how hard you tap the bottles!" Ms. Logan moderates the conversation but lets students confront each other's ideas and give their own arguments.

The next day, Ms. Logan teaches a lesson on sound. She explains how sound causes waves in the air, and how the waves cause the eardrum to vibrate, transmitting sound information to the brain. She has two students come to the front of the class with a Slinky and uses the Slinky to illustrate how sound waves travel. She asks many questions of students, both to see if they are understanding and to get them to take the next mental step. She then explains how sound waves in a tube become lower in pitch the longer the tube. To illustrate this she plays a flute and a piccolo. Light bulbs are starting to click on in the students' minds, and Ms. Logan can tell from the responses to her questions that the students are starting to get the idea. They get back into their groups to discuss what they have learned and to try to apply their knowledge to the bottle problem.

When the students come into class on the third day of the sound lesson they are buzzing with excitement. They rush to their lab stations and start madly filling and tapping bottles to test out the theories they came up with the day before. Ms. Logan walks among the groups listening in on their conversations. "It's not the amount of *water*, it's the amount of *air*," she hears one student say. "It's not the *bottle*; it's the *air*," says a student in another group. She helps one group that is still floundering to get on track. Finally, Ms. Logan calls the class together to discuss their findings and conclusions. Representatives of some of the groups demonstrate the experiments they used to show how it was the amount of air in each bottle that determined the sound.

"How could we make one elegant demonstration that it's *only* the amount of air that controls the sound?" asks Ms. Logan. The students buzz among themselves, and then assemble all their bottles into one experiment. They make one line of identical bottles with different amounts of water. Then to show that it is the air, not the water, that matters, they put the same amount of water in bottles of different sizes. Sure enough, in either case the more air space left in the bottle, the lower the sound.

Ms. Logan ends the period with a homework assignment to read a chapter on sound in a textbook. She tells the students they will have an opportunity to work in their groups to make certain that every group member understands everything in the sound lesson, and then there will be a quiz in which students will have to individually show that they can apply their new knowledge. She reminds them that their groups can be "Superteams" only if everyone knows the material.

The bell rings, and students pour into the hallway, still talking excitedly about what they have learned. Some groupmates promise to call each other that evening to prepare for the group study the next day.

On the fourth day of the sound lesson, Ms. Logan passes out worksheets containing questions relating to sound and hearing. She gives each team two copies of the worksheets, and reminds the students to work together to make sure that everyone on the team can answer the questions correctly. The students get right to work. They discuss answers to the questions, argue about their understandings, and finally come to consensus. If they run into trouble, they call on Ms. Logan, but she will only respond to four hands raised, indicating that everyone on the team has the same problem and they've already tried to solve it themselves.

As the teams work they are trying to make sure that every teammate is learning. They quiz each other and encourage each team member to explain his or her current understandings so that others can correct misunderstandings and profit from each other's thinking processes. For example, the "Fizziks" are struggling with a question about how sound travels through different substances.

"I think sound should go faster through air than through water or wood, because water and wood are thicker," says Jennifer. Matthew and Rosa nod in agreement, but Thomas is doubtful.

"I know that sounds right," he says, "but don't you remember that experiment where they put a watch on a table and you could hear it tick through the wood?"

Rosa frowns, but then brightens up. "Now I remember! And remember how Indians and trackers put their ears on the ground or on the railroad tracks to hear sounds far away?"

"OK, you guys are right. Let's go on to the next question," says Matthew.

"Not so fast." Jennifer objects. "Let's make sure everyone has this. Matthew, can you say how sound travels through different substances?"

"Sure. You can hear a watch tick through a table even though it's thicker than air."

"I think you're on the right track," says Thomas, "but let's work on it."

"How about this," asks Rosa. "Sound goes through different substances at different speeds. It goes faster through wood, water, steel, and ground than it does through air. Does that help you, Matthew?"

"Yeah, I got it," he says. "Like sound waves work better in thick stuff than in air."

"Great!" says Jennifer. "I'm pretty sure I've got it too. Sound waves go real fast in solids and water, but slow in air. That's why you see a kid hit a ball before you hear the sound."

"I think we've got that one. Now let's go on," says Thomas.

After all the teams have studied their worksheets most of the period, Ms. Logan has them put their worksheets away and take a quiz on sound. On the quiz the students are not allowed to help each other. The next day, Ms. Logan announces the team scores and gives fancy certificates to teams that met a high standard of excellence.[1]

The most important goal of cooperative learning is to provide students with the-

---

1 Adapted from Slavin, 1994.

knowledge, concepts, skills, and understandings they need to become happy and contributing members of our society. From early on, research on cooperative learning has shown how these strategies can enhance student achievement. However, this research also has identified many of the reasons that cooperative learning enhances achievement and, most importantly, the elements of cooperative learning that must be in place if it is to have a maximum effect on achievement.

Think back to Ms. Logan's class. Which of the strategies she is using contribute most to student learning? Is it the opportunity she gave students to share their thinking processes as they discovered the principles of sound? Is it the more structured interaction within the teams when they worked to master the worksheets? Is it the intrinsic motivation she aroused by making students curious about sound? Is it the team recognition she provided based on the learning of all team members? Could Ms. Logan have gotten the same results by just telling students to work together, without any tests or team recognition? Do the answers to these questions depend on the types of objectives being pursued, the assessments being used, or the age or achievement level of the students? These and many related questions have been studied in research on cooperative learning presented in this chapter.

What Makes Group Work Work?

Why should students who work in cooperative groups learn more than those in traditionally organized classes? Researchers investigating this question have suggested a wide range of theoretical models to explain the superiority of cooperative learning (see Slavin, 1992, 1993). The theories fall into two major categories, *motivational* and *cognitive.*

Motivational Theories

Motivational perspectives on cooperative learning focus primarily on the reward or goal structures under which students operate (see Slavin,1993). Deutsch (1949) identified three goal structures: cooperative, where each individual's goal-oriented efforts contribute to others' goal attainment; competitive, where each individual's goal-oriented efforts frustrate others' goal attainment; and individualistic, where individuals' goal-oriented efforts have no consequences for others' goal attainment. From a motivational perspective (such as those of Johnson et al., 1981, and Slavin, 1983a), cooperative goal structures create a situation in which the only way group members can attain their own personal goals is if the group is successful. Therefore, to meet their personal goals, group members must help their groupmates to do whatever helps the group to succeed, and, perhaps more important, encourage their groupmates to exert maximum effort. In other words, rewarding groups based on group performance (or the sum of individual performances) creates an interpersonal reward structure in which group members will give or withhold social reinforcers (such as praise and encouragement) in response to groupmates' task-related efforts (see Slavin, 1983a).

The critique of traditional classroom organization made by motivational theorists is that the competitive grading and informal reward system of the classroom create peer norms that oppose academic efforts (see Coleman,1961). Since one student's success decreases the chances that others will succeed, students are likely to express norms that high achievement is for "nerds" or teachers' pets. Such work-restriction norms are familiar in industry, where the "rate buster" is scorned by fellow workers (Vroom,

1969). But when students work together toward a common goal, as they do when a cooperative reward structure is in place, their learning efforts help their groupmates succeed. Students therefore encourage one another's learning, reinforce one another's academic efforts, and express norms favoring academic achievement.

Several studies have found that when students work together to accomplish a group goal, they come to express norms in favor of doing whatever is necessary for the group to succeed (Deutsch, 1949; Thomas, 1957). In a cooperative classroom, a student who tries hard, attends class regularly, and helps others to learn is praised and encouraged by groupmates, much in contrast with the situation in a traditional class. For example, Hulten and DeVries (1976), Madden and Slavin (1983a), and Slavin (1978b) all found that students in cooperative learning classes felt that their classmates wanted them to learn. In cooperative groups, learning becomes an activity that gets students ahead in their peer group. Slavin (1975) and Slavin, DeVries, and Hulten (1975) found that students in cooperative groups who gained in achievement improved their social status in the classroom, whereas in traditional classes such students lost status. These changes in the social consequences of academic success can be very important. Coleman (1961) found that bright students in high schools in which academic achievement helped a student to be accepted by the "leading crowd" turned their efforts more toward learning than bright students in schools in which athletic or social achievement was considered more important. Brookover, Beady, Flood, Schweitzer, and Wisenbaker (1979) found that students' support for academic goals was the most important predictor of their achievement (controlling for ability and social class).

Clearly, cooperative goals create proacademic norms among students, and proacademic norms have important effects on student achievement.

Cognitive Theories

Whereas motivational theories of cooperative learning emphasize the degree to which cooperative goals change students' incentives to do academic work, cognitive theories emphasize the effects of working together in itself (whether or not the groups are trying to achieve a group goal). There are several different cognitive theories, which fall into two major categories: developmental theories and cognitive elaboration theories.

Developmental Theories

The fundamental assumption of the developmental theories is that interaction among children around appropriate tasks increases their mastery of critical concepts (Damon, 1984; Murray, 1982). Vygotsky (1978, p. 86) defines the zone of proximal development as "the distance between the actual developmental level as determined by independent problem solving and the level of potential development as determined through problem solving under adult guidance *or in collaboration with more capable peers*" (emphasis added). In his view, collaborative activity among children promotes growth because children of similar ages are likely to be operating within one another's proximal zones of development, modeling in the collaborating group behaviors more advanced than those they could perform as individuals. Vygotsky (1978, p. 47) describes the influence of collaborative activity on learning as follows: "Functions are first formed in the collective in the form of relations among children and then become mental functions for the individual. . . Research shows that reflection is spawned from argument."

Similarly, Piaget (1926) held that social-arbitrary knowledge – language, values, rules, morality, and symbol systems (such as reading and math) – can be learned only in interactions with others. Research in the Piagetian tradition has focused on *conser-*

vation, the ability to recognize that certain characteristics of objects remain the same when others change. For example, a child who has not yet learned the conservation principle will watch an experimenter pour a liquid from a wide jar into a tall, narrow one and say that the tall jar contains more liquid, or will believe that a ball of clay has a different weight when it is flattened. Most children attain the principal of conservation between the ages of five and seven.

There is a great deal of support for the idea that peer interaction can help non-conservers become conservers. Many studies have shown that when conservers and nonconservers of about the same age work collaboratively on tasks requiring conservation, the nonconservers generally develop and maintain conservation concepts (Bell, Grossen, and Perret-Clermont, 1985; Murray, 1982; Perret-Clermont, 1980). In fact, a few studies (for example, Ames and Murray, 1982; Mugny and Doise, 1978) have found that *both* members of pairs of disagreeing nonconservers who had to come to consensus on conservation problems gained in conservation. The importance of peers operating in one another's proximal zones of development was demonstrated by Kuhn (1972), who found that a small difference in cognitive level between a child and a social model was more conducive to cognitive growth than a larger difference.

On the basis of these and other findings, many Piagetians (such as Damon, 1984; Murray, 1982; Wadsworth, 1984) have called for an increased use of cooperative activities in schools. They argue that interaction among students on learning tasks will lead in *itself* to improved student achievement. Students will learn from one another because in their discussions of the content, cognitive conflicts will arise, inadequate reasoning will be exposed, and higher-quality understandings will emerge.

Cognitive Elaboration Theories
What we might call the cognitive elaboration perspective is quite different from the developmental viewpoint. Research in cognitive psychology has found that if information is to be retained in memory and related to information already in memory, the learner must engage in some sort of cognitive restructuring, or elaboration, of the material (Wittrock, 1978). For example, writing a summary or outline of a lecture is a better study aid than simply taking notes, because the summary or outline requires the student to reorganize the material and sort out what is important in it (Brown, Bransford, Ferrara, and Campione, 1983; Hidi and Anderson, 1986).

One of the most effective means of elaboration is explaining the material to someone else. Research on peer tutoring has long found achievement benefits for the tutor as well as the tutee (Devin-Sheehan, Feldman, and Allen, 1976). More recently, Donald Dansereau and his colleagues have found in a series of studies that college students working on structured "cooperative scripts" can learn technical material or procedures far better than can students working alone (Dansereau,1985). In this method, students take roles as recaller and listener. They read a section of text, and then the recaller summarizes the information while the listener corrects any errors, fills in any omitted material, and thinks of ways both students can remember the main ideas. On the next section, the students switch roles. Dansereau (1988) has found that while both recaller and listener learned more than did students working alone, the recaller learned more. This mirrors both the peer-tutoring findings and the findings of Noreen Webb (1985), who discovered that the students who gained the most from cooperative activities were those who provided elaborated explanations to others. In this research as well as in Dansereau's, students who received elaborated explanations learned-more than those who worked alone, but not as much as those who served as explainers.

Pitfalls of Cooperative Learning

While both the motivational and the cognitive theories support the achievement benefits of cooperative learning, there is one important pitfall that must be avoided if cooperative learning is to be instructionally effective. If not properly constructed, cooperative learning methods can allow for the "free rider" effect, in which some group members do all or most of the work (and learning) while others go along for the ride. The free-rider effect is most likely to occur when the group has a single task, as when they are asked to hand in a single report, complete a single worksheet, or produce one project. Such assignments can also create a situation in which students who are perceived to be less skillful are ignored by other group members. For example, if a group's assignment is to solve a complex math problem, the ideas or contributions of students believed to be poor at math could be ignored or brushed off, and there is little incentive for the more active participants in the problem-solving activity to take time to explain what they are doing to the less active group members.

This problem, which we might call "diffusion of responsibility" (Slavin, 1983a), can be detrimental to the achievement effects of cooperative learning. Diffusion of responsibility can be eliminated in cooperative learning in two principal ways. One is to make each group member responsible for a unique part of the group's task, as in Jigsaw, Group Investigation, and related methods. The danger of task specialization, however, is that students may learn a great deal about the portion of the task they worked on themselves but not about the rest of the content.

The second means of eliminating diffusion of responsibility is to have students be individually accountable for their learning. For example in the Student Team Learning methods (Slavin, 1986a), groups are rewarded based on the sum of their members' individual quiz scores or other individual performances. In this way, the group's task is to make sure that everyone has learned the content. No one can be a free rider, and it would be foolish for a group to ignore any of its members.

Classroom Research on the Achievement Effects of Cooperative Learning

There is a strong theoretical basis for predicting that cooperative learning methods that use group goals and individual accountability will increase student achievement. Even so, it is critical that cooperative methods be assessed in actual classrooms over realistic time periods to determine if they have an impact on measures of school achievement. Fortunately, cooperative learning is one of the most extensively evaluated of all instructional innovations. The remainder of this chapter reviews the research on practical applications of cooperative learning methods in elementary and secondary schools.

Review Methods

This research review uses an abbreviated form of best-evidence synthesis (Slavin, 1986b). The literature-search procedures, statistical methods, and study-inclusion criteria are essentially identical to those used in earlier reviews of research on group-based mastery learning (Slavin, 1987a), ability grouping (Slavin, 1987b, 1990a), and cooperative learning (Slavin, 1990b). The study-inclusion criteria were slightly adapted to the characteristics of the cooperative learning literature. They were as follows.

Germaneness Criteria

To be included in this review, studies had to evaluate forms of cooperative learning in which small groups of elementary or secondary students worked together to learn. Studies of cross-age peer tutoring, in which an older student teaches a younger one, were excluded.

Methodological Criteria

1. Studies had to compare cooperative learning with control groups studying the same material. This excluded a few studies that used time-series designs (for instance, Lew, Mesch, Johnson, and Johnson, 1986; Delquadri et al., 1986; Fantuzzo, Polite, & Grayson, 1990), and a few comparisons within studies in which control groups were not studying the same materials as experimental groups (for example, Vedder, 1985). In a few studies (such as Johnson, Johnson, and Scott, 1978), cooperative learning students could help one another on the test used as the outcome measure while individualistic or competitive students could not. Comparisons involving these "congruent tests" or "daily achievement" were excluded (see Slavin, 1984a, for more on this). Studies comparing alternate forms of cooperative learning (but not control groups) are reviewed in a separate section.
2. Evidence had to be given that experimental and control groups were initially equivalent. Studies had to either use random assignment of students to conditions or present evidence that classes were initially within 50 percent of a standard deviation of one another and used statistical controls for pretest differences. This excluded a few studies with large pretest differences (Ziegler, 1981; Okebukola, 1986c; Oishi, Slavin, and Madden, 1983) and studies that failed to present evidence of initial equality (e.g., Stokes, 1990).
3. Study duration had to be at least four weeks (twenty hours). This caused by far the largest number of exclusions. For example, of seventeen achievement studies that used control groups at the elementary or secondary level cited by Johnson and Johnson (1985) in a review of their own work, only one (just barely) met the four-week requirement. The median duration of all seventeen studies was ten days. Such brief studies are useful for theory building, but are too short to serve as evidence of the likely achievement effects of cooperative learning as a principal mode of classroom instruction. Studies of such limited duration are often also quite artificial.
4. Achievement measures had to assess objectives taught in experimental as well as control classes. If experimental and control classes were not studying precisely the same materials, then standardized or other broad-based tests had to be used to assess objectives pursued by all classes.

Computation of Effect Sizes and Medians

This research review uses effect size as the measure of the impact of cooperative learning on student achievement. An *effect size* is the proportion of a standard deviation by which an experimental group exceeds a control group (see Glass, McGaw, and Smith, 1981). An effect size of +1.0, for example, would be the equivalent of an experimental-control difference of 100 points on the SAT Verbal or Quantitative scales, two stanines, or fifteen IQ points. With the exception of one-to-one tutoring, few educational interventions produce effect sizes this large; effect sizes of +.20 or +.25 are generally considered educationally significant. Effect sizes provide a convenient measure of program impact, but for any given study they should be interpreted cautiously, for they can be greatly influenced by study characteristics.

Research on the achievement effects of cooperative learning in comparison to control groups is summarized in Tables 2-1 to 2-9. A total of ninety studies met the inclusion requirements. Since studies that compared multiple cooperative learning methods to control groups are listed more than once, the tables present ninety-nine separate comparisons of cooperative learning and control methods. Effect sizes could be computed for seventy-seven of these comparisons. Outcomes in the remaining twenty-two are characterized in the tables as significantly favoring cooperative learning (+), no significant difference (0), or significantly favoring control groups (–).

Overall, the effects of cooperative learning on achievement are clearly positive. Sixty-three (64%) of the ninety-nine experimental-control comparisons significantly favored cooperative learning. Only five (5%) significantly favored control groups. However, it is clear from looking across the tables that achievement effects vary widely, depending on methods and many other factors.

Table 2-10 summarizes the outcomes of cooperative learning according to several criteria. The first column lists the median effect sizes for the seventy-seven studies that included enough data to enable computation of effect sizes (e.g., means and standard deviations, or exact t, F, or p values). The second column lists effect sizes for standardized tests for the Student Team Learning methods (no other method had more than one study that used standardized measures). The third section shows the percentage of studies whose outcomes significantly favored cooperative learning, showed no differences, or significantly favored control groups. If studies used multiple achievement measures, outcomes were considered significantly positive if at least half of the measures showed significantly positive effects and there were no negative effects. This column includes all studies that met the inclusion criteria. The top part of Table 2-10 summarizes outcomes by method, and the lower part shows the outcomes according to their use of group goals and individual accountability.

STAD

 More than half of all experimental-control comparisons involved the Student Team Learning methods, and more than half of these evaluated STAD (or combinations of STAD and TGT, which are included in Table 2-1). The median effect size for all STAD studies is +.32 on all tests and +.21 on standardized measures. Studies of STAD have implemented this model in language arts, math, spelling, social studies, science, and other subjects; they have taken place throughout the United States and in Israel and Nigeria, in inner-city, suburban, and rural schools; they have involved students in grades 3-11. Effects of STAD have been consistently positive in all subjects, with the surprising exception of spelling, where three studies failed to find experimental-control differences. They have been equally positive with younger and older students, and with students in different types of schools. Ten of the twenty-nine STAD evaluations were done by the developers at Johns Hopkins University, and it is interesting to note that effect sizes obtained by the developers (median ES = +.20) were substantially lower than for studies done by other researchers (median ES = +.52). This may be due in part to more extensive use of standardized tests in the Hopkins research; standardized measures have characteristically lower effect sizes than are found on tests more directly adapted to the content taught in experimental and control groups. In addition, the Hopkins studies generally involved much larger numbers of students and classes than non-Hopkins studies, and large sample size is negatively related to effect sizes. The largest effect sizes in all categories are almost invariably from studies involving only one to three experimental classes. In small studies it is usually much easier to ensure that treatments are conscientiously implemented, and this may account for their larger

Table 2-1 Student Teams-Achievement Divisions

Article	Grades	Location	Sample Size	Duration (weeks)	Design	Subjects	Effect Size by Subgroup/Measure	ES Standardized	ES Total
Slavin (1980b)	4	Hagerstown, MD	424	12	STAD classes compared with matched controls.	Language Arts	Hoyum-Sanders	+.18*	+.18*
Madden and Slavin (1983b)	3,4,6	Baltimore, MD	183	7	Teachers taught 1 STAD, 1 focused-instruction class, randomly assigned.	Math	Nonhandicapped +.12* Handicapped +.18	+.13*	+.13*
Slavin and Karweit (1981)	4-5	Hagerstown, MD	486 (17 classes)	16	Cooperative learning classes used STAD in language arts, TGT in math, Jigsaw in social studies. Compared with matched controls.	Language Arts	CTBS Language Mechanics +.12* CTBS Language Expression +.12*	+.12*	+.12*
Stevens, Slavin, Farnish, and Madden (1988)	3-4	Harrisburg, PA	30 classes	4	Classes randomly assigned to STAD, control.	Reading Comprehension	Main Ideas +.24* Inferences +.11*		+.18*
Slavin (1978b)	7	Frederick, MD	205 (8 classes)	10	Teachers taught 1 STAD, 1 focused-instruction class, randomly assigned.	Language Arts		(0)	(0)
Slavin (1977c)	7	Baltimore, MD	62 (2 classes)	10	Teachers taught 1 STAD, 1 focused-instruction class, randomly assigned.	Language Arts	Hoyum-Sanders +.76* Experimenter-Made Test +.36*	+.76	+.56*
Slavin (1979)	7-8	Baltimore, MD	424	12	Teachers taught 1 STAD, 1 focused-instruction class, randomly assigned.	Language Arts	Hoyum-Sanders	(0)	(0)
Slavin and Oickle (1981)	6-8	Rural MD	230	12	Teachers taught 1 STAD, 1 focused-instruction class, randomly assigned.	Language Arts	Hoyum-Sanders: Blacks +.72* Whites +.14	+.33*	+.33*
Slavin and Karweit (1984)	9	Philadelphia, PA Low ach.	588	30	Teachers randomly assigned to STAD, mastery learning, combined, and control.	General Math	CTBS: STAD vs. control +.19* STAD + ML vs. control +.23*	+.21*	+.21*
Tomblin and Davis (1985)	4-6	San Diego, CA	509 (8 classes)	8	Classes randomly assigned to STAD, untreated control.	Spelling			+.07
Frantz (1979)	4-5	Rural VA	48	6	Classes randomly assigned to STAD control.	Reading			+.27
Kagan, Zahn, Widaman, Schwarzwald, and Tyrell (1985)	2-6	Riverside, CA	600 (25 classes)	6	Student teachers randomly assigned to STAD, TGT, untreated control.	Spelling			(0)

Table 2-1 (Continued)

Article	Grades	Location	Sample Size	Duration (weeks)	Design	Subjects	Effect Size by Subgroup/Measure	ES Standardized	ES Total
Perrault (1982):									
Study 1	7	Suburban MD	88 (4 classes)	6	Teacher taught 2 STAD, 2 control classes, randomly assigned.	Drafting	Achievement Test +.28* Drawing +.81 *		+.55*
Study 2	7	Suburban MD	48 (4 classes)	6	Teacher taught 2 STAD, 2 control classes, randomly assigned.	Drafting	Achievement Test +.53* Drawing (0)		+.26*
Sherman and Thomas (1986)	10	Ohio	38 (2 classes)	5	Classes randomly assigned to STAD, individualistic.	General Math			+1.20*
Allen and Van Sickle (1984)	9	Rural GA	51 (2 classes)	6	Teacher taught 1 STAD, 1 control class, randomly assigned.	Geography			+.94*
Peck (1991)	5-6	Muncie, IN	135 (6 classes)	7	STAD classes compared to control on spelling test, controlling for CAT scores.	Spelling			-.04
Kosters (1990)	11	Midwest	52 (2 classes)	1 semester	STAD and control classes taught by same teacher compared on history achievement test controlling for pretests, ability measure.	U.S. History			+.32
Stevens, et al. (1991)	3-4	Harrisburg, PA	319 (20 classes)	4	Teachers randomly assigned to STAD and direct instruction classes studying same materials.	Reading (Main Idea)			+.32*
Sharan et al. (1984)	7	Israel	Eng: 470 Lit: 538	16	Teachers randomly assigned to STAD, Group Investigation, or control.	English as a second Language +.14* English literature -.08		+.03	
Mevarech (1985b)	5	Israel	134	15	Students randomly assigned to STAD, ML, combined, and control.	Math	STAD vs. control +.19* STAD + ML vs. control +.28*		+.24*
Mevarech (1985b)	9	Israel	113	18	Students randomly assigned to STAD ML, combined, and control.	Consumer Math			+1.04*
Mevarech (1991)	3	Israel	117 (4 classes)	3 mos.	Matched classes using STAD, ML, combined, or control compared.	Math	STAD vs. control +.48* STAD + ML vs. control +.16*		+.62*

Table 2-1 (Continued)

Article	Grades	Location	Sample Size	Duration (weeks)	Design	Subjects	Effect Size by Subgroup/Measure	ES Standardized	ES Total
Okebukola (1985)	8	Nigeria	358	6	Student teachers randomly assigned to STAD, TGT, Jigsow, IT, competitive, or control.	Science	STAD vs. control +2.52* STAD vs. competitive +1.79*		+2.15*
Okebukola (1986a)	7	Nigeria	99	24	Students randomly assigned to STAD, LT, competitive, control, all taught by same teacher.	Science	STAD vs. control +5.14* STAD vs. competitive +2.72*		+3.93*
Kinney (1989)	9	Alexandria, VA	82 (6 classes)	1 yr	Compared students in STAD/TGT combination to students taught by same teacher in the previous year, controlling for pre-tests.	Biology	Chapter Tests +.99* Blacks +1.74* Whites +.74* Males +.66* Females +1.43* Stdized Tests +.24 Blacks +.33 Whites +.12 Males +.37 Females +.10	+.24	+.62*
Chapman (1991)	9	Rural NE Alabama (honors classes)	62	1 yr	STAD/TGT class compared to matched control class, controlling for PSAT Verbal, pretest.	English (verbal learning)	PSAT-Verbal	+.49*	+.49*
Hawkins et al. (1988)	7	Seattle (low ach.)	149 (5 schools)	1 yr	Teachers randomly assigned to STAD/TGT or control in 3 schools, 2 ad'l schools matched on CAT. Only low achieving subsample assessed.	Lang. Arts Math Social Studies	CAT Total +.15 Reading +.22 Language -.01 Math +.14	+.15	+.15
Sherman and Thomas (1986)	10	Ohio	38 (2 classes)	5	Classes randomly assigned to STAD/TGT, control.	General Math			.+.120*

Key: * Statistically Significant.
(+) Significant effect favoring cooperative learning, ES unknown.
(0) No significant difference, ES unknown.

effect sizes (within categories of methods). However, the large sample sizes contribute to the very high proportion of statistically significant positive outcomes found for all of the Student Team Learning methods. Twenty of the 29 STAD studies (69%) found significant positive effects, and none were negative. Across all five STL methods, 40 of 52 studies (77%) found significantly positive effects.

TGT

Table 2-2 summarizes research on TGT. TGT is identical to STAD except in its use of academic games instead of quizzes, and its effects are similar to those found for STAD. The median effect size across seven studies was +.38, and was +.40 in the four studies that used standardized measures.

CIRC

Six studies of CIRC are presented in Table 2-3. These studies are much larger (median = 842 students) and much longer (median duration = 1 year) than the studies in any other category, including the other Student Team Learning methods. Also, all six CIRC evaluations primarily used norm-referenced standardized tests. These factors make the CIRC studies particularly high in generalizability to large-scale application in schools.

Due to the nature of the program, CIRC studies are limited to the elementary and middle grades (2-8) and to reading, language arts, and writing. All eight studies found significantly positive effects, with a median effect size of +.29. With the exception of writing samples, all measures in five of the studies were standardized California Achievement Tests. Hertz-Lazarowitz, Ivory, and Calderón (1993) used CIRC in bilingual classes in El Paso, and assessed their performance on the Spanish Texas Assessment of Academic Skills (TAAS) in second grades and the English Norm Referenced Assessment Program for Texas (NAPT) in third and fourth grades. In two studies in Israel, Hertz-Lazarowitz, Lerner, Schaedel, Walk, & Sarid (in press) and Schaedel, Hertz-Lazarowitz, Walk, Lerner, Juberan, & Sarid (in press) evaluated responses to questions about a fable. They did two studies in schools near Haifa, Israel, one in Jewish schools and one in Arab schools.

The CIRC model is also used in grades 2 through 6 in a comprehensive school change program called Success for All (Slavin, Madden, Karweit, Dolan, & Wasik, 1992). Success for All has been evaluated in multi-year assessments in seven school districts, involving thousands of students in high-poverty schools. Effect sizes on individually administered standardized tests of reading have averaged +.65 in second grade and +.53 in third (Slavin, Madden, Dolan, Wasik, Ross, & Smith, 1994). These effects may be seen as evaluations of CIRC, but the same students also received innovative programs and, in some cases, one-to-one tutoring and family support services before second grade, so the unique contribution of the CIRC elements cannot be determined. However, in the first year-long evaluation of Success for All (Slavin, Madden, Karweit, Livermon, & Dolan, 1990), second and third graders had not received the early intenention elements of the program and still scored substantially better than controls on individually administered standardized reading tests (ES = +.62) and on group-administered California Achievement Tests (ES = +.52). These outcomes can be primarily attributed to the CIRC elements.

TAI

The outcomes of six studies of TAI are summarized in Table 2-4. Every one of the studies found statistically significant positive effects of TAI on Comprehensive Test of Basic Skills or California Achievement Test mathematics scales, with a median effect size of +.15. Effects were substantially higher for computations than for concepts and applications.

Table 2-2 Teams-Games-Tournaments

Article	Grades	Location	Sample Size	Duration (weeks)	Design	Subjects	Effect Size by Subgroup/Measure	ES Standardized	ES Total
DeVries and Mescon (1975)	3	Syracuse, NY	60 (2 classes)	6	Students randomly assigned to TGT or control classes.	Language Arts	Hoyum-Sanders +.19* Experimenter-made test +.57*	+.19*	+.38*
DeVries, Mescon, and Shackman (1975a)	3	Syracuse, NY	53 (2 classes)	5	Students randomly assigned to TGT or control classes.	Verbal Analogies	Gates-McGinitie +.60* Experimenter-made tests +.85*	+.60*	+.73*
DeVries, Mescon, and Shackman (1975b)	3	Syracuse, NY	54 (2 classes)	6	Students randomly assigned to TGT or control classes.	Language Arts	Hoyum-Sanders +.64* Experimenter-made tests +.80*	+.64*	+.72*
Edwards, DeVries, and Snyder (1972)	7	Baltimore, MD	96 (4 classes)	9	Classes randomly assigned to TGT or control all taught by same teacher.	Math	Stonford: +.50 GE* Experimenter-made tests (+)*	(+)*	(+)*
Edwards and DeVries (1972)	7	Baltimore, MD	117 (4 classes)	4	Students randomly assigned to TGT or control classes.	Math			(0)
Edwards and DeVries (1974)	7	Baltimore, MD	128 (4 classes)	12	Students randomly assigned to TGT or control classes.	Math (+) Social Studies (0)			(+)*
DeVries, Edwards, and Wells (1974)	10-12	Suburban FL	191 (6 classes)	12	Classes randomly assigned to TGT, control.	American History			+.29*
Hulten and DeVries (1976)	7	Suburban MD	299	10	Classes randomly assigned to TGT, control.	Math	Stanford	+.33	+.33*
DeVries, Lucasse, and Shackman (1980)	7-8	Grand Rapids, MI	1742	10	Teachers randomly assigned to TGT or individualized instruction.	Language Arts	Hoyum-Sanders (0) Experimenter-made tests (+)*	(0)	(+)*
Slavin and Karweit (1981)	4-5	Hagerstown, MD	465 (17 classes)	16	Coop learning classes used TGT in math, STAD in language arts, Jigsaw in social studies. Compared with matched controls.	Math	CTBS: Math computations .00 Math concepts -.10	-.05	-.05
Kagan, Zahn, Widaman, Schwarzwald, and Tyrell (1985)	2-6	Riverside, CA	600 (25 classes)	6	Student teachers randomly assigned to STAD, TGT, or control.	Spelling			(0)
Okebukola (1985)	8	Nigeria	359	6	Student teachers randomly assigned to TGT, STAD, Jigsaw, LT, competitive, or control.	Science	TGT vs. control +2.41* TGT vs. competitive +1.69*		+2.05*

Key: * Statistically significant.
(+) Significant effect favoring cooperative learning, ES unknown.
(0) No significant difference, ES unknown.

Table 2-3 Cooperative Integraded Reading and Composition

Article	Grades	Location	Sample Size	Duration (weeks)	Design	Subjects	Effect Size by Subgroup/Measure	ES Standardized	ES Total
Stevens, Madden, Slavin, and Farnish (1987): Study 1	3-4	Suburban MD	461	12	CIRC classes matched with control.	Reading	CAT ReadingComprehension +.19* CAT Reoding Vocabulary +.18*	+.20*	+.21*
						Language	CAT Language Expression +.24* CAT Language Mechanics +.12 *		
						Spelling Writing	CAT Spelling +.29* Writing Samples +.25		
Study 2	3-4	Suburban MD	450	24	CIRC classes matched with control.	Reading	CAT Reading Comprehension +.35* CAT Reeding Vocabulary +.12* Durrell IRI's +.60*	+.33*	+.32*
						Language	CAT Language Expression +.29* CAT Language Mechanics +.30* Writing Samples +.23*		
Stevens and Slavin (1995).	2-6	Suburban MD	1299 (63 classes)	2 yrs	Classes in 4 schools matched on reading achievement, assigned to CIRC or control, compared on CAT, controlling for prior CAT's.	Writing Reading Language	CAT Reading Vocab. +.20* Reading Comp. +.26* Lang. Mech. +.13 Lang. Exp. +.26* Index of Reading Awareness +.40* Special Ed:	+.21 *	+.25*
							CAT Reading Vocab. +.37* Reading Comp. +.32* Lang. Mech. +.28* Lang. Exp. +.36*	+.33 *	

Table 2-3 (Continued)

Article	Grades	Location	Sample Size	Duration (weeks)	Design	Subjects	Effect Size by Subgroup/Measure	ES Standardized	ES Total
Stevens and Durkin (1992a)	6	Baltimore, MD	1223 (54 classes)	1 yr	CIRC classes compared to control matched on prior CAT's.	Reading	CAT Reading Vocab. .00 Reading Comp. +.13* Special Ed: CAT Reading Vocab. +.28* Reading Comp. +.61*	+.07* +.45*	+.07*
Stevens and Durkin (1992b)	6-8	Baltimore, MD	3986	1 yr	Two CIRC schools compared with three control schools matched on CAT's.	Reading Writing	CAT Reading Vocab. +.33* Reading Comp. +.25* Lang. Mech. .00 Lang. Exp. +.38*	+.24*	+.24*
Hertz-Lazarowitz, Ivory, and Calderón (1993)	2-4	El Paso, TX (Spanish bilingual classes)	364 (24 classes)	Gr2:1yr Gr3:2yrs Gr4:3yrs	CIRC classes compared to control matched on Spanish and English pretests. Most instruction in Spanish in Grade 2, transition to English by Grade 4.	Reading Language	Spanish TAAS-Gr. 2 Reading +.43* Writing +.47* NAPT (English) Gr. 3 Reading +.59* Language +.29* NAPT (English) Gr. 4 Reading +.19 Language -.02	+.45* +.44* +.09	+.33*
Hertz-Lazarowitz, et al. (in press); Schaedel et al. (in press) Study 1 (Jewish Schools)	3-4	Israel	244	1 yr	CIRC schools compared to matched control schools.	Reading Writing			+.30*
Study 2 (Arab Schools)	4-5	Israel	199	1 yr	CIRC schools compared to matched control schools.	Reading Writing (Hebrew as a Second Language)			+.47*

* Statistically significant.

Table 2-4 Team-Assisted Individualization

Article	Grades	Location	Sample Size	Duration (weeks)	Design	Subjects	Effect Size by Subgroup/Measure	ES Standardized	ES Total
Slavin, Leavey, and Madden (1984):									
Study 1	3-5	Suburban MD	506	8	Schools randomly assigned to TAI, control.	Math	CTBS Computations	+.09*	+.09*
Study 2	4-6	Suburban MD	320	10	TAI classes compared with matched control.	Math	CTBS Computations	+.11*	+.11*
Slavin, Madden, and Leavey(1984)	3-5	Suburban MD	1371	24	TAI classes compared with matched control.	Math	CTBS Computations +.18* CTBS Concepts +.10*	+.14*	+.14*
Slavin and Karweit (1985):									
Experiment 1	4-6	Wilmington, DE	212	18	Classes randomly assigned to TAI, control.	Math	CTBS Computations +.77* CTBS Concepts +.00*	+.39*	+.39*
Experiment 2	3-5	Hagerstown, MD	220	16	Classes randomly assigned to TAI, control.	Math	CTBS Computations +.58* CTBS Concepts +.04	+.31*	+.31*
Stevens & Slavin (1993)	2-6	Suburban MD	1112 (5 sch.)	2 yrs	Cooperative schools using TAI matched with control schools.	Math	CAT Computations +.27* CAT Applications +.05	+.15*	+.15*

* Statistically significant.

Learning Together

David and Roger Johnsons' (1994) Learning Together models of cooperative learning are probably the most widely used of all cooperative methods, and have been evaluated in the largest number of studies. However, the great majority of their studies have taken place over periods of less than four weeks (many as little as three days). Such brief experiments are often quite artificial and have little generalizabiliy to classroom applications intended to last many months or years. For this reason, studies of less than four weeks' duration did not meet the inclusion requirements established in this review. Also, many of the Johnsons' studies included achievement measures (called "congruent measures" or "daily achievement") on which cooperative students were able to work together while children in individualistic and competitive conditions had to work by themselves. These measures also did not meet the inclusion criteria, as they give an obvious advantage to the students in the cooperative conditions.

Studies of the Johnsons' methods that did meet the inclusion criteria are summarized in Table 2-5. The studies are divided into two sections. The first lists studies in which groups were "praised and rewarded" based on the quality of a common group product, such as a worksheet. This method is low in individual accountability; in theory, one student could do all the work or tell others the answers. In the second category, students were given grades based on the average performance of all group members on tests or quizzes they took without groupmate help, or based on the performance of students who were randomly chosen to represent their groups. These procedures add individual accountability to the Learning Together procedures.

Studies of the Learning Together models without individual accountability produced highly variable results. One study by Johnson, Johnson, & Scott (1978) found significant differences favoring the individualistic group, while another by Johnson, Johnson, Scott, & Ramolae (1985) found no differences. A series of studies in Nigeria by Peter Okebukola (1984, 1985, 1986a,b) found some positive and some negative effects in comparison to individualistic and competitive conditions.

In contrast, studies of Learning Together models that did incorporate individual accountability were fairly consistent in showing significant positive effects. Three of four such studies found that cooperative learning with grades or stars based on the individual learning of group members produced greater learning than individualistic or control methods. Another study that narrowly missed the four-week duration requirement (Yager, Johnson, & Johnson, 1985) also found a positive effect of the Learning Together model with individual accountability on student achievement.

Jigsaw

Table 2-6 shows the outcomes of thirteen studies of Jigsaw. Eleven of these evaluated the original program (Aronson et al., 1978), and two evaluated Jigsaw II (Slavin, 1986). The findings of both sets of studies are highly variable. Two studies (Moskowitz et al., 1985; Tomblin & Davis, 1985) found significantly higher achievement in control groups than in Jigsaw. On the other hand, studies in Israel by Reuven Lazarowitz (1990; Lazarowitz & Karsenty,1990) in high school biology classes showed significantly positive effects of Jigsaw, as did a study in Nigeria (Okebukola, 1985). The Lazarowitz studies made important modifications in Jigsaw, including having expert groups working together on discovery-oriented science activities that group members then brought back to teach to their home groups. Of two studies of Jigsaw II, one (Phelps, 1990) found no significant differences in achievement, while the other (Mattingly & Van Sickle, 1991) found substantial positive effects of this method on geography achievement.

Table 2-5 Learning Together (Johnsons' Methods)

Learning Together Models Lacking Individual Accountability

Article	Grades	Location	Sample Size	Duration (weeks)	Design	Subjects	Effect Size by Subgroup/Measure	ES Standardized	ES Total
Johnson, Johnson, and Scott (1978)	5-6	Minnesota, High ach.	30	10	Students randomly assigned to LT or individual instruction.	Math	Posttests -.71* Retention -.81* (2 months)		-.71*
Johnson, Johnson, Scott, and Ramolae (1985)	5-6	Suburban MN	154	4	Students randomly assigned to LT or individualistic. No teacher instruction.	Science			.00
Robertson (1982)	2-3	Suburban NJ	166	6	Teachers randomly assigned to LT, individualistic/competitive, control.	Math	LT vs. individualistic/competitive -.02 LT vs. control +.16		+.07
Martinez (1990)	3	Southern California (bilingual, Mexican-American)	60	1 yr	Students randomly assigned to LT, control classes, compared on CAT. Bilingual classes taught primarily in English.	Reading Language Spelling Math		+.04 +.37 -.23 +.17	+.12
Okebukola (1985)	8	Nigeria	356	6	Student teachers randomly assigned to LT, STAD, TGT, Jigsaw, competitive, control.	Science	LT vs. control +.95* LT vs. competitive +.49*		+.72*
Okebukola (1986a)	7	Nigeria	97	24	Students randomly assigned to LT, STAD, competitive, or control, all taught by same teacher.	Science	LT vs. control +3.43* LT vs. competitive +1.20*		+2.32*
Okebukola (1984)	9	Nigeria	720	11	Student teachers randomly assigned to LT, competitive, control.	Biology	LT vs. individualistic +.27 LT vs. competitive -.89*		-.31*
Okebukola (1986b)	9	Nigeria	493	6	Student teachers randomly assigned to LT or competitive.	Biology	Prefer cooperative +1.85* Prefer competitive -1.97*		-.06

Table 2-5 (Continued)

Learning Together Models with Individual Accountability

Article	Grades	Location	Sample Size	Duration (weeks)	Design	Subjects	Effect Size by Subgroup/Measure	ES Standardized	ES Total
Humphreys, Johnson, and Johnson (1982)	9	Suburban MN	44	6	Compared LT, competitive, individualistic, controlling for pretests. Grades in LT based on group averages. No teacher instruction.	Physical Science	Posttest (+)* Retention (+)* (1 week)		(+)*
Yager, Johnson, Johnson, and Snider (1986)	3	Suburban MN	88	5	Students randomly assigned to LT or individualistic. Grades in LT based on group averages. No teacher instruction.	Transportation	Posttest (+)* Retention (+)*		(+)*
Kambiss (1990)	4	Urban Middle-Class	51 (2 classes)	12	LT class compared to matched control class. Groups received stars if randomly chosen member could answer questions correctly.	Spelling			(+)*
Meadows (1988)	7	Suburban Washington State	138 (5 classes)	6	LT compared to matched individual and teacher-directed controls. Groups received points based on behavior, randomly chosen worksheets, games.	Geography	Non-LD (0) LD (O)		(0)

Key: * Statistically significant.
(+) Significant effect favoring cooperative learning, ES unknown.
(0) No significant difference, ES unknown.

Table 2-6 Jigsaw

Article	Grades	Location	Sample Size	Duration (weeks)	Design	Subjects	Effect Size by Subgroup/Measure	ES Standardized	ES Total
Original Jigsaw									
Gonzales (1981)	3-4	Hollister, CA (bilingual)	99	20	Bilingual Jigsaw classes compared with matched controls.	Reading Language Arts Math	CAT Reading CAT Language	(0) (0)	(0)
Moskowitz, Malvin, Schaeffer, and Schaps (1983)	5-6	Suburban CA	261	24	Schools randomly assigned to Jigsaw, control.	Reading Math	CAT Math	(0) (0) (0)	(0)
Moskowitz, Malvin, Schaeffer, and Schaps (1985)	5	Suburban CA	480	30	Jigsaw classes compared with matched controls.	Reading Math	Stanford Reading Stanford Math	(0) (-)*	(-)
Tomblin and Davis (1985)	Jr. High	San Diego, CA	90	8	Classes randomly assigned to Jigsaw, control.	English			-.51*
Lazarowitz, Baird, Hertz-Lazarowitz, and Jenkins (1985)	10-12	Suburban UT	113 (4 classes)	6	Jigsaw classes compared with matched individualized instruction.	Biology			(0)
Lazarowitz, Hertz-Lazarowitz, and Jenkins, in press.	11-12	Utah	120 (5 classes)	5	Classes randomly assigned to Jigsaw with mastery testing or individual mastery learning.	Earth Science	Achievement Test +.18 Writing Sample -.15		+.02
Hertz-Lazarowitz, Sapir, and Sharan	8	Israel	68	5	Same teacher taught matched Jigsaw, GI control classes.	Arabic language and culture			+.22
Rich, Amir, and Slavin (1986)	7	Israel	339 (9 classes)	12	Classes randomly assigned to Jigsaw, control.	Literature -.32* History +.04			-.14*

Table 2-6 (Continued)

Article	Grades	Location	Sample Size	Duration (weeks)	Design	Subjects	Effect Size by Subgroup/Measure	ES Standardized	ES Total
Original Jigsaw (cont.)									
Okebukola (1985)	8	Nigeria	359	6	Student teachers randomly assigned to Jigsaw, STAD, TGT, LT, competitive, control.	Science	Jigsaw vs. control +1.41 * Jigsaw vs. competitive +.87*		+1.14*
Lazarowitz (1991)	9	Israel	201 (6 classes)	6	Jigsaw classes compared with matched controls.	Biology (Mitosis Meiosis)			(+)*
Lazarowitz & Karsenty (1990)	10	Israel	708 (10 schools)	12	Jigsaw classes compared with matched controls.	Biology (Photosynthesis)			+.51*
Jigsaw II Phelps (1990)	8	Rural Indiana	107 (4 classes)	9	Jigsaw II classes compared with matched controls.	U.S. History			-.32
Mattingly & Van Sickle (1991)	9	Germany (U.S. Military Dependents)	45 (2 classes)	9	Jigsaw II classes compared with matched controls.	Geography			+.80*

Key: * Statistically significant.
(+) Significant effect favoring cooperative learning, ES unknown.
(0) No significant difference. ES unknown.
(-) Significant effect favoring control group, ES unknown.

Group Investigation
Evaluations of Group Investigation, summarized in Table 2-7, mostly show small effects on achievement measures. However, one remarkable study, by Sharan & Shachar (1988), found very large positive effects. In this study teachers used the Group Investigation methods for several months before the study began, so this may be the one study in which the teachers were fully competent to implement this complex method from the beginning of the study.

Structured Dyads
Studies of four types of cooperative methods involving highly structured interaction within dyads are presented in Table 2-8. Both Reciprocal Peer Tutoring (RPT) and Classwide Peer Tutoring (CWPT) are structured dyadic methods that incorporate group goals and individual accountability; students earn points based on quiz scores that they can use to purchase special classroom privileges. Three studies compared these methods to control groups and found substantially higher achievement. One remarkable study of CWPT followed children in six schools from first through fourth grades. Students who participated in CWPT throughout that period scored significantly better on Metropolitan Achievement Tests in reading, language, and math than did control students (Greenwood, Delquadri, & Hall, 1989). A two-year follow-up found that the experimental students maintained most of their gains. In addition to these experimental-control studies, several studies using reversal and multiple-baseline designs have also supported the effectiveness of the RPT and CWPT strategies (e.g., Wolfe, Fantuzzo, & Wolfe, 1986; Maheady, Mallette, Harper, & Sacca, 1991).

Dansereau (1988) and his colleagues have conducted dozens of brief laboratory studies involving college students which demonstrate strong positive effects of a scripted form of dyadic learning on student achievement. This method does not use any group goals or individual accountability, but does exactly prescribe how dyads are to work together. Only one study, by Berg (1993), has implemented this method long enough to meet the duration criteria required in this review. This study found that students in a scripted pair learning program achieved significantly better than similar students the previous year.

Finally, Van Oudenhoven and his colleagues in the Netherlands conducted two studies in which student pairs were assigned to study spelling lists in a structured way. This procedure was significantly more effective than control treatments whether or not group goals (shared feedback) were used.

Other Cooperative Methods
The final category of cooperative learning studies, presented in Table 2-9, consists of methods that did not fit into any other category, but there are also important commonalties among them. Two studies by Lizbeth Johnson (1985; Johnson & Waxman, 1985) evaluated Marilyn Burns' (1981) Groups of Four method for teaching mathematics. These studies found some improvement in problem solving offset by losses in other aspects of mathematics. Groups of Four involves students in four-member learning groups that work together to solve common problems; that is, there is no group goal and no individual accountability.

Solomon et al. (1988) evaluated a form of cooperative learning similar to Groups of Four over a four-year period. This program, called the Child Development Project (CDP), strongly discourages teachers from using any form of group rewards. Students work together on projects and other common activities. Comparisons of CDP with matched control schools found several affective benefits of the program but no achievement differences in any area.

Two Dutch studies did find positive effects of cooperative programs lacking group goals and individual accountability. Lamberigts & Diepenbroek (1992) found positive

Table 2-7 Group Investigation and Related Methods

Article	Grades	Location	Sample Size	Duration (weeks)	Design	Subjects	Effect Size by Subgroup/Measure	ES Standardized	ES Total
Sharan et al. (1984)	7	Israel	English: 504 Literature: 465	18	Teachers randomly assigned to GI, STAD, or control.	English as a second language Literature	+.10* +.14*		+.12*
Hertz-Lazarowitz, Sapir, and Sharan (1981)	8	Israel	67	5	Same teachers taught matched Jigsaw, GI, control classes.	Arabic language and culture			.00
Sharan and Shachar (1988)	8	Israel	351 (11 classes)	18	Classes randomly assigned to GI, control.	Geography +1.41* History +1.45*			+1.43*
Sherman and Zimmerman (1986)	10	Ohio	46 (2 classes)	7	Compared GI and matched competitive class taught by same teacher	Biology			-.15
Talmage, Pascarella, and Ford (1984)	2-6	Elgin, IL	493	3 yrs.	Teachers using a form of GI compared with matched controls.	Reading	SAT Reading +.18* SAT Language Arts +.11	+.14	+.14*
Sherman (1988)	High school	Rural Midwest	46 (2 classes)	7	GI class compared to matched control class in which students did same activities as in GI individually, controlling for pretests.	Biology (Biomes)			-.16

* Statistically significant.

Table 2-8 Structured Dyadic Methods

Article	Grades	Location	Sample Size	Duration (weeks)	Design	Subjects	Effect Size by Subgroup/Measure	ES Standardized	ES Total
Reciprocal Peer Tutoring									
Fantuzzo, King, & Heller (1992)	4-5	Philadelphia (at-risk)	64	20	Students randomly assigned to four classes: structure + Reward (RPT), structure, reward, and control.	Math	Reward + Structure +1.42* / Reward Only +.21 / Structure Only -.26		+1.42*
Heller & Fantuzzo (in press)	4-5	Philadelphia (at-risk)	84	8 mo.	Students randomly assigned to three classes: RPT, RPT + parent involvement, and control.	Math	Curric.-based Test RPT +1.31* / RPT-P +2.08* / SDMT / RPT +1.08* / RPT-P +1.40*	+1.08*	+1.20*
Classwide Peer Tutoring									
Greenwood, Del-quadri, & Hall (1989)	1-4	Kansas City, KS (Chapter 1 Schools)	123 (6 schools)	4 yrs.	Schools randomly assigned to CWPT or control condition. First grade cohorts then remained in program through grade 4. Compared on MAT in grade 4, controlling for grade 1 MAT.	Reading / Language / Math / Followup / Reading / Language / Math / Science / Social Studies	MAT / MAT / MAT / Grade 6 / CTBS / CTBS / CTBS / CTBS / CTBS → +.57* / +.37* / +.60* / / +.38* / +.31* / +.52* / +.42* / +.38*		+.51*
Scripted Pair Learning									
Berg (1993)	11	Honolulu (Lab School)	56 (2 classes)	8	Students in scripted pair learning class compared to matched students in some class previous year.	Math	During treatment +.25* / Followup +.73* (8 weeks)		+.25*

Table 2-8 (Continued)

Article	Grades	Location	Sample Size	Duration (weeks)	Design	Subjects	Effect Size by Subgroup/Measure	ES Standardized	ES Total
Paired Spelling									
Van Oudenhoven, Van Berkum and Swen-Koopmans (1987)	3	Netherlands	218 (14 classes)	12	Classes randomly assigned to pair learning with individual feedback, pair learning with shared feedback, or control.	Spelling	Pairs with individual feedback: (+) Pairs with shared feedback: (+)		(+)*
Van Oudenhoven, Wiersma, and Van Yperen (1987)	3	Netherlands	261	15	Classes randomly assigned to pair learning, individual learning with feedback from classmate, or individual learning with no feedback.	Spelling	High Average Achievement (0) Low Achievement (+)		(+)*

Key: * Statistically significant.
(+) Significant effect favoring cooperative learning, ES unknown.
(0) No significant difference, ES unknown.

Table 9-2 Other Cooperative Learning Methods

Methods Lacking Group Goals and Individual Accountability

Article	Grades	Location	Sample Size	Duration (weeks)	Design	Subjects	Effect Size by Subgroup/Measure	ES Standardized	ES Total
L. Johnson (1985)	4-5	Suburban Houston, TX	859 (51 classes)	27	Teachers trained in Marilyn Burns "Groups of Four" method compared to matched controls.	Math	Romberg-Wearne: Comprehension -.08 Application +.01 Problem Solving +.22	+.04	+.04
L. Johnson and Waxman (1985)	8	Houston, TX	150	1 yr	Teachers trained in Marilyn Burns "Groups of Four" method compared to matched controls.	Math	SAT Problem Solving: High Achievement (0) Moderate Achievement (0) Low Achievement (+)	(0)	(0)
Vedder (1985)	4	Netherlands	191	4	Classes randomly assigned to pair learning or individual learning.	Geometry	Posttests +.06 Follow-up +.05 (3 weeks)		+.06
Solomon et al. (1988)	K-4	Suburban California	350 (6 schools)	5 yrs.	Compared Child Development Project schools to matched control schools.	Reading Language Spelling Science Social Studies	CAT CAT CAT CAT CAT	(0) (0) (0) (0) (0)	(0)
Lamberigts & Diepenbroek (1992)	High school	Netherlands	27 (2 classes)	4	Compared cooperative learning method emphasizing learning strategies to direct instruction, controlling for ability, motivation.	Reading Comprehension Spelling			(+)*

Table 2-9 (Continued)

Article	Grades	Location	Sample Size	Duration (weeks)	Design	Subjects	Effect Size by Subgroup/Measure	ES Standardized	ES Total
Methods Lacking Group Goals and Individual Accountability (cont.)									
Terwel et al. (in press) school	Middle	Netherlands (23 classes)	600		Compared students in Adaptive Learning and Cooperative Learning (AGO) model to students in matched schools, controlling for pretests.	Math			+.68*
Methods with Group Goals but No Individual Accountability									
Arzt (1983)	9-11	Suburban New York	304	20	Teachers taught experimental and control classes. Cooperative groups in competition based on one group worksheet.	Math			+.13

Key: * Statistically significant
(+) Significant effect favoring cooperative learning
(0) No significant difference

effects of a method emphasizing metacognitive learning strategies as well as cooperative learning. Terwel et al. (in press) investigated a mathematics teaching method called Adaptive Learning and Cooperative Learning (AGO in Dutch). Strong positive effects of AGO were found in comparison to control groups.

Two of the most widely used cooperative learning methods have not been evaluated in studies that met the inclusion criteria established for this review. Studies of Complex Instruction have established that classes in which students talked and worked together more also gained more in mean math achievement than Complex Instruction classes in which students talked and worked together less frequently (Cohen, Lotan, & Leechor, 1989). However, this correlational study cannot be seen as an evaluation of the method in comparison to traditional practices. Spencer Kagan's (1992) methods have been evaluated in a dissertation by Stokes (1990), but this study failed to present data indicating that experimental and control groups were equivalent before the study began.

What Factors Contribute to Achievement Effects of Cooperative Learning?

Research on the achievement effects of cooperative learning has focused not only on documenting the effects of these methods on achievement but also on understanding the conditions under which positive effects are most likely to be seen. There are two primary ways to learn about factors that contribute to the effectiveness of cooperative learning. One is to compare the outcomes of studies of alternative methods, as in Table 2-10. For example, if programs that incorporated group rewards produced stronger or more consistent positive effects (in comparison to control groups) than programs that did not, then this would provide one kind of evidence that group rewards enhance the outcomes of cooperative learning. The problem with such comparisons is that the studies being compared usually differ in measures, durations, subjects, and many other factors that could explain differing outcomes. Better evidence is provided by studies that compared alternative forms of cooperative learning. In such studies, most factors other than the ones being studied can be held constant. The following sections discuss both types of studies to explore factors that contribute to the effectiveness of cooperative learning for increasing achievement.

Group Goals and Individual Accountability

From early on, reviewers of the cooperative learning literature have concluded that cooperative learning has its greatest effects on student learning when groups are recognized or rewarded based on the individual learning of their members (Slavin, 1983a, 1983b, 1989, 1990, 1993; Ellis & Fouts, 1993; Newmann & Thompson, 1987; Manning & Lucking, 1991; Davidson, 1985; Mergendollar & Packer, 1989). For example, methods of this type may give groups certificates based on the average of individual quiz scores of group members, where group members could not help each other on the quizzes. Alternatively, group members might be chosen at random to represent the group, and the whole group might be rewarded based on the selected member's performance. In contrast, methods lacking group goals give students only individual grades or other individual feedback, and there is no group consequence for doing well as a group. Methods lacking individual accountability might reward groups for doing well, but the basis for this reward would be a single project, worksheet, quiz, or other product that could theoretically have been done by only one group member.

The importance of group goals and individual accountability is in providing students with an incentive to help each other and to encourage each other to put forth

Table 2-10 Breakdown of Effect Sizes by Characteristics of Cooperative Methods

	Mean ES	Mean ES, Standardized Tests	Percentage of Studies:			Total Studies
			Significantly Positive	No Difference	Significantly Negative	
All Studies	+.26 (77)		64 (63)	31 (31)	5 (5)	99
Student Team Learning:						
STAD	+.32 (26)	+.21 (9)	69 (20)	31 (9)	0 (0)	29
TGT	+.38 (7)	+.40 (4)	75 (9)	25 (3)	0 (0)	12
CIRC	+.29 (8)	+.23 (6)	100 (8)	0 (0)	0 (0)	8
TAI	+.15 (6)	+.15 (6)	100 (6)	0 (0)	0 (0)	6
All STL	+.32 (47)	+.21 (25)	77 (40)	23 (12)	0 (0)	52
Learning Together	+.04 (8)		42 (5)	42 (5)	17 (2)	12
Jigsaw	+.12 (8)		31 (4)	46 (6)	23 (3)	13
Group Investigation	+.06 (6)		50 (3)	50 (3)	0 (0)	6
Structured Dyads	+.86 (4)		100 (6)	0 (0)	0 (0)	6
Other	+.10 (4)		29 (2)	71 (5)	0 (0)	7
Group Goals and Individual Accountability	+.32 (52)		78 (50)	22 (14)	0 (0)	64
Group Goals Only	+.07 (9)		22 (2)	56 (5)	22 (2)	9
Individual Accountability (Task Specialization)	+.07 (12)		35 (6)	47 (8)	18 (3)	17
No Group Goals or Individual Accountability	+.16 (4)		56 (5)	44 (4)	0 (0)	9

Note: Numbers in parentheses are total numbers of studies in each category.

maximum effort (Slavin, 1993). If students value doing well as a group, and the group can succeed only by ensuring that all group members have learned the material, then group members will be motivated to teach each other. Studies of behaviors within groups that relate most to achievement gains consistently show that students who give each other elaborated explanations (and, less consistently, those who receive such explanations) are the students who learn the most in cooperative learning. Giving or receiving answers without explanation generally reduces achievement (Webb, 1989, 1992). At least in theory, group goals and individual accountability should motivate students to engage in the behaviors that increase achievement and avoid those that reduce it. If a group member wants her group to be successful, she must teach her groupmates (and learn the material herself). If she simply tells her groupmates the answers, they will fail the quiz that they must take individually. If she ignores a group-mate who is not understanding the material, the groupmate will fail and the group will fail as well. In groups lacking individual accountability, one or two students may do the group's work, while others engage in "social loafing" (Latane, Williams, & Harkins, 1979). For example, in a group asked to complete a single project or solve a single problem, some students may actively participate while others watch. Worse, students felt to have little to contribute to the group or who are lower in status or less aggressive may be discouraged from participating. A group trying to complete a com-mon project or solve a common problem may not want to stop and explain what is going on to a groupmate who doesn't understand, or may feel it is useless or counterproductive to try to involve certain groupmates.

The evidence from research on cooperative learning strongly supports the impor-tance of group goals that can be achieved only by ensuring the learning of all group members. Table 2-10 provides one kind of evidence to support this conclusion. Stud-ies of methods that incorporated group goals and individual accountability produced a much higher median effect size than did studies of other methods. The median effect size across fifty-two such studies was +.32, compared to a median of only +.07 across twenty-five studies that did not incorporate group goals and individual accountability. Seventy-eight percent of studies of methods using group goals and individual accountability found significantly positive effects, and there were no significantly negative effects. In methods lacking these elements only 37% of studies found significantly positive effects, and 14% found significantly negative effects.

A comparison of Learning Together studies also supports the same conclusion. Across eight studies of Learning Together methods in which students were rewarded based on a single worksheet or product, the median effect size was near zero (+.035). However, among four studies that evaluated forms of the program in which students were graded based on the average performance of all group members on individual assessments, three found significantly positive effects.

Finally, comparisons within the same studies consistently support the importance of group goals and individual accountability. For example, Fantuzzo, King, & Heller (1992) conducted a component analysis of Reciprocal Peer Tutoring (RPT). They compared four conditions in which students worked in dyads to learn math. In one, students were rewarded with opportunities to engage in special activities of their choice if the sum of the dyad's scores on daily quizzes exceeded a criterion. In another, students were taught a structured method of tutoring each other, correcting errors, and alternating tutor-tutee roles. A third condition involved a combination of rewards and structure, and a fourth was a control condition in which students worked in pairs but were given neither rewards nor structure. The results showed that the reward + structure condition had by far the largest effects on math achievement (ES = + 1.42), but rewards had much larger effects than structure. The reward + structure condition exceeded structure only by an effect size of + 1.88, and the reward-only group exceeded control by an effect size of +.21 (the structure-only group performed less well than the control group).

Other studies also found greater achievement for cooperative methods using group goals and individual accountability than for those that do not. Huber, Bogatzki, & Winter (1982) compared a form of STAD to traditional group work lacking group goals or individual accountability. The STAD group scored significantly better on a math test (ES = +.23). In a study of TAI, Cavanaugh (1984) found that students who received group recognition based on the number of units accurately completed by all group members both learned more (ES = +.24) and completed more units (ES = +.25) than did students who received individual recognition only. Okebukola (1985), studying science in Nigeria, found substantially greater achievement in STAD and TGT, methods using group goals and individual accountability, than in forms of Jigsaw and the Johnsons' methods that did not. In another study, Okebukola found much higher achievement in classes that used a method combining cooperation and group competition (one form of group reward) than in a "pure" cooperative method that did not use group rewards of any kind (ES = + 1.28).

A few reviewers (e.g., Damon, 1984; Kohn, 1986) have recommended against the use of group rewards, fearing that they may undermine long-term motivation. There is no evidence that they do so, and they certainly do not undermine long-term achievement. Among multi-year studies, methods that incorporate group rewards based on individual learning performance have consistently shown continued or enhanced achievement gains over time (Stevens & Slavin, in press; Hertz-Lazarowitz et al., 1993; Greenwood et al., 1989). In contrast, multi-year studies of methods lacking group rewards found few achievement effects in the short or long-term (Solomon et al., 1988; Talmage et al., 1984).

Cohen (1994) raises the possibility that while group rewards and individual accountability may be necessary for lower-level skills, they may not be for higher-level ones. As evidence of this she cites a study by Sharan et al. (1984) that compared STAD and Group Investigation. In this study STAD and GI students performed equally well (and better than controls) on a test of English as a foreign language, and STAD students did significantly better than GI on "lower level" (knowledge) items (ES =. 38). On "high level" items, GI students performed non-significantly higher than STAD students, with a difference of less than half of a point on a 15-point test. Otherwise, there is no evidence that group rewards are less important for higher-order skills, although the possibility is intriguing.

Structuring Group Interactions

While it is clear that, all other things being equal, group rewards and individual accountability greatly enhance the achievement outcomes of cooperative learning, there is some evidence that carefully structuring the interactions among students in cooperative groups also can be effective, even in the absence of group rewards. For example Meloth & Deering (1992) compared students working in two cooperative conditions. In one, students were taught specific reading comprehension strategies and given "think sheets" to remind them to use these strategies (e.g., prediction, summarization, character mapping). In the other group, students earned team scores if their members improved each week on quizzes. A comparison of the two groups on a reading comprehension test found greater gains for the strategy group (also see Meloth & Deering, in press). As noted earlier, Berg (1993) found positive effects of scripted dyadic methods that did not use group rewards, and Van Oudenhoven et al. (1987) found positive effects of structured pair learning whether feedback was given to the pairs or only to individuals.

Research on Reciprocal Teaching (Palincsar & Brown, 1984) also shows how direct strategy instruction can enhance the effects of a technique related to cooperative learning. In this method, the teacher works with small groups of students and models such cognitive strategies as question generation and summarization. The teacher then gradually turns over responsibility to the students to carry on these activities with each other. Studies of Reciprocal Teaching have generally found positive effects of this method on reading comprehension (Lysynchuk et al., 1990; Palincsar & Brown, 1984; Palincsar et al., 1987; Rosenshine & Meister, 1992). The effects of group rewards based on the individual learning of all group members are dearly indirect; they only motivate students to engage in certain behaviors, such as providing each other with elaborated explanation. The research by Meloth and Deering (1992), Berg (1993), and others suggests that students can be directly taught to engage in cognitive and interpersonal behaviors that lead to higher achievement, without the need for group rewards.

However, there is a growing body of evidence to suggest that a combination of group rewards and strategy training produces much better outcomes than either alone. First, the Fantuzzo et al. (1992) study, cited earlier, directly made a comparison between rewards alone, strategy alone, and a combination, and found the combination to be by far the most effective. Further, the outcomes of the RPT and CWPT dyadic learning methods, which use group rewards as well as strategy instruction; produced some of the largest positive effects of any cooperative methods, much larger than those found in the Berg (1993) study that provided groups with structure but not rewards. The consistent positive findings for CIRC, which uses both group rewards and strategy instruction, also argue for this combination.

Which Students Gain Most from Cooperative Learning?

Several studies have focused on the question of which students gain the most from cooperative learning. One particularly important question relates to whether cooperative learning is beneficial to students at all levels of prior achievement. It would be possible to argue (see, for example, Allan, 1991; Robinson, 1990) that high achievers could be held back by having to explain material to their low-achieving groupmates. However, it would be equally possible to argue that because students who give elaborated explanations typically learn more than those who receive them (Webb, 1989), high achievers should be the students who benefit most from cooperative learning because they give the most frequent elaborated explanations.

The evidence from experimental studies that met the inclusion criteria for this review support neither position. A few studies found better outcomes for high achievers than for low (e.g., Edwards & DeVries, 1972) and a few found that low achievers gained the most (e.g., Edwards et al., 1972; Johnson & Waxman, 1985; Van Oudenhoven et al., 1987). Most, however, found equal benefits for high, average, and low achievers in comparison to their counterparts in control groups (e.g., Sharan et al., 1984). One two-year study of schools using cooperative learning most of their instructional day found that high, average, and low achievers all achieved better than controls at similar achievement levels. However, a separate analysis of the very highest achievers, those in the top 10% and top 5% of their classes at pretest, found particularly large positive effects of cooperative learning on these students (Slavin, 1991; Stevens & Slavin, 1993).

A few studies have looked for possible differences in the effects of cooperative learning on students of different ethnicities. Several have found particularly large effects for black students (e.g., Slavin & Oickle, 1981; Slavin, 1977). However, other studies have found equal effects of cooperative learning for students of different ethnic backgrounds (Slavin & Karweit, 1984; Edwards et al., 1972; Slavin, Leavey, & Madden, 1984; Sharan & Shachar, 1988).

Other studies have examined a variety of factors that might interact with achievement gain in cooperative learning. Okebukola (1986b) and Wheeler & Ryan (1973) found that students who preferred cooperative learning learned more in cooperative methods than those who preferred competition. Chambers & Abrami (1991) found that students on successful teams learned more than those on less successful teams.

Finally, a small number of studies have compared variations in cooperative procedures. Moody & Gifford (1990) found that while there was no difference in achievement gains of homogeneous and heterogeneous groups, pairs produced greater science achievement than groups of four and gender-homogeneous groups performed better than mixed groups. Foyle, Lyman, Tompkins, Perne, & Foyle (1993) found that cooperative learning classes assigned daily homework achieved more than those not assigned homework. Kaminski (1991) and Rich et al. (1986) found that explicit teaching of collaborative skills had no effect on student achievement. Jones (1990) compared cooperative learning using group competition to an otherwise identical method that compared groups to a set standard (as in STAD). There were no achievement differences, but a few attitude differences favored the group competition.

Conclusion

Research on the achievement effects of cooperative learning has progressed substantially since the first edition of this book (Slavin, 1990). In the earlier edition, sixty-eight experimental-control studies qualified for inclusion; only five years later, there are ninety-nine that qualify, and many more that compared alternative cooperative approaches. The main conclusion of this review is similar to that of the first edition and of other reviewers (e.g., Davidson, 1985; Ellis & Fouts, 1993; Newmann & Thompson, 1987). Group rewards based on the individual learning of all group members are extremely important in producing positive achievement outcomes in cooperative learning. What recent research has added to this conclusion, however, is the possibility that it is possible to create conditions leading to positive achievement outcomes by directly teaching students structured methods of working with each other (especially in pairs) or teaching them learning strategies closely related to the instructional objective (especially for teaching reading comprehension skills).

The possibility that effective learning strategies could be directly taught to cooperative groups fits well within a theoretical framework described earlier (Slavin, 1989, 1993). This framework is diagrammed in Figure 2.1.

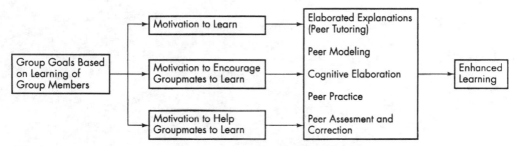

Figure 2.1 Model of factors influencing learning gains in cooperative learning.

The theory outlined in Figure 2.1 assumes that it is behaviors within cooperative groups, such as cognitive elaboration, peer tutoring, peer modeling, and mutual assessment, that lead to enhanced achievement. Group rewards based on individual learning performances are hypothesized to motivate students to engage in these behaviors, but have no direct impact on learning. Clearly, if the behaviors can be taught and maintained directly, then no group rewards are needed. However, it is likely that, especially over the long run, students need some kind of group goal based on group members' learning if they are to continue to spend significant time and effort helping each other learn, assessing each other's progress, encouraging each other's efforts, and so on. This may explain why combinations of group rewards and explicit strategy instruction have produced some of the strongest effects of cooperative learning.

The research on cooperative learning is remarkable in its breadth and its quality. There is still much more to be learned about how, why, and under what conditions cooperative learning enhances student achievement, but it is clear that under certain well-defined circumstances, cooperative learning can have consistent and important effects on the learning of all students.

Chapter 3

The Cooperative Elementary School

Background and Commentary

By 1985, our group at Johns Hopkins University had amassed considerable evidence about the effectiveness of cooperative learning methods that incorporated group goals and individual accountability, and we had developed and successfully evaluated comprehensive cooperative learning methods capable of entirely replacing traditional instruction in the 3 R's: TAI-Mathematics, and Cooperative Integrated Reading and Composition. However, up to this point individual teachers were the unit of change. In our dissemination efforts, for example, it didn't matter whether we worked with twenty teachers from one school or twenty teachers from twenty schools. Teachers could make an individual decision to use cooperative learning, and usually did so with little involvement or assistance from colleagues or administrators.

After a decade of working in this way, we began to see its limitations. We saw teachers leaving our workshops highly motivated and apparently able to use cooperative methods, but failing to actually do so in their classroom. We saw others try and fail to use cooperative methods for reasons that seemed trivial or because of problems that the teacher in the next-door classroom was solving quite effectively. We saw teachers who had used cooperative learning methods effectively, sometimes for many years, failing to continue them when they changed schools, or even when they began teaching a different grade in the same school.

Further, we saw that some of the goals of cooperative learning were difficult to achieve on a teacher-by-teacher basis. In particular, the use of cooperative learning as a means of facilitating mainstreaming or integration of academically handicapped children in regular classes was difficult when the whole school (or major sections of it) was not involved.

It occurred to us that to improve the quality and longevity of cooperative learning, we needed to make it the *school's* program, not just individual teachers' programs. In 1986, we began to experiment with what we called the cooperative elementary school, in which a variety of cooperative learning programs were phased in over a period of time, in which teachers and administrators worked together to help with the quality of implementation in all classes, and in which special education and general education teachers would work collaboratively to meet the needs of academically handicapped students in the least restrictive environment possible. The research on the cooperative elementary school, led by my colleague Robert Stevens, is presented in the article reprinted in this chapter, from the *American Educational Research Journal*.

Despite the success of the cooperative school on most measures, this idea has not caught on among educators on any significant scale. Part of the problem was that we did not disseminate it very vigorously; shortly after completing early work on the cooperative school we turned our intention to Success for All, which incorporates all aspects of the cooperative school but adds to it early childhood and first grade programs, tutors, facilitators, and family support programs (see Chapter 4). However, many schools have implemented versions of the cooperative elementary school, usually building around schoolwide adoptions of CIRC and adding other curricula and school-level changes over time. The cooperative school gave us invaluable experience in the use of peer coaching among teachers, increasing administrator involvement in innovation, and other means of mobilizing whole school staffs for change that we have subsequent put to good use in all of our dissemination programs.

In addition to demonstrating how school staffs could work together to infuse cooperative learning through most of the school day and year, the cooperative school study added two key findings to research on cooperative learning. First it utterly laid to rest the often-expressed concern that students would get tired of cooperative learning, or that the effects of cooperative learning might be due to a novelty or Hawthorne effect. These students had cooperative learning in most of their classes every day for two years or more, and there was no indication that they got tired of it or that the effects wore off over time. Second, this was the first study to analyze outcomes separately for highly able students, (those in the top 10% and top 5% of their classes). The finding that cooperative learning (in heterogeneous classes) was more successful than other approaches with these very high achieving students has become important in the broader debate about the need for separate classes for gifted students (Slavin, 1991).

This article reports the results of a 2-year study of the cooperative elementary school model which used cooperation as an overarching philosophy to change school and classroom organization and instructional processes. The components of the model include: using cooperative learning across a variety of content areas, full-scale mainstreaming of academically handicapped students, teachers using peer coaching, teachers planning cooperatively, and parent involvement in the school. After the first year of implementation, students in cooperative elementary schools had significantly higher achievement in reading vocabulary. After the second year, students had significantly higher achievement in reading vocabulary, reading comprehension, language expression, and math computation than did their peers in traditional schools. After 2 years, academically handicapped students in cooperative elementary schools had significantly higher achievement in reading vocabulary, reading comprehension, language expression, math computation, and math application in comparison with similar students in comparison schools. There also were better social relations in cooperative elementary schools, and handicapped students were more accepted socially by their nonhandicapped peers than were similar students in traditional schools with pull-out remedial programs. The results also suggest that gifted students in heterogeneous cooperative learning classes had significantly higher achievement than their peers in enrichment programs without cooperative learning.

Cooperative learning methods have been researched for more than 20 years and have been rapidly increasing in use over the past decade. However, until recently almost all applications of cooperative learning methods have been implemented teacher by teacher, usually in one subject at a time, and few studies have involved durations of more than a semester (see Slavin, 1994). Cooperative learning methods can certainly be justified as additions to teachers' instructional repertoires in traditionally organized schools, but the potential for cooperative approaches to serve as the organizing principle for the entire school and as the core instructional element has not been assessed. Could cooperative learning be used on a broad scale in many subjects and over extended time periods to fundamentally change the organization of schools and classrooms? Would cooperative learning methods still be effective if they became the primary mode of instruction in schools, and would they maintain their effectiveness over time? Would schoolwide use of cooperative principles enhance the school's potential to successfully mainstream learning disabled students? These were the principal questions addressed in the study reported here.

The goal of this research was to use cooperative learning as a way to change classrooms and learning activities. Cooperative learning was seen as a vehicle for increasing students' active involvement in learning activities and having students take more responsibility for managing their own instruction, thus allowing the teacher more opportunities to have academic interactions with individuals and small groups.

At the same time, we attempted to use cooperation as a philosophy to change the school and classroom organization and the way members of the school community interacted. Peer coaching was used as a form of cooperation where teachers would help one another improve their teaching (Joyce, Hersh, & McKibbin, 1983). Similarly, teachers and the principal were encouraged to collaborate in planning instruction and to develop interdependence within the faculty so they would take collective responsibility for the learning of all students.

Research on Cooperative Learning

In a review of the research on cooperative learning, Slavin (1994) focused on studies of at least 4 weeks' duration that compared cooperative learning and control classes on objectives pursued equally by both over periods of at least 4 weeks. The review concluded that for cooperative learning to have an effect on achievement two conditions were essential: group goals and individual accountability. Group goals motivate students to help their group mates learn. They develop positive interdependence between the individuals in the group, giving them a reason to cooperate in a meaningful fashion (Deutsch, 1949; Johnson & Johnson, 1989; Stevens, 1994). Individual accountability, a measure of each students' learning, increases the probability that all students will learn and reduces the potential for a *free rider effect*, where a student does little and relies on others in the group to accomplish the goal (Johnson & Johnson, 1989; Slavin, 1994; Stevens, 1994).

However, these two elements must be linked to make cooperative learning processes effective in increasing learning. The attainment of the group goal must depend on the individual learning of all group members. In this way, the group's success depends on successful learning by everyone and produces positive interdependence. When this kind of interdependence is developed, students perceive their role to be to help their groupmates *learn*, not to simply help them complete a single group task. The research of Webb (1985) and others has shown that achievement gains in cooperative learning depend on the giving and receiving of elaborate explanations of concepts and skills and that giving and receiving answers without explanations are negatively related to achievement gain. Explanations force students to evaluate, integrate, and elaborate on knowledge and thus improve their understanding or depth of processing of what is being learned (Brown & Campione, 1986; Wittrock, 1986). It is hypothesized that the use of group goals based on individual accountability motivates students to provide high-quality explanations to one another and to avoid simply giving answers. Reviewers of the extensive literature on cooperative learning and student achievement have generally agreed that group goals based on individual accountability measures are essential for achievement effects (Davidson, 1985; Johnson & Johnson, 1989; Newmann & Thompson, 1987; Slavin, 1994). Cooperative learning programs that incorporate these elements consistently increase student achievement more than control methods. Slavin (1995) reports that a total of 52 such studies had a median effect size on achievement measures of +.32. In contrast, the median effect size for methods lacking group goals and individual accountability was only +.07 across 25 studies.

Cooperative learning has been found to have a positive impact on many variables other than achievement, including intergroup relations, self-esteem, attitudes toward class and school, and ability to work collaboratively with others (Johnson & Johnson, 1989; Slavin, 1994). These noncognitive outcomes, however, do not seem to depend on the use of group goals based on individual accountability measures.

There has been some question about whether the effects of cooperative learning can be maintained in longer implementations and when cooperative learning is used throughout the school day rather than in one subject at a time. Critics (e.g., Bossert, 1989) have sometimes questioned whether the effects of cooperative learning may be due to novelty or Hawthorne effects and whether intensive, long-term use of cooperative learning programs might be less effective. On the other hand, it may be that, as students experience cooperative methods in many subjects and for longer periods, their increasing skills in working cooperatively and a perception that cooperation leads to success could make widespread and long-term use of cooperative learning more effective than more time-limited implementations.

Year-long studies of cooperative learning programs using group goals based on individual accountability have generally been as successful as shorter studies in enhancing student achievement (e.g., Slavin & Karweit, 1984; Slavin, Leavey, & Madden, 1984; Stevens, Madden, Slavin, & Farnish, 1987). The few multiyear studies that have been done have found effects that vary according to the methods used. In a 2-year study of Cooperative Integrated Reading and Composition (CIRC), Stevens and Slavin (1995) found that a cooperative learning reading and language arts program that used group goals based on individual accountability had significant positive effects on students' achievement in reading and language arts. Similarly, a 3-year study of CIRC in bilingual classes by Hertz-Lazarowitz, Ivory, and Calderon (1993) found positive effects on reading and language performance. Several multiyear studies of Success for All, an elementary school restructuring model that makes extensive use of cooperative learning methods, have also found substantial positive effects on students' reading performance (Slavin et al., in press).

In contrast, a multiple-year study by Talmage, Pascarella, and Ford (1984) studied a 3-year staff development emphasizing cooperative learning methods developed by Sharan and Sharan (1976) in which students were instructed to break group tasks into subtasks and were graded based on an evaluation of the group's product. This program produced small positive effects on student achievement in reading and language arts. A study by Solomon Watson, Schaps, Battistich, and Solomon (1990) implemented cooperative learning methods over a 5-year period, starting with children in kindergarten and continuing with the same cohort through the fourth grade. The study found no effects on achievement as compared to control schools, but there were positive effects on prosocial behavior, social problem solving, and adherence to democratic values. It should be noted, however, that the latter two studies used forms of cooperative learning that emphasized the production of a group product. While the models in these studies used individual accountability (e.g., grades), *group* success was not dependent on individual learning performance, and there were no formal group goals. The group products did not require each student to participate equally, and there was no way to hold each student accountable for learning during the process.

Given the quantity of research that documents the potential for cooperative learning to have a positive impact on students' achievement in a variety of subject areas, as well as the potential social and attitudinal benefits, it seems that cooperative learning may be a viable theme for changing the organization of an entire school. Therefore this study attempted to reorganize the school and classrooms in the school by using cooperative learning across the curriculum and using it as a theme to better integrate instruction between special and regular education.

The Cooperative Elementary School

The cooperative elementary school model uses cooperation as a philosophical and practical approach to changing classroom and school organization, classroom processes, and learning activities to provide students with more active learning experiences, equal access to learning by all students, and a more supportive social environment for the students and teachers (Slavin 1987). The elements were the following:

1. Widespread use of cooperative learning in academic classes.
2. Mainstreaming learning disabled students in regular education.
3. Teachers coaching one another.
4. Teachers collaborating in instructional planning.
5. Principal and teachers collaborating on school planning and decision making.
6. Principal and teachers encouraging active involvement of parents.

We describe each of the elements and the theoretical and research rationale for them below.

Widespread Use of Cooperative Learning in Academic Classes

A major emphasis of the cooperative elementary school design is the use of cooperative learning methods in many subjects and grade levels. Cooperative learning changes the classroom organization and processes to make learning more active for students and increase students' responsibility for their learning activities. Similarly, a cooperative learning classroom organization allows teachers more opportunities to meet instructionally with small groups and individual students because it frees teachers from many of their management responsibilities, which are assumed by the students and their learning groups (Johnson & Johnson, 1981; Slavin, 1994).

Teachers in cooperative elementary schools are given training, materials, follow-up, and assistance to use the Johns Hopkins cooperative learning models (Slavin, 1986). For the present study, particular emphasis was placed on the two comprehensive programs designed to accommodate student diversity: Cooperative Integrated Reading and Composition, or CIRC, (Hertz-Lazarowitz et al., 1993; Stevens et al., 1987; Stevens & Slavin, 1995) and Team Assisted Individualization-Mathematics, or TAI (Slavin et al., 1984, 1986). These programs included instructional materials, teachers' manuals, and other supports to enable teachers of Grades 2-6 to use them as their primary approaches to the teaching of reading, language arts/writing, and mathematics, respectively. Research on TAI (e.g., Slavin et al., 1984, 1986; Slavin & Karweit, 1985) and CIRC (Stevens et al., 1987; Stevens & Slavin, 1995) has found these methods to be among the most effective of all cooperative learning programs for accelerating student achievement.

Mainstreaming Learning Disabled Students in Regular Education

Another key element of the cooperative school model is the intention to mainstream learning disabled students into regular education. There has been a great deal of discussion and research on mainstreaming special education students since the passage of PL 94-142, the Education of All Handicapped Act in 1975. Yet the question of the efficacy of special education and remedial pull-out programs for mildly handicapped students is certainly not a new one (Lloyd & Gambatese, 1991). Separate special education programs have been depicted as expensive, ineffective, and poorly integrated with regular education instruction (cf. Johnston, Allington, & Afflerbach, 1985; Stainback & Stainback, 1984).

Current research on mainstreaming has been strongly influenced by advocates of educating students with handicaps in the least restrictive environment possible (Reynolds, 1989; Will, 1986). However, the research on mainstreaming to date is less than conclusive and has often raised more questions than answers (Fuchs & Fuchs, 1991; Lloyd, Crowley, Kohler, & Strain, 1988). Research has suggested that simply placing students with learning disabilities in regular classes is not sufficient to produce significantly better achievement (Marston, 1987). Similarly, regular classroom teachers often feel unprepared for or incapable of providing adequate instruction for learning disabled students who are added to their classes (Gersten, Walker, & Darch, 1988). It seems that effective mainstreaming may require restructuring schools to provide additional support for the instructional needs of learning disabled students in the regular classroom setting (Stainback & Stainback, 1991).

How can schools effectively mainstream academically handicapped students? One alternative may be cooperative learning. Previous research has shown the potential benefits of cooperative learning processes for helping teachers mainstream special education students and for improving their social relations and academic achievement (Johnson, Johnson, Waring, & Maruyama, 1986; Madden & Slavin, 1983; Slavin, 1984; Slavin, Stevens, & Madden, 1988; Slavin & Stevens, 1991; Stevens & Slavin, 1991, 1995). However, there is a need for more research; not all of the programs that mainstreamed learning disabled students through cooperative learning have shown significant effects on students' achievement (Fuchs & Fuchs, 1991; Jenkins et al., 1994; Lloyd et al., 1988; Stevens & Slavin, 1991). Furthermore, questions remain about the long-term academic and social effects of mainstreaming through cooperative learning and about the impact of using cooperative learning throughout the day in a variety of subjects as a means of mainstreaming learning disabled students.

The previous research that effectively used cooperative learning to mainstream learning disabled students typically discontinued remedial pull-out programs. Instead, the special education teacher team taught in the regular classroom along with the regular education teacher (e.g., Stevens et al., 1987; Stevens & Slavin, 1995). In this way, it was possible to give the learning disabled students the additional instructional support they required.

In the cooperative elementary school model, learning disabled students receive all of their instruction in the regular classroom. During reading and/or math periods, the special education teacher team teaches with the regular classroom teacher using CIRC or TAI. In both of these methods, students work in heterogeneous learning teams, but receive instruction in relatively homogeneous teaching groups drawn from the various teams. The teaching groups are often composed of both learning disabled and nondisabled students. Typically, the regular education teacher provides the initial instruction for the students. The special education teacher provides follow-up on and extensions of the regular instruction. For example, the special education teacher often provides more support for the learning disabled students by giving them additional instruction related to what was covered in the teaching group or by guiding them and giving them feedback during their initial practice activities.

Mainstreamed learning disabled students are integrated into the cooperative learning teams within the class, distributing the learning disabled students among the teams with the nonhandicapped students. The mainstreamed students interact with other students on academic tasks as they work in cooperative learning teams, giving the learning disabled students all of the potential cognitive advantages of cooperative learning described previously. The cooperative learning processes also provide a way for special education students to become socially mainstreamed through academic interactions with their nonhandicapped classmates.

Organizational Support

Other elements of the cooperative elementary school model are used to create more support within the schools for the changes being made in the school and classroom organization aspects of the model. In order to fully implement widespread use of cooperative learning and full-scale mainstreaming of learning disabled students, certain school environment elements are also considered. Those elements include: peer coaching, teacher collaboration in instructional planning, principal-teacher collaboration in administrative planning, and parent involvement.

Peer coaching. One major difficulty in using an innovative program that changes how classrooms are organized and how teachers teach is the quality of the implementation (Showers, 1987). While some implementation issues may be addressed through the specificity of the training and periodic follow-up, implementation continues to be problematic. Peer coaching, where teachers observe and give feedback to one another, is an approach that has successfully produced higher implementation of innovative programs (Joyce & Showers, 1980, 1981; Showers, 1987; Williamson & Russell, 1990) and specific teacher behaviors (Sparks, 1986).

Peer coaching provides more opportunities for classroom teachers to get specific and nonthreatening feedback as they use new programs in their classrooms. Through coaching one another, teachers develop expertise within the school, rather than simply being the recipients of external expertise. This helps them develop more ownership of the innovation and more confidence as they try the new strategies involved (Sparks & Burder, 1987). At the same time, peer coaching provides a vehicle for professional dialogue about teaching that also seems to increase feelings of collegiality within the school (Sparks & Burder, 1987).

In the cooperative elementary school model, teachers are given many opportunities to visit one another's classes and provide support and feedback to one another. Teachers who have successfully implemented a given cooperative learning method serve as peer coaches for those who will receive training later. The principals encourage peer coaching by offering to teach classes for the coaches when they are observing their colleagues.

Teacher collaboration in instructional planning. As part of the cooperative elementary school's philosophical approach to using cooperation as the central element in school change, we wanted teachers to collaborate with one another on substantive issues related to instructional planning and implementation. Little (1982) found that continuous professional development can be found in schools that promote substantive dialogue among teachers about teaching and curriculum planning and development. Similarly, Johnson and Johnson (1987) suggested that teacher collaboration can help improve teachers' competence in implementing innovations.

In the cooperative elementary schools, we promoted collaborative teacher dialogue within the context of implementing the elements of the model. Teachers were given time to plan common strategies, instructional content, and activities in the forum of grade-level meetings. This provided the teachers an opportunity to coordinate academic content across classes and facilitated the sharing of instructional materials. Teachers also used their planning time to develop classroom strategies for covering content effectively and resolving management problems.

Principal and teacher collaboration in administrative planning. Another key element of the cooperative elementary school model was to develop collaboration between the administrators and teachers to further support the process of change. The building steering committee was made up of the principal, representatives from each of the grade levels, special services, and other faculty members. The goal of the committee was to develop goals for the school and to discuss progress toward the goals periodically. The committee provided a forum for discussion of broad issues related to school management and policy as well as integration of instructional goals and practices across grade levels. While the principal remained the instructional leader in the school, and was ultimately responsible for making schoolwide decisions, the steering committee provided valuable input in the decision-making process.

Parent involvement. The cooperative elementary school model encouraged parent participation in school activities to develop a sense of joint responsibility for children's success in school. Parents and the community were introduced to the project as it was initiated. They were given a theoretical rationale for the project and an explanation of each of the components of the cooperative elementary school. The parents were encouraged to monitor students' educational progress at home. The parents were also presented with the school's expectations concerning homework and independent reading, and the parent's role in making sure that those were accomplished. The cooperative elementary schools also maintained regular contact with parents through regularly scheduled teacher-parent conferences, PTA meetings, and a school newsletter. The latter kept parents informed of events at the school, changes in school policy or programs, and notable successes of students, classes, or teachers.

Method

Sample and Design

The sample consisted of 1,012 students in second through sixth grades in five elementary schools of a suburban Maryland school district. Twenty-one classes in the two treatment schools were matched with 24 classes in the three comparison schools on mean California Achievement Test scores for Total Reading, Total Language, and Total Mathematics. There was also an attempt to control for ethnic and socioeconomic background of the students by selecting comparison schools from the same or similar neighborhoods (see below). The student populations of each school ranged from 4% to 15% minority students (M = 7.3%), and from 2% to 20% (M = 10.2%) disadvantaged students (i.e., those receiving free or reduced-price lunch). The schools were all located in predominantly working-class neighborhoods. Approximately 9.3% of the five schools' student populations were identified as learning disabled, ranging from 7% to 12% in each school.

Selection of Schools

Treatment schools
Because the cooperative elementary school required schoolwide change and encouraged teacher participation in decision making, we felt it was important to have teach-

ers involved in determining whether their school would participate in this project. Prior to beginning the project, we approached four schools in the Maryland school district that agreed to participate in the project. At each school, we presented a brief overview of the project, the duration, and the potential benefits. After the authors answered the teachers' questions and concerns, the staff voted on participating in the proposed project. Only if 75% of the faculty agreed to participate in all elements of the project was the school selected; of the four schools to whom presentations were made, two voted overwhelmingly to participate.

Comparison schools
The comparison schools were selected from school district records of average student achievement, ethnicity, and socioeconomic background. These data were used to select comparison schools that were well matched with the two treatment schools on these independent and demographic variables. Schools that were approached for the study but did not volunteer to participate were not included as comparison schools.

Treatments

Treatment schools
The treatment schools adopted the cooperative elementary school model, as described previously. Each component was gradually phased in during the first year of the study. Prior to implementation of a component, the faculty was given training or an introduction to the component that included the rationale and goals as well as detailed teachers' manuals for each instructional program. During the school year, members of the research staff observed classes, participated in meetings with teachers, and observed steering committee meetings to facilitate implementation of the program components.

Comparison schools
The comparison schools continued using their regular teaching methods and curriculum. In reading, these methods consisted of the use of two or three ability-based reading groups. A basal series was used with workbooks, worksheets, and teacher-prepared materials as follow-up activities. The teachers also integrated literature into reading instruction by reading two or three novels a year. In language arts, the teachers used whole-class instruction and usually used a published language arts series as well as teacher-made activities in writing and language usage. In mathematics, the teachers used a district-adopted mathematics text. Typically mathematics involved whole-class instruction with follow-up activities from the text or generated by the teacher. While the comparison teachers occasionally used various kinds of group work, none of them used structured cooperative learning that involved group goals based on group members' individual learning performances.

The comparison schools did not implement most of the other components of the cooperative elementary school model. Specifically, the comparison schools did not use an in-class model for mainstreaming special education students. They did not use peer coaching or regularly use cooperative planning between teachers. The comparison schools did have school improvement teams made up of administrators and faculty, much like the cooperative elementary school's building-level steering committee, but typically the school improvement team met much less frequently (e.g., 2 times a semester rather than 2 times a month).

Allocated time and curricula
Both the treatment and comparison schools allocated the same amount of time to reading, language arts and mathematics instruction daily, in compliance with school district guidelines In reading, all of the schools used similar basal reading programs selected from an adoption list provided by the school district. However, the treatment schools did not use adjunct materials (i.e., workbooks) provided by the basal publisher, whereas the comparison schools typically did use them. Similarly in language arts both the treatment and comparison schools used a district-adopted language arts program, but in the cooperative elementary schools this program was used as supplemental material during CIRC writing rather than as the central curriculum. In mathematics, the comparison schools used the district-adopted textbook, whereas the treatment schools used the TAI curriculum, using the district text for application and problem solving activities only.

Implementation of the Model Components

Steering committee. The first component of the cooperative elementary school to be implemented was the building-level steering committee. It was important to establish this committee because it would act as the main vehicle for communication between teachers, school administration, and research staff. The steering committee was organized prior to the start of the school so that it could help resolve problems that might occur as we began implementing the other components of the model.

Cooperative learning programs. We gradually phased in the implementation of the cooperative learning programs during the first year of the project. The goal was to have cooperative learning used by all of the teachers across a variety of curriculum areas, with specific emphasis on the CIRC program in reading and language arts and the TAI program in mathematics. The teachers also were trained in Jigsaw II, Teams-Games-Tournaments (TGT), and Student Teams Achievement Division (STAD), which are generic cooperative learning models that can be used in a variety of content areas, such as social studies, spelling, or science (see Slavin, 1986). The training for each program included explanation of the processes and the rationale behind them, and a detailed manual on how to use the program in a classroom. Trainers also provided teachers with simulated demonstration lessons for each program.

The initial training occurred in August, with training on subsequent cooperative learning programs taking place at roughly 2-month intervals so the teachers were not overwhelmed by too many new programs at the same time. Initially, the teachers were trained in the CIRC program's reading component because we felt it would be easier to start the year with a new reading program than it would be to reorganize reading instruction during the year. Also, CIRC was considered the most complex program, so teachers would have the most time to master it.

Following training, the research staff observed the teachers as they used the programs in their classrooms. We gave the teachers feedback on their implementation to improve the quality of their instruction and maintain fidelity to the program. Teachers also coached one another as they started using the cooperative learning programs.

Subsequent training was planned in collaboration with the building steering committee. They provided valuable feedback on whether the faculty was ready to take on another new program and which program they preferred to implement next. By March of the first year of the study, all of the curricular elements of the program were in place. All of the teachers were using CIRC and TAI as their reading, language arts, and mathematics curricula. In a typical week, the classes used the instructional procedures prescribed in those models for least 4 days. (Teachers occasionally used one day a week for other kinds of activities like performing a play, watching a movie; engaging in reading, writing, or mathematics activities related to other subjects like social studies and science, or engaging in enrichment activities like field trips.)

All of the teachers were trained in the more generic cooperative learning models (Jigsaw II, TGT, and STAD). These model are designed and used for specific activities in a variety of subjects like spelling, social studies, and science. Unlike the curriculum-specific models of cooperative learning, the generic models are designed to be an adjunct to instruction, and, to a great extent, they are different ways to accomplish the same goal. As such, the teachers used them much more irregularly and may not have used each of them. However, all of the teachers were observed a number of times using at least one of the generic models during instructional activities in subjects other than reading, language arts, and mathematics.

One aspect of the cooperative elementary school, mainstreaming learning disabled students, was only partially implemented the first year. Due to scheduling difficulties, the schools were able to mainstream only about 60% of their learning disabled students.

Elements of the cooperative elementary school
During the second year of the study, all the elements of the cooperative elementary school were in use. The teachers were using cooperative learning in a variety of subject areas, and the students were involved in cooperative learning on a daily basis. Special education pull-out remediation for learning disabled students had been discontinued, and special education teachers provided remedial instruction to them in the regular classroom in team-teaching arrangements. The learning disabled students were also integrated with nonlearning disabled students on heterogeneous learning teams. The teachers within and sometimes across grades collaborated in planning instruction, and teachers at a grade level met at least weekly for instructional planning. The building steering committee met at least monthly to plan and evaluate the school's progress in meeting their goals. Finally, a biweekly newsletter was published by the school to keep the parents and community informed about the activities at the school, and parents were provided with periodic parent-faculty meetings about the nature of and events in the cooperative elementary school.

Measures

Achievement pretests
We used standardized test scores from district records to match classes on entering ability and as a covariate in the statistical analyses for the entire population. The scores used were the Total Reading Total Language, and Total Mathematics scores from the California Achievement Test, Form C, which were given in the fall of the first year of the study. The pretest scores were transformed to z scores separately for each grade so that the data could be analyzed by grade level and collapsed across grade levels if there were no grade-by-treatment interactions.

Achievement posttests
In the spring of the first and second years of the study, teachers administered the reading, language, and mathematics subtests of the California Achievement Test, Form E. The substests of California Achievement Test have reliabilities (internal consistency) ranging from .80 to .95, with a median .91 reliability (CTB/McGraw-Hill, 1986). Raw scores from the subtests were transformed to z scores separately for each grade to enable us to combine scores across grades if there were no grade-by-treatment interactions.

Attitude measures

The students were given an attitude measure as a pre- and posttest. Pretests were given in the fall of the first year of the project, and posttests were given in the spring of the second year. The attitude measure asked students to rate their attitude toward and their perceived ability in reading, language arts, and mathematics. For each subject, the students were also asked to rate their interest and their ability on a 3-point scale ("I like it a lot"/"I don't like it" and "I am really good at this"/"I am not very good at this"). The attitude measure had alpha reliabilities of .35 for the premeasure and .33 for the postmeasure. For the perceived ability measure, the alpha reliabilities were .32 and .37, respectively.

Social relations measure

The students were given a measure of their social relations as a pre- and postmeasure at approximately the same time they were given the other pre- and posttests. The social relations measure asked students to list the names of their friends in the class (Moreno, 1953). The results were analyzed by comparing the average number of friends listed by treatment and control students. The test-retest reliability for the number of friends was .57 for this population. For learning disabled students, the social relations measures were reanalyzed to determine the number of times each learning disabled student was selected as a friend by his/her nonhandicapped peers. This second analysis provided a measure of social acceptance of mainstreamed learning disabled students.

Results

Analyses

The analyses were somewhat complicated by the use of intact classes and schools. The students were nested within classes, classes were nested within schools, and schools were nested within treatment. Because of the nested nature of the design, each level of the hierarchy potentially influences the processes at the lower levels (Raudenbush & Bryk, 1988). In an attempt to disentangle the effects of schools or teachers from the effects of the treatments on the student outcomes, we used the hierarchical linear model (HLM; Bryk, Raudenbush, Seltzer, & Congdon, 1988). Raudenbush and Bryk (1988) provide a thorough discussion of the use of HLM with multilevel data.

Since the study lasted 2 years, the posttest data were analyzed and reported separately for each year. The initial sample of the five schools included 1,012 students. Of those, 873 students had pretest data available in the district's records and were in the study for the entire 2 years. Approximately 13.7% of the original students moved or otherwise did not stay in the participating school for 2 years. (In the treatment schools, 11% of the students remained in the school for the entire 2 years.)

Standardized test scores that were converted into z scores were used in the achievement analyses, with the appropriate pretest (Total Reading, Total Language, or Total Mathematics) used as a covariate to increase the power of the analyses. The test of the homogeneity of regressions indicated no significant violations of that assumption, all $t < 1.0$. Initially, grade-by-treatment analyses were conducted to determine if the treatment had differential effects on achievement, attitudes, or social relations at different grade levels. No grade-by-treatment analyses were significant, so the data were collapsed across grade levels to simplify the analyses and discussion of the results.

Achievement pretests. As noted previously, the treatment and comparison classes were matched on initial achievement on the California Achievement Test. There were no significant pretest differences on Total Reading, $t < 1.0$, or Total Language, $t < 1.0$ (see

Table 1). There were, however, significant pretest differences on Total Mathematics, t = 2.02, p = .05, favoring the comparison students.

Posttests after the first year. The results from the first-year analyses indicated significant differences in favor of the treatment group on reading vocabulary, t = 2.14, p < .05 (see Table 1). The magnitude of the effect, however, was relatively small; the effect size was +.17 standard deviations. There were no significant differences on the measures of reading comprehension, t = 1.63; language mechanics, t < 1.0; language expression, t < 1.0; math computation, t = 1.34; and math application, t < 1.0. The means, HLM-fitted means, standard deviations, and effect sizes are presented in Table 1. The HLM-fitted means are equivalent to adjusted means and are computed from the base and gamma produced by the HLM analyses and the orthogonally coded comparison coefficient. The HLM mean = base + (G X C), where G is the gamma and C is the comparison coefficient.

It should be noted that the HLM analysis is analgous to a conservative class-level analysis. Individual-level analyses showed significantly positive effects of the model on reading vocabulary, reading comprehension, and math computations.

Posttests after the second year

The second year HLM analyses yielded significant effects favoring the treatment group on reading vocabulary, t = 3.04, p < .01; reading comprehension, t = 3.62, p < .01; language expression, t = 2.93, p < .01; and math computation, t = 3.77, p < .01 (see Table

Table 1   Students' Achievement: Means, Standard Deviations, Analyses, and Effect Sizes (Collapsed Across Grades)

Measure	CES			Comparison				
	M	(SD)	HLM fitted	M	(SD)	HLM fitted M	t	Effect size[1]
Pretests								
Reading	−.03	(.99)		.04	(1.00)		<1.0	−.07
Language	−.01	(.98)		.00	(1.00)		<1.0	−.01
Mathematics	−.09	(.97)		.08	(1.02)		2.02*	−.17
			Year 1					
Posttests								
Read. voc.	.04	(.99)	.08	−.14	(1.01)	−.09	2.14*	+.17
Read. comp.	.03	(1.01)	.08	−.03	(.99)	−.05	1.63	+.13
Lang. mech.	−.01	(.99)	.00	−.02	(1.00)	.01	<1.0	−.01
Lang. expr.	.04	(1.01)	.04	−.04	(.99)	−.04	<1.0	+.08
Math comp.	.02	(.99)	.06	−.01	(1.00)	−.06	1.34	+.12
Math appl.	−.07	(.98)	−.02	.08	(1.01)	.03	<1.0	−.05
			Year 2					
Posttests								
Read. voc.	.05	(.98)	.10	−.04	(1.01)	−.11	3.04*	+.21
Read. comp.	.08	(.98)	.15	−.07	(1.01)	−.13	3.62**	+.28
Lang. mech.	.03	(.97)	.05	−.02	(1.02)	−.05	1.16	+.10
Lang. expr.	.10	(.96)	.11	−.09	(1.03)	−.10	2.93**	+.21
Math comp.	.03	(1.00)	.15	−.04	(.99)	−.14	.377**	+.29
Math appl.	−.03	(1.01)	.04	.01	(1.01)	−.06	1.24	+.10
N (students)	411			462				
Number of classes	21			24				

1   Effect size equals the difference of the HLM-fitted means divided by the control group standard deviation.
* p < .05.
** p < .01.

1). There were no significant effects on language mechanics, $t = 1.16$, or math applications, $t = 1.24$. The magnitude of the significant effects ranged from .21 to .29 standard deviations.

Attitude measures
There were no significant pre- or postmeasure differences on students' attitudes toward reading, language arts, or mathematics. Similarly, there were no premeasure differences on students' perceived ability in reading, language arts, or mathematics. After 2 years, students in the cooperative elementary school did have higher perceived ability in reading ($t = 2.02$, $p < .05$) and language arts ($t = 2.99$, $p < .01$). The magnitude of these significant effects was +.20 and +.26, respectively (see Table 2).

Social relations
The social relations measure examined the number of friends listed by students in the treatment conditions at the beginning of the study and the number listed 2 years later at the end of the study. There were no significant differences on the premeasure, $t < 1.0$. On the postmeasure, the students in the treatment schools listed significantly more friends than did students in the comparison schools, $t = 3.92$, $p < .01$ (see Table 3). The magnitude of the significant effect was +.42 standard deviations.

Effects for learning disabled students after the first year
The pretests indicated that the learning disabled students in the cooperative elementary schools and in the comparison schools had similar levels of initial achievement. There were no significant pretest differences on standardized measures of their academic ability in reading, $F = 1.33$; language arts, $F = 3.19$; or mathematics, $F < 1.0$. After the first year, the achievement results for learning disabled students were somewhat similar to those of the whole population. The analysis of variance results indicated no significant differences between the achievement of learning disabled students in the cooperative elementary school and those in the comparison schools; however, learning disabled students in the cooperative elementary school did have slightly better achievement scores on all measures (see Table 4).

Effects for learning disabled students after the second year
After the second year, the differences in achievement for learning disabled students again followed the pattern of the differences found in the entire population. There were significant posttest differences in favor of the learning disabled students in the cooperative elementary school on reading vocabulary, $F = 13.48$, $p < .01$; reading comprehension, $F = 14.39$, $p < .01$; language expression, 11.41, $p < .01$; math computation, $F = 10.77$, $p < .01$; and math application, $F = 3.75$, $p < .05$ (see Table 4). There were no significant differences in language mechanics, $F < 1.0$. The effect sizes for the significant effects were much larger than those found for nonhandicapped students, ranging from +.35 to +.85.

Initially, there were no significant differences between the treatment groups on learning disabled students' social relations, $F < 1.0$, as measured by the number of friends they selected. On the postmeasure, learning disabled students in the treatment schools listed significantly more friends than did their counterparts in the comparison schools, $t = 3.42$, $p < .01$ (see Table 3). This was a substantial difference, with an effect size of +.86. The social relations measure was also used to assess the level of the social acceptance of mainstreamed learning disabled students, as measured by the number of times learning disabled students were selected as friends by their nonhandicapped classmates. Initially, there were no significant differences, $t < 1.0$. On the post-

Table 2 Students Attitude, Means, Standard Deviations, and Effects

	CES		Comparison		*t*	Effect size[1]
Attitude toward reading						
Pretest	1.47	(.61)	1.55	(.63)	1.79	-.12
Posttest	1.70	(.69)	1.71	(.68)	<1.0	-.01
Attitude toward language arts						
Pretest	1.80	(.69)	1.74	(.74)	1.44	+.08
Posttest	1.82	(.76)	1.78	(.74)	<1.0	+.05
Attitude toward mathematics						
Pretest	1.50	(.69)	1.43	(.61)	1.52	+.11
Posttest	1.48	(.65)	1.42	(.69)	1.37	+.09
Perceived reading ability						
Pretest	1.58	(.56)	1.62	(.60)	<1.0	-.07
Posttest	1.72	(.59)	1.61	(.55)	2.02*	+.20
Perceived language arts ability						
Pretest	1.73	(.65)	1.68	(.65)	1.31	+.08
Posttest	1.74	(.65)	1.58	(.62)	2.99**	+.26
Perceived math ability						
Pretest	1.39	(.66)	1.47	(.62)	1.22	-.12
Posttest	1.45	(.63)	1.51	(.67)	1.11	-.07
N (students)	411		462			
Number of classes	21		24			

1 Effect size equals the difference in treatment means divided by the control standard deviation
* $p < .05$.
** $p < .01$.

Table 3 Classroom Social Relations, Means, Standard Deviations, and Effect Sizes

	CES		Comparison		*t*	Effect size[1]
Number of friends						
Premeasure	5.91	(3.67)	5.89	(3.59)	<1.0	+.01
Postmeasure	8.71	(3.82)	7.12	(3.62)	3.92**	+.42
N (students)	411		462			
Gifted students						
Number of friends						
Premeasure	5.74	(2.23)	5.66	(2.54)	<1.0	+.03
Postmeasure	8.92	(3.01)	7.59	(2.89)	2.64**	+.46
N (students)	46		61			
Special education students						
Number of friends						
Premeasure	5.03	(2.81)	5.10	(3.04)	<1.0	–.02
Postmeasure	9.11	(4.27)	6.29	(3.29)	3.42**	+.86
Number of times picked by nonspecial education students						
Premeasure	2.83	(1.76)	2.78	(1.54)	<1.0	+.03
Postmeasure	5.83	(2.39)	3.81	(2.24)	433**	+.90
N (students)	40		36			

1 Effect size equals the difference in treatment means divided by the control standard deviation.
* $p < .05$.
** $p < .01$.

measure, learning disabled students in the treatment schools were selected as friends by nonhandicapped students significantly more often than were their counterparts in the comparison schools, $t = 4.33$, $p < .01$. This comparison had an effect size of $+.90$, nearly a full standard deviation difference (see Table 3).

There were no initial differences between the cooperative elementary school's and the comparison school's learning disabled students' attitudes toward or perceived ability in reading, language arts, or mathematics. Learning disabled students in the cooperative elementary school had higher postmeasures of their perceived ability in reading ($t = 2.01$, $p < .05$) and language arts ($t = 2.19$, $p < .05$). The effects sizes for these significant effects are $+.26$ and $+.33$, respectively (see Table 5).

Effects for gifted students
Our initial hypotheses did not explicitly predict effects of the cooperative elementary school on gifted students, but, in response to recent discussions about the effects of cooperative learning's effects on gifted students (e.g., Allan, 1991; Feldhusen & Moon, 1992; Slavin, 1991), we also investigated the effects in these data post hoc. Gifted students were identified as the top 10% of students on the standardized achievement pretests. Typically, they were more than a year above grade level. In the comparison schools, the gifted students participated in a twice-weekly pullout program that focused on enrichment rather than acceleration. The goal of the enrichment program was to extend gifted students' learning to higher levels of understanding through self-directed investigations and problem-solving activities. In the cooperative elementary school, the gifted students did not participate in enrichment activities; instead, they remained in the regular classroom and participated on heterogeneous cooperative learning teams.

The gifted students in the treatment and comparison schools were not significantly different from one another on their initial reading ($F = 1.21$) and language achievement ($F = 2.11$). However, the gifted students in the comparison schools had significantly higher initial math achievement than those in the treatment schools ($F = 4.28$, $p < .05$). Due to the initial differences, an analyses of covariance (ANCOVA) was used to analyze the posttest data, controlling for pretests in reading, language, or mathematics.

The first-year results indicate no significant differences on achievement for gifted students in the cooperative elementary schools as compared to those in the comparison schools. After 2 years, the gifted students in the treatment schools had significantly higher achievement in reading vocabulary, $F = 11.06$, $p < .01$; reading comprehension, $F = 12.13$, $p < .01$; language expression, $F = 6.09$, $p < .05$; and math computation, $F = 9.16$, $p < .05$, than did their counterparts in the comparison schools who participated in enrichment pull-out programs (see Table 6). There were no significant differences on measures of language mechanics or math applications. The effect sizes for the significant main effects ranged from $+.48$ to $+.68$ standard deviations, larger than the effects for students in general, but similar to those found for special education students.

Initially there were no significant differences between gifted students in the cooperative elementary school and those in the comparison schools on the number of friends they listed. After 2 years in the cooperative elementary school, gifted students reported significantly more friends than did the gifted students in the comparison schools ($t = 2.64$, $p < .01$). The gifted students in the cooperative elementary school averaged 1.5 more friends, an effect size of $+.46$ (see Table 3).

The attitude measures indicated no significant pretest differences between gifted students in the cooperative elementary school and those in the comparison schools on their attitudes toward reading, language arts, or mathematics, or their perceived

Table 4 Special Education Students: Standardized Achievement Measures, Means, Standard Deviations, and Effect Sizes (Collapsed Across Grades)

Measure	CES		Comparison		F	Effect size[1]
	M	*(SD)*	*M*	*(SD)*		
Pretests						
Reading	−1.31	(.78)	−1.29	(.70)	1.33	−.03
Language	−1.27	(.81)	−1.30	(.65)	3.19	+.05
Mathematics	−.98	(.73)	−.98	(.80)	<1.0	.00
Year 1						
Posttests						
Read. voc.	−1.20	(.81)	−1.38	(.69)	1.28	+.26
Read. comp.	−1.21	(.83)	−1.40	(.65)	1.59	+.29
Lang. mech.	−1.16	(.87)	−1.25	(.79)	<1.0	+.11
Lang. expr.	−1.19	(.85)	−1.45	(.70)	2.59	+.37
Math comp.	−.45	(.70)	−.71	(.78)	2.06	+.33
Math appl.	−.71	(.66)	−.83	(.75)	<1.0	+.16
Year 2						
Posttests						
Read. voc.	−.98	(.86)	−1.48	(.66)	13.83**	+.76
Read. comp.	−1.00	(.96)	−1.53	(.62)	14.39**	+.85
Lang. mech.	−.99	(.96)	−1.21	(.85)	<1.0	+.25
Lang. expr.	−.97	(.97)	−1.55	(.78)	11.41**	+.74
Math comp.	−.10	(.65)	−.54	(.75)	10.77**	+.59
Math appl.	−.28	(.58)	−.53	(.71)	3.98*	+.35
N (students)	40		36			

1 Effect size equals the difference in treatment means divided by the control standard deviation.
* $p < .05$.
** $p < .01$.

Table 5 Special Education Students Attitudes, Means, Standard Deviations, and Effects

	CES	Comparison	t	Effect size[1]
Attitude toward reading				
Pretest	1.53 (.64)	1.67 (.67)	1.34	−.21
Posttest	1.56 (.68)	1.63 (.70)	1.21	−.10
Attitude toward language arts				
Pretest	2.07 (.70)	2.05 (.80)	<1.0	+.03
Posttest	1.89 (.81)	1.83 (.74)	<1.0	+.08
Attitude toward mathematics				
Pretest	1.30 (.74)	1.27 (.54)	<1.0	+.06
Posttest	1.43 (.75)	1.44 (.59)	<1.0	−.02
Perceived reading ability				
Pretest	1.80 (.68)	1.84 (.77)	<1.0	−.05
Posttest	1.95 (.50)	1.79 (.54)	2.01*	+.26
Perceived language arts ability				
Pretest	1.95 (.70)	2.00 (.77)	<1.0	−.06
Posttest	2.00 (.77)	1.79 (.63)	2.19*	+.33
Perceived math ability				
Pretest	1.73 (.71)	1.67 (.67)	1.48	+.09
Posttest	1.71 (.85)	1.63 (.70)	1.30	+.11
N (students)	40	36		

1 Effect size equals the difference in treatment means divided by the control standard deviation.
* $p < .05$.
** $p < .01$.

Table 6 Gifted Students: Standardized Achievement Measures, Means, Standard Deviations, and Effect Sizes (Collapsed Across Grades)

	CES			Comparison				
Measure	M	(SD)	Adjusted M	M	(SD)	Adjusted M	F	Effect size[1]
Pretests								
Reading	1.47	(.48)		1.59	(.60)		1.21	−.20
Language	1.13	(.53)		1.31	(.68)		2.11	−.26
Mathematics	1.38	(.64)		1.68	(.55)		4.28*	−.54
			Year 1					
Posttests								
Read. voc.	1.43	(.39)	1.47	1.30	(.64)	1.25	3.03	+.34
Read. comp.	1.34	(.40)	1.38	1.35	(.57)	1.30	<1.0	+.14
Lang. mech.	1.10	(.51)	1.10	1.14	(.62)	1.13	<1.0	-.05
Lang. expr.	1.12	(.46)	1.13	1.10	(.57)	1.09	<1.0	+.07
Math comp.	1.32	(.63)	1.37	1.24	(.56)	2.69	9.16**	+.32
Math appl.	1.34	(.59)	1.39	1.48	(.55)	1.42	<1.0	−.05
			Year 2					
Posttests								
Read. voc.	1.38	(.33)	1.41	1.01	(.66)	.99	11.06**	+.65
Read. comp.	1.14	(.24)	1.15	.86	(.44)	.85	12.13**	+.68
Lang. mech.	1.06	(.48)	1.12	1.10	(.53)	1.06	<1.0	+.11
Lang. expr.	1.16	(.31)	1.22	1.07	(.40)	1.03	6.09	+.48
Math comp.	1.24	(.61)	1.33	1.05	(.58)	.99	9.16**	+.59
Math appl.	1.30	(.43)	1.37	1.33	(.54)	1.27	<1.0	+.19
N (students)	46			61				

1 Effect size equals the difference in adjusted means divided by the control standard deviation.
* $p < .05$.
** $p < .01$.

Table 7 Gifted Students Attitudes, Means, Standard Deviations, and Effects

	CES		Comparison		t	Effect size[1]
Attitude toward reading						
Pretest	1.60	(75)	1.69	(.62)	1.33	−.14
Posttest	1.95	(.69)	1.90	(.65)	<1.0	+.08
Attitude toward language arts						
Pretest	1.78	(.70)	1.81	(.75)	<1.0	−.04
Posttest	2.05	(.83)	1.67	(.79)	2.09*	+.48
Attitude toward mathematics						
Pretest	1.45	(.61)	1.49	(.64)	<1.0	−.06
Posttest	1.52	(.54)	1.50	(.58)	<1.0	+.03
Perceived reading ability						
Pretest	1.51	(.59)	1.55	(.61)	<1.0	−.07
Posttest	1.52	(.61)	1.50	(.51)	<1.0	+.04
Perceived language arts ability						
Pretest	1.75	(.55)	1.79	(.54)	<1.0	−.07
Posttest	1.85	(.75)	1.33	(.77)	3.55**	+.68
Perceived math ability						
Pretest	1.30	(.67)	1.28	(.54)	<1.0	+.04
Posttest	1.26	(.58)	1.27	(.51)	<1.0	−.02
N (students)	46		61			

1 Effect size equals the difference in treatment means divided by the control standard deviation.
* $p < .05$.
** $p < .01$.

ability in those subjects. After 2 years, gifted students in the cooperative elementary school had significantly higher attitudes toward language arts ($t = 2.09$, $p < .05$) and higher perceived ability in language arts ($t = 3.55$, $p < .01$). Their attitudes toward school subjects and perceived ability were rated nearly a half to two-thirds of a standard deviation higher on these two measures (effects sizes = +.48 and +.68 respectively; see Table 7).

Discussion

Because the experimental program had so many components, it is difficult to ascribe the program's outcomes to any single element. However, the results of this study support the hypothesis that cooperative learning can be the primary mode of instruction and – when integrated with effective instruction in reading, language arts, and mathematics and with changes in school organization, peer coaching, mainstreaming, and other elements – can be effective in producing higher student achievement. After the first year, the effects were relatively small, and the only significant differences were on measures of reading vocabulary achievement, which indicated that students in the cooperative elementary school outperformed those in the traditional schools. The first year results in reading, to some degree, replicate earlier findings of the positive effects for the CIRC program (see Stevens et al. 1987). The lack of significant positive effects in math and language arts is likely to be due to the later implementation of these components, which were gradually phased in over the course of the year. As a result, the students received the treatment in language and math for much less than a year, and it is possible that students have to experience the programs longer for them to have a strong impact on standardized achievement test scores.

After the second year, the students in the cooperative elementary schools clearly outperformed their peers in the more traditional schools on standardized measures of reading vocabulary, reading comprehension, language expression, and mathematics computation. The effect sizes, which ranged from +.21 to +.29 standard deviations, suggest that these differences were educationally significant – particularly, considering that the differences were obtained on standardized achievement tests, which typically are difficult to influence due to their limited overlap with the curriculum and their strong correlation with general ability. The pattern and the magnitude of these effects replicate the findings in earlier studies for the CIRC and TAI programs (see Slavin et al., 1984; Stevens et al., 1987; Stevens & Slavin, 1995).

These results are important because they are the result of a long-term implementation of cooperative learning. The positive effects suggest that multifaceted cooperative learning programs like TAI and CIRC can maintain their effectiveness over time and that the effectiveness is not likely to be due to the novelty of cooperative learning or to the Hawthorne effect, as has been suggested (Bossert, 1989). This shows that multifaceted instructional programs that use state-of-the-art instruction and curriculum with cooperative learning classroom processes can produce and maintain significantly higher achievement.

The students in the cooperative elementary school listed significantly more friends than did students in the more traditional schools, indicating that there were better peer relations in the cooperative elementary schools, a result that replicates that of previous research in cooperative learning (cf. Johnson & Johnson, 1989; Sharan, 1980; Slavin, 1994). Unlike typical classroom instruction, during which students tend to have only casual and rather superficial contact, cooperative learning processes have students working together to achieve a common goal. This produces more meaningful interactions between the students and a sense of positive interdependence. As these work groups change over time, the students collaborate with a variety of students in the class, leading to better peer relations in the class and an increase in students' friendships.

Typically, cooperative learning results in more positive student attitudes in the subjects where cooperative learning processes are used (cf. Slavin 1994). Unlike the previous research, this study did not find that students had enhanced attitudes toward subjects where cooperative learning was used. The lack of significant findings on students' attitudes and on most of the measures of their perceived ability may be due to errors in measurement. The attitude and perceived ability measures had few (3) items, and alpha reliabilities were less than .40. However, students in the cooperative elementary school did have more positive perceptions of their ability in reading and language arts than did the students in the more traditional schools. It is difficult to determine which component of the programs produced this effect on students' perceptions of their ability. The result may in part be due to the cooperative learning processes that reduce the competition in the classroom, for competition has been found to lower most students' perceptions of their ability (Deutsch, 1949). It may be due to the support structure that cooperative learning provides, which may keep students from being frustrated and hence feeling less successful when they engage in their learning activities. The result may also be due to the curricula used in reading and language arts. Students may perceive their ability as being higher because the instructional activities build competence the students can easily see. The best example of this would be in the case of the writing, where students spent much of their time writing, revising their writing, and critiquing their peers' writing. It is possible that the students more readily perceived the development of their writing ability through the frequent writing activities and that this perception was reflected in their ratings of their perceived ability. Finally, the positive results may be due to some combination of these and other elements of the model.

Mainstreamed learning disabled students
The first-year achievement data show small and nonsignificant advantages for learning disabled students in the cooperative elementary schools. This may in part be due to the difficulties encountered in scheduling fully mainstreamed services for many of these students during the first year. Before the beginning of the second school year, we collaborated with the principals and special education teachers to schedule special education services for learning disabled students when the entire school schedule was being developed. This alleviated what had previously been a roadblock to mainstreaming the learning disabled students and special education teachers into regular classes for instruction in reading and mathematics.

The results after the second year of the study support the hypothesis that the well-structured, cooperative learning programs used in the cooperative elementary school can be vehicles to enhance a school's ability to successfully mainstream learning disabled students into the regular classroom. The achievement effect sizes for learning disabled students, which ranged from a third to more than three quarters of a standard deviation, suggest that learning disabled students have much to gain by being mainstreamed in these kinds of programs. However, these results are dependent on (a) the learning disabled students' being integrated into heterogeneous learning teams in classrooms, (b) the cooperative learning programs' using group goals based on individual accountability, and (c) the special education teachers' being scheduled to provide additional instruction and support to the learning disabled students in the regular classroom.

One advantage of cooperative learning is that it provides a structure that allows students to help manage the classroom, which evolves from the positive interdependence created within the learning teams. When the students take more responsibility, it frees teachers from time that must be spent on managerial tasks so they can provide more individualized instruction to students who need it. This gives teachers more instructional flexibility to accommodate the increased classwide heterogeneity inherent in mainstreaming (Johnson & Johnson, 1989; Slavin, 1994). Additionally, students become an instructional and motivational resource in the classroom so that students who need help can rely on support and feedback from their peers, providing another mechanism for accommodating students with diverse abilities.

The instructional support for learning disabled students in the cooperative elementary school is further enhanced by the use of an in-class model of support services, where the special education teacher goes into the regular education classroom and team teaches with the classroom teacher. In this way, the learning disabled student continues to receive the additional instruction from the special education teacher while at the same time getting the social and motivational benefits of being in the regular classroom.

The students in the cooperative elementary schools also displayed much better social acceptance of mainstreamed learning disabled students. A learning disabled student in the cooperative elementary school was likely to be chosen as a friend by a nondisabled student 50% more often than were his or her peers in the more traditional schools. This is a particularly important result because one of the goals of mainstreaming is to promote more social acceptance of students who have special needs. However, research has suggested that the opposite effect may often be true. Mainstreamed handicapped students typically are not well integrated into the class socially or academically and as a result are poorly accepted by their peers and often ostracized (Bryan, Bay, Lopez-Reyna & Donahue, 1991; Gottlieb, Corman, & Curci, 1984; Reid, 1984).

Perhaps the critical element in the social acceptance of mainstreamed students is that cooperative learning creates an environment of positive interdependence within the teams, where students depend on one another and where all must succeed in order for the group and any one member to succeed. This makes contact between learning disabled and nondisabled students on the cooperative learning teams meaningful and of more equal status. Allport (1954) found that prejudice actually increased when different racial groups had more frequent superficial contact, much like what may be the case in some mainstreamed classrooms. On the other hand, prejudice decreased when individuals had equal status contact in situations where they could develop a feeling of positive interdependence (Allport, 1954) like that found in well-structured, cooperative learning. The results of this study suggest that true mainstreaming goes beyond academics and that disabled and nondisabled students also need classroom environments that will promote social development and acceptance in an environment of meaningful contact focused on attaining a common goal.

Cooperative learning and gifted students
While we had no a priori hypotheses about the effects of the cooperative elementary school model on gifted students, there was the implicit expectation that the elements of the model would be beneficial for all students, including the gifted. In particular, recent discussions in journals suggest that the effect of cooperative learning on gifted students is an important question. Critics of cooperative learning have suggested that when gifted students are in cooperative learning and take time out of their academic tasks to explain content to their classmates it reduces the achievement, motivation, and attitudes of gifted students (Allan, 1991; Feldhusen & Moon, 1992). Cooperative learning is claimed to result in a slower instructional pace for gifted students, which in turn contributes to lower achievement than that they might have obtained other-

wise (Robinson, 1990). Similarly, Linnemeyer (1992) suggests that in heterogeneous cooperative groups the gifted students have no one who can adequately give them the critical feedback assumed to produce higher achievement. On the other hand, researchers have suggested that cooperative learning in heterogeneous groups can be beneficial for gifted students, particularly if teachers are well-trained and the process is implemented well (Gallagher, 1991; Mills & Durden, 1992; Slavin, 1991).

The results from this study suggest that well-structured cooperative learning is not detrimental to the achievement of gifted students and, in fact, can produce signifi- cant and substantial positive effects on gifted students' achievement. Gifted students in this study who worked in heterogeneous cooperative groups in a variety of content areas had much higher achievement than similar students in the comparison schools who received enrichment programs twice a week and did not participate in coopera- tive learning. These results corroborate previous evidence about the benefits of coop- erative learning for gifted students (Johnson & Johnson, 1989; Slavin, 1994; Webb, 1985). Gifted students seem to benefit from being the one who provides elaborate explanations to classmates, a process that is important to learning both at a theoret- ical level (Brown & Palincsar, 1989; Collins, Brown, & Newman, 1989; Vygotsky, 1978) and in practice (Webb, 1985).

Limitations and Future Research

This study did not evaluate all of the components of the cooperative learning pro- gram. Instead, it focused on using cooperative learning in a variety of content areas and mainstreaming learning disabled students. Components, such as peer coaching and cooperative planning among teachers and between the teachers and principal, were not specifically addressed. The authors felt that, for the purposes of this study, these factors primarily served a facilitative role in promoting the changes in curric- ula, instructional processes, and school and classroom organization. It is impossible to determine the impact of any of these components from the data presented. Similarly, each of the cooperative learning programs (e.g., TAI, CIRC) changed group processes, curricula, and reward structures, and the relative effects of parts cannot be disentangled in this study. Rather, such effects need to be the goal of future compo- nent analysis studies (cf. Stevens, Slavin, & Farnish, 1991).

The study's assignment procedures raise the possibility of selection bias. The exper- imental schools were selected only if their teachers voted to implement the program. Selection bias would have been a more serious problem if we had compared these schools to schools that considered and rejected the program. In fact, there were only two schools we approached that did not choose to participate, so it is reasonable to assume that, given the chance, the control schools would quite likely have chosen to participate.

The only way to have removed selection bias would have been to randomly assign schools to treatments, something that, as a practical matter, was simply infeasible. Our use of HLM gives us a statistical penalty for having students grouped within schools whose characteristics may influence outcomes. Also if better schools had been selected for the experimental group, it would presumably have been reflected in the premea- sures, which, in fact, favored the control schools.

Another potential limitation of this study is that we did not quantitatively mea- sure the implementation of the various cooperative learning programs or other ele- ments of the cooperative elementary school. Instead, our goal was to coach teachers to improve the quality of their implementation. Clearly there was a significant amount of natural variation across teachers in the cooperative elementary school, as there is in all schools and all programs, which to some degree limited the generalizability of these results. Similarly the teachers in the cooperative elementary school were volunteers, which suggests that these results cannot be generalized to situations where the model is mandated.

The fact that standardized achievement tests were our only measure of student learning is somewhat limiting. The stability, multiple-choice format, and lack of overlap with the curricula taught inherent in these tests were disadvantages. However, we felt that the disadvantages were more than balanced by the advantages of higher reliability and the fact that the tests were generic. They did not measure specific curriculum content, hence, no specific curricular advantage was given to either the cooperative elementary school or the comparison schools.

Finally, the comparison of gifted students in the cooperative elementary schools to students in gifted enrichment programs may not be generalizable to all forms of gifted education. Clearly, enrichment programs differ from accelerated programs for gifted students, as well as from one another, making generalization to other programs limited. However, the results do show that well-structured, cooperative learning can be beneficial for gifted students.

Any study of a multifaceted program leaves many questions unanswered, and this one is no exception. Perhaps the most evident question revolves around the relative impact of each of the elements of the cooperative elementary school in producing the various effects on achievement, social relations, and students' attitudes. Only through component analyses of this model can one begin to disentangle the elements and their effects. Similarly, there remain questions about the nature of the student dialogue and classroom processes in the cooperative learning setting. It may be particularly important to understand these processes in terms of how learning disabled or other below-average students participate in collaborative dialogue and how that dialogue can help or hinder their development.

The duration of the study greatly reduces the potential for the Hawthorne effect to have produced the observed differences in students' achievement. Hawthorne effects are primarily novelty and attention effects. In a brief study these effects could be consequential. However, it does not seem likely that novelty and attention could sustain outstanding efforts in all major academic subjects for 2 years, as in this study. If the outcomes were due to Hawthorne effects, the entire effect would be expected to be seen during the first months. Yet in this study more gains were seen in the second year than in the first (see Table 1). This is more consistent with the argument that teachers were getting better at a new method than that students were profiting from a short-term burst of enthusiasm on the part of their teachers. For the students themselves, cooperative learning would have been seen as standard operating procedure long before the end of the 2 years.

While further research is needed to understand the effects of this program, the results of this study do suggest that cooperative learning can serve as a basis for school restructuring and can produce important benefits for a wide range of students. This study is the first and only evaluation of a cooperative elementary school. It is not merely another study of cooperative learning; it is the only study to evaluate cooperative learning as the focus of schoolwide change, the only study to evaluate cooperative learning in many subjects at once, and one of the few to show the effects of cooperative learning over a multiyear period.

The authors would like to thank the faculty and students of the Anne Arundel (MD) County Schools for their participation in this project. We would also like to thank the editors and the three anonymous reviewers for their helpful comments on the drafts of the article. This research was conducted by the Center for Research on Effective Schooling of Disadvantaged Students and supported by funds from the Office of Educational Research and Improvement (Grant No. R117 R90002) and the Office of Special Education and Rehabilitative Services (Grant No. G008730141-89), U.S. Department of Education. The opinions expressed in this report do not necessarily reflect the position or policy of OERI, OSERS, or the Department of Education, and no official endorsement should be inferred.
1Effect size is the mean difference between the experimental and the control groups divided by the control group's standard deviation.

CHAPTER 4

Success for All

Background and Commentary

Without any doubt, the development and evaluation of Success for All is the most important thing my colleagues and I have done to move the practice of education toward the idea of building a high floor under the achievement of all students. Not only has research on this program found consistent and powerful effects on the achievement of students in high-poverty schools, but the program has been enormously successful in dissemination. By fall, 1996, more than 400 schools in 26 U.S. states and three foreign countries will be using Success for All, reaching a total of more than 200,000 children.

Success for All come into being as a result of a request made to our group at Johns Hopkins in 1986 by a former Maryland Secretary of Human Resources, Kalman Hettleman, and the then Superintendent (Alice Pinderhughes) and Board President (Robert Embry) of the Baltimore City Public Schools. They asked us to design a program for students in inner-city schools that would not only raise students' average achievement levels, but would ensure that all students would succeed, at least in reading. We eagerly took on this challenge; we were finishing a book, *Effective Programs for Students At Risk* (Slavin, Karweit, & Madden, 1989), and were very interested in applying what we had learned to high-poverty schools

After a series of discussions with district leaders we worked with the district's elementary division to apply everything known about how to organize schools and classrooms to ensure that all children will learn. We began to work with one Baltimore elementary school in 1987-1988, and then added four more (plus one school in Philadelphia) in 1988-89. As research on the early implementers began to show strong positive effects on student reading performance and other outcomes, we began a national dissemination program, continuing to add schools to our evaluation design as we grow. At present, we are following schools in ten districts.

Most recently, we have begun a very large study of the program in Houston. Fifty-three schools are being compared to twenty-three matched comparison schools. More than half of the schools are using the Spanish version of the program, *Lee Conmigo*, as well as the English version. In this study we are comparing the effects of the program in schools implementing only the reading program, reading plus tutoring, or the full model. We are also learning how to maintain program quality in such a large implementation, and will be interested to see whether program outcomes can be as positive as they are in our more typical implementations involving many fewer schools. Early evaluation evidence is showing positive effects in schools that were able to get their programs off the ground in the early months of implementation, especially those implementing the full model. Complete data on program outcomes will not be collected until spring, 1996.

At this writing, Success for All is continuing to expand to many additional schools and districts, and our organization is learning new ways to provide quality professional development to ensure faithful, flexible, and thoughtful implementation. We are continuing our development of the program, especially of the Spanish version used in bilingual schools and the additional program elements that make a Success for All school into a Roots and Wings school (see Chapter 5).

Beyond the children and families our schools touch directly, the research on Success for All is important in demonstrating again and again that children placed at risk of school failure due to poverty or other factors can succeed throughout elementary school. This finding undermines the excuses so often given for the failure of children from poor communities. Implementing Success for All is certainly not the only way to ensure the success of the most at-risk children, but the program's consistent effects and replicability make it no longer possible to pretend that there is no practical solution to the school failure of so many children. Further, research on Success for All shows that it is possible to create school-by-school change on a large scale without compromising on the effectiveness of that change, and that it is possible to have fairly prescriptive school reform models introduced in schools without having to have each school staff reinvent the program. This demonstration also has importance beyond the program itself.

Ms. Martin's kindergarten class has some of the brightest, happiest, friendliest, and most optimistic kids you'll ever meet. Students in her class are glad to be in school, proud of their accomplishments, certain that they will succeed at whatever the school has to offer. Every one of them is a natural scientist, a storyteller, a creative thinker, a curious seeker of knowledge. Ms. Martin's class could be anywhere, in suburb or ghetto, small town or barrio, it doesn't matter. Kindergartners everywhere are just as bright, enthusiastic, and confident as her kids are.

Only a few years from now, many of these same children will have lost the spark they all started with. Some will have failed a grade. Some will be in special education. Some will be in long term remediation, such as Title I or other remedial programs. Some will be bored or anxious or unmotivated. Many will see school as a chore rather than a pleasure and will no longer expect to excel. In a very brief span of time, Ms. Martin's children will have defined themselves as successes or failures in school. All too often, only a few will still have a sense of excitement and positive self-expectations about learning. We cannot predict very well which of Ms. Martin's students will

This research was supported by grants from the Office of Educational Research and Improvement, U.S. Department of Education (Nos. OERI-R-117-R90002 and R-117-D40005), from the Carnegie Corporation, the Pew Charitable Trusts, and the Abell Foundation. However, any opinions expressed do not necessarily represent the positions or policies of our funders.

We would like to thank the following individuals for their help in research on Success for All: Matthew Riley, Lawrence Howe, and Fred Cottman of the Baltimore City Public Schools; Katherine Conner, Allie Mulvihill, and Inez Hill of the Philadelphia Public Schools; Claire Eadon, Cheryl Boan, and Patsy Griffing of the Charleston County (SC) Public Schools; Charles Welch, Cornelia Shideler, and Sharon Mukes of the Ft. Wayne (IN) Community Schools; Janis Hull of the Vallivue (ID) School District; Carolyn Burks of the Montgomery County (AL) Public Schools; Jerry Frye, Mark Lewis, and Dionicio Cruz of Modesto (CA) City Schools; Carley Ochoa and Susan Toscano of the Riverside Unified School District; Nancy Karweit, Gretta Gordy, Renee Kling, Alta Shaw, Robert Petza, Mary Alice Bond, and Barbara Haxby of the Johns Hopkins University; and Jason Casey, Brenda Johnson, Carole Bond, Anne Faulks, Ann Crawford, and Michele Shapiro of the University of Memphis.

succeed and which will fail, but we can predict based on the past that if nothing changes, far too many will fail. This is especially true if Ms. Martin's kindergarten happens to be located in a high-poverty neighborhood, in which there are typically fewer resources in the school to provide top-quality instruction to every child, fewer forms of rescue if children run into academic difficulties, and fewer supports for learning at home. Preventable failures occur in all schools, but in high poverty schools failure can be endemic, so widespread that it makes it difficult to treat each child at risk of failure as a person of value in need of emergency assistance to get back on track. Instead, many such schools do their best to provide the greatest benefit to the greatest number of children possible, but have an unfortunately well-founded expectation that a certain percentage of students will fall by the wayside during the elementary years.

Any discussion of school reform should begin with Ms. Martin's kindergartners. The first goal of reform should be to ensure that every child, regardless of home background, home language, or learning style, achieves the success that he or she so confidently expected in kindergarten, that all children maintain their motivation, enthusiasm, and optimism because they are objectively succeeding at the school's tasks. Any reform that does less than this is hollow and self-defeating.

What does it mean to succeed in the early grades? The elementary school's definition of success, and therefore the parents' and children's definition as well, is overwhelmingly success in reading. Very few children who are reading adequately are retained, assigned to special education, or given long-term remedial services. Other subjects are important, of course, but reading and language arts form the core of what school success means in the early grades.

When a child fails to read well in the early grades, he or she begins a downward progression. In first grade, some children begin to notice that they are not reading adequately. They may fail first grade or be assigned to long term remediation. As they proceed through the elementary grades, many students begin to see that they are failing at their full-time jobs. When this happens, things begin to unravel. Failing students begin to have poor motivation and poor self-expectations, which lead to continued poor achievement, in a declining spiral that ultimately leads to despair, delinquency, and dropout.

Remediating learning deficits after they are already well established is extremely difficult. Children who have already failed to learn to read, for example, are now anxious about reading, and doubt their ability to learn it. Their motivation to read may be low. They may ultimately learn to read but it will always be a chore, not a pleasure. Clearly, the time to provide additional help to children who are at risk is early, when children are still motivated and confident and when any learning deficits are relatively small and remediable. The most important goal in educational programming for students at risk of school failure is to try to make certain that we do not squander the greatest resource we have: the enthusiasm and positive self-expectations of young children themselves.

In practical terms, what this perspective implies is that schools, and especially Title 1, special education, and other services for at-risk children, must be shifted from an emphasis on remediation to an emphasis on prevention and early intervention. Prevention means providing developmentally appropriate preschool and kindergarten programs so that students will enter first grade ready to succeed, and it means providing regular classroom teachers with effective instructional programs, curricula, and professional development to enable them to see that most students are successful the first time they are taught. Early intervention means that supplementary instructional services are provided early in students' schooling and that they are intensive enough to bring at-risk students quickly to a level at which they can profit from good quality classroom instruction.

The purpose of this paper is to describe the current state of research on Success for All, a program built around the idea that every child can and must succeed in the early grades, no matter what this takes. The idea behind Success for All is to use everything we know about effective instruction for students at risk to direct all aspects of school and classroom organization toward the goal of preventing academic deficits from appearing in the first place; recognizing and intensively intervening with any deficits that do appear; and providing students with a rich and full curriculum to enable them to build on their firm foundation in basic skills. The commitment of Success for All is to do whatever it takes to see that every child becomes a skilled, strategic, and enthusiastic reader as they progress through the elementary grades.

Program Description

Success for All is built around the assumption that every child can read. We mean this not as wishful thinking or as philosophical statement, but as a practical, attainable reality. In particular, every child without organic retardation can learn to read. Some children need more help than others and may need different approaches than those needed by others, but one way or another every child can become a successful reader.

The first requirement for the success of every child is *prevention*. This means providing excellent preschool and kindergarten programs, improving curriculum, instruction, and classroom management throughout the grades, assessing students frequently to make sure they are making adequate progress, and establishing cooperative relationships with parents so they can support students learning at home.

Top-quality curriculum and instruction from age four on will ensure the success of most students, but not all of them. The next requirement for the success of *all* students is *intensive early intervention*. This means one-to-one tutoring by certified teachers for first graders having reading problems. It means being able to work with parents and social service agencies to be sure that all students attend school, have medical services or eyeglasses if they need them, have help with behavior problems, and so on.

The most important idea in Success for All is that the school must relentlessly stick with every child until that child is succeeding. If prevention is not enough the child may need tutoring. If this is not enough he or she may need help with behavior or attendance or eyeglasses. If this is not enough he may need a modified approach to reading. The school does not merely provide services to children, it constantly assesses the results of the services it provides and keeps varying or adding services until every child is successful.

Success for All began in one Baltimore elementary school in 1987-1988, and since then has expanded each year to additional schools. As of fall, 1995, it is in about 300 schools in 70 districts in 23 states throughout the U.S. The districts range from some of the largest in the country, such as Baltimore, Houston, Memphis, Philadelphia, Cincinnati, Cleveland, Chicago, New York, and Miami, to such middle-sized districts as Montgomery, Alabama; Rockford, Illinois; and Modesto and Riverside, California, to tiny rural districts, including two on the Navajo reservation in Arizona. Success for All reading curricula in Spanish have been developed and researched and are used in bilingual programs in California, Texas, Arizona, Florida, Illinois, New York, New Jersey, and Philadelphia. Almost all Success for All schools are high-poverty Title I schools, and the great majority are schoolwide projects. Otherwise, the schools vary widely.

Overview of Success for All Components

Success for All has somewhat different components at different sites, depending on the school's needs and resources available to implement the program (adapted from Slavin et al., 1992). However, there is a common set of elements characteristic of all.

Reading Program

Success for All uses a reading curriculum based on research and effective practices in beginning reading (e.g., Adams, 1990), and on effective use of cooperative learning (Slavin, 1995; Stevens, Madden, Slavin, and Farnish, 1987).

Reading teachers at every grade level begin the reading time by reading children's literature to students and engaging them in a discussion of the story to enhance their understanding of the story, listening and speaking vocabulary, and knowledge of story structure. In kindergarten and first grade, the program emphasizes the development of oral language and pre-reading skills through the use of thematically-based units which incorporate areas such as language, art, and writing under a science or social studies topic. A component called Story Telling and Retelling (STaR) involves the students in listening to, retelling, and dramatizing children's literature. Big books as well as oral and written composing activities allow students to develop concepts of print as they also develop knowledge of story structure. There is also a strong emphasis on phonetic awareness activities which help develop auditory discrimination and supports the development of reading readiness strategies.

Reading Roots is typically introduced in the second semester of kindergarten or in first grade. This K-1 beginning reading program uses as its base a series of phonetically regular but meaningful and interesting minibooks and emphasizes repeated oral reading to partners as well as to the teacher. The minibooks begin with a set of "shared stories," in which part of a story is written in small type (read by the teacher) and part is written in large type (read by the students). The student portion uses a phonetically controlled vocabulary. Taken together, the teacher and student portions create interesting, worthwhile stories. Over time, the teacher portion diminishes and the student portion lengthens, until students are reading the entire book. This scaffolding allows students to read interesting literature when they only have a few letter sounds.

Letters and letter sounds are introduced in an active, engaging set of activities that begins with oral language and moves into written symbols. Individual sounds are integrated into a context of words, sentences and stories. Instruction is provided in story structure, specific comprehension skills, metacognitive strategies for self-assessment and self-correction, and integration of reading and writing.

Spanish bilingual programs use an adaptation of *Reading Roots* called *Lee Conmigo* ("Read With Me"). *Lee Conmigo* uses the same instructional strategies as *Reading Roots,* but is built around the Macmillan *Campanitas de Oro* series.

When students reach the primer reading level, they use a program called *Beyond the Basics,* an adaptation of Cooperative Integrated Reading and Composition (CIRC) (Stevens, Madden, Slavin, & Farnish, 1987). *Beyond the Basics* uses cooperative learning activities built around story structure, prediction, summarization, vocabulary building, decoding practice, and story-related writing. Students engage in partner reading and structured discussion of stories or novels, and work toward mastery of the vocabulary and content of the story in teams. Story-related writing is also shared within teams. Cooperative learning both increases students' motivation and engages students in cognitive activities known to contribute to reading comprehension, such as elaboration, summarization, and rephrasing (see Slavin, 1995). Research on CIRC has found it to significantly increase students' reading comprehension and language skills (Stevens et al., 1987).

In addition to these story-related activities, teachers provide direct instruction in reading comprehension skills, and students practice these skills in their teams. Classroom libraries of trade books at students' reading levels are provided for each teacher, and students read books of their choice for homework for 20 minutes each night. Home readings are shared via presentations, summaries, puppet shows, and other formats twice a week during "book club" sessions.

Materials to support *Beyond the Basics* through the sixth grade (or beyond) exist in English and Spanish. The English materials are build around children's literature and around the most widely used basal series and anthologies. Supportive materials have been developed for more than 100 children's novels and for most current basal series. Spanish materials are similarly built around Spanish-language novels and the *Campanitas* basal program.

Beginning in the second semester of program implementation, Success for All schools usually implement a writing/ language arts program based primarily on cooperative learning principles (see Slavin, Madden, & Stevens, 1989/90).

Students in grades one to three (and sometimes 4 to 5 or 4 to 6) are regrouped for reading. The students are assigned to heterogeneous, age-grouped classes most of the day, but during a regular 90-minute reading period they are regrouped by reading performance levels into reading classes of students all at the same level. For example, a 2-1 reading class might contain first, second, and third grade students all reading at the same level. The reading classes are smaller than homerooms because tutors and other certified staff (such as librarians or art teachers) teach reading during this common reading period. Regrouping allows teachers to teach the whole reading class without having to break the class into reading groups. This greatly reduces the time spent in seatwork and increases direct instruction time, eliminating workbooks, dittos, or other follow-up activities which are needed in classes that have multiple reading groups. The regrouping is a form of the Joplin Plan, which has been found to increase reading achievement in the elementary grades (Slavin, 1987).

Eight-Week Reading Assessments

At eight week intervals, reading teachers assess student progress through the reading program. The results of the assessments are used to determine who is to receive tutoring, to change students' reading groups, to suggest other adaptations in students' programs, and to identify students who need other types of assistance, such as family interventions or screening for vision and hearing problems. The assessments are curriculum-based measures that include teacher observations and judgments as well as more formal measures of reading comprehension.

Reading Tutors

One of the most important elements of the Success for All model is the use of tutors to promote students' success in reading. One-to-one tutoring is the most effective form of instruction known (see Wasik & Slavin, 1993). The tutors are certified teachers with experience teaching Title 1, special education, and/or primary reading. Often, well-qualified paraprofessionals also tutor children with less severe reading problems. In this case, a certified tutor monitors their work and assists with the diagnostic assessment and intervention strategies. Tutors work one-on-one with students who are having difficulties keeping up with their reading groups. The tutoring occurs in 20-minute sessions during times other than reading or math periods.

In general, tutors support students' success in the regular reading curriculum, rather than teaching different objectives. For example, the tutor will work with a student on the same story and concepts being read and taught in the regular reading class. However, tutors seek to identify learning problems and use different strategies to teach the same skills. They also teach metacognitive skills beyond those taught in the class room program (Wasik & Madden, 1995). Schools may have as many as six or more teachers serving as tutors depending on school size, need for tutoring, and other factors.

During daily 90-minute reading periods, certified tutors serve as additional reading teachers to reduce class size for reading. Reading teachers and tutors use brief forms to communicate about students' specific problems and needs and meet at regular times to coordinate their approaches with individual children.

Initial decisions about reading group placement and the need for tutoring are based on informal reading inventories that the tutors give to each child. Subsequent reading group placements and tutoring assignments are made using the curriculum-based assessments described above. First graders receive priority for tutoring, on the assumption that the primary function of the tutors is to help all students be successful in reading the first time, before they fail and become remedial readers.

Preschool and Kindergarten

Most Success for All schools provide a half-day preschool and/or a full-day kindergarten for eligible students. The preschool and kindergarten programs focus on providing a balanced and developmentally appropriate learning experience for young children. The curriculum emphasizes the development and use of language. It provides a balance of academic readiness and nonacademic music, art, and movement activities in a series of thematic units. Readiness activities include use of the Peabody Language Development Kits and Story Telling and Retelling (STaR) in which students retell stories read by the teachers. Pre-reading activities begin during the second semester of kindergarten.

Family Support Team

Parents are an essential part of the formula for success in Success for All. A Family Support Team works in each school, serving to make families feel comfortable in the school and become active supporters of their child's education as well as providing specific services. The Family Support Team consists of the Title I parent liaison, vice-principal (if any), counselor (if any), facilitator, and any other appropriate staff already present in the school or added to the school staff.

The Family Support Team first works toward good relations with parents and to increase involvement in the schools. Family Support Team members may complete "welcome" visits for new families. They organize many attractive programs in the school, such as parenting skills workshops. Many schools use a program called "Raising Readers" in which parents are given strategies to use in reading with their own children.

The Family Support Team also intervenes to solve problems. For example, they may contact parents whose children are frequently absent to see what resources can be provided to assist the family in getting their child to school. Family support staff, teachers, and parents work together to solve school behavior problems. Also, family support staff are called on to provide assistance when students seem to be working at less than their full potential because of problems at home. Families of students who are not receiving adequate sleep or nutrition, need glasses, are not attending school regularly, or are exhibiting serious behavior problems, may receive family support assistance.

The Family Support Team is strongly integrated into the academic program of the school. It receives referrals from teachers and tutors regarding children who are not making adequate academic progress, and thereby constitutes an additional stage of intervention for students in need above and beyond that provided by the classroom teacher or tutor. The Family Support Team also encourages and trains the parents to fulfill numerous volunteer roles within the school, ranging from providing a listening ear to emerging readers to helping in the school cafeteria.

Program Facilitator

A program facilitator works at each school to oversee (with the principal) the operation of the Success for All model. The facilitator helps plan the Success for All program, helps the principal with scheduling, and visits classes and tutoring sessions frequently to help teachers and tutors with individual problems. He or she works directly with the teachers on implementation of the curriculum, classroom management, and other issues, helps teachers and tutors deal with any behavior problems or other special problems, and coordinates the activities of the Family Support Team with those of the instructional staff.

Teachers and Teacher Training

The teachers and tutors are regular certified teachers. They receive detailed teacher's manuals supplemented by three days of inservice at the beginning of the school year. For classroom teachers of grades 1-3 and for reading tutors, these training sessions focus on implementation of the reading program, and their detailed teachers' manuals cover general teaching strategies as well as specific lessons. Preschool and kindergarten teachers and aides are trained in use of the STaR and Peabody programs, thematic units, and other aspects of the preschool and kindergarten models. Tutors later receive two additional days of training on tutoring strategies and reading assessment.

Throughout the year, additional inservice presentations are made by the facilitators and other project staff on such topics as classroom management, instructional pace, and cooperative learning. Facilitators also organize many informal sessions to allow teachers to share problems and problem solutions, suggest changes, and discuss individual children. The staff development model used in Success for All emphasizes relatively brief initial training with extensive classroom follow-up, coaching, and group discussion.

Advisory Committee

An advisory committee composed of the building principal, program facilitator, teacher representatives, parent representatives, and family support staff meets regularly to review the progress of the program and to identify and solve any problems that arise. In most schools existing site-based management teams are adapted to fulfill this function. In addition, grade level teams and the Family Support Team meet regularly to discuss common problems and solutions and to make decisions in their areas of responsibility.

Special Education

Every effort is made to deal with students' learning problems within the context of the regular classroom, as supplemented by tutors. Tutors evaluate students' strengths and weaknesses and develop strategies to teach in the most effective way. In some schools, special education teachers work as tutors and reading teachers with students identified as learning disabled as well as other students experiencing learning problems who are at risk for special education placement. One major goal of Success for All is to keep students with learning problems out of special education if at all possible, and to serve any students who do qualify for special education in a way that does not disrupt their regular classroom experience (see Slavin, Madden, Karweit, Dolan, Wasik, Shaw, Mainzer, & Haxby, 1991).

Relentlessness

While the particular elements of Success for All may vary from school to school, there is one feature we try to make consistent in all: a relentless focus on the success of every child. It would be entirely possible to have tutoring and curriculum change and family support and other services yet still not ensure the success of at-risk children. Success does not come from piling on additional services but from coordinating human resources around a well-defined goal, constantly assessing progress toward that goal, and never giving up until success is achieved.

None of the elements of Success for All are completely new or unique to this program. What is most distinctive about the program is its schoolwide, coordinated, and proactive plan for translating positive expectations into concrete success for all children. Every child can complete elementary school reading confidently, strategically, and joyfully and can maintain the enthusiasm and positive self-expectations with which they came to first grade. The purpose of Success for All is to see that this vision can become a practical reality in every school.

Research on Success for All

From the very beginning, there has been a strong focus in Success for All on research and evaluation. We began longitudinal evaluations of the program in its earliest sites, six schools in Baltimore and Philadelphia and one in Charleston, SC. Later, third-party evaluators at the University of Memphis, Steve Ross, Lana Smith, and their colleagues, added evaluations in Memphis, Montgomery (AL), Ft. Wayne (IN), and Caldwell (ID). Most recently, studies focusing on English language learners in California have been conducted in Modesto and Riverside by Marcie Dianda of the Southwest Regional Laboratory. Each of these evaluations has compared Success for All schools to matched comparison schools on measures of reading performance, starting with cohorts in kindergarten or in first grade and continuing to follow these students as long as possible (details of the evaluation design appear below). Vaguaries of funding and other local problems have ended some evaluations prematurely, but most have been able to follow Success for All schools for many years. As of this writing, there are seven years of continuous data from the six original schools in Baltimore and Philadelphia, and varying numbers of years of data from seven other districts, a total of nineteen schools (and their matched control schools). Table 1 lists the districts and characteristics of the schools.

Earlier evaluations of Success for All schools have found almost uniformly positive outcomes for all schools on all reading measures (see Slavin et al., 1990; Slavin et al., 1993; Madden et al., 1994). Smaller special-purpose studies have also found positive effects of Success for All on such outcomes as attendance and reduced special education placement and referrals (Slavin et al., 1992, 1994).

In order to summarize the outcomes from all schools and all years involved in experimental control comparisons, this paper uses a method of analysis, called a multi-site replicated experiment (Slavin, 1993), in which each grade level *cohort* (students in all classes in that grade in a given year) in each school is considered a replication. In other words, if three first grades have proceeded through school X, each first grade cohort (compared to its control group) produces an effect size representing the experimental-control difference in student achievement that year. For example, across 19 schools ever involved in Success for All evaluations, there are a total of 55 first grade cohorts from which experimental and control achievement data have been collected. This procedure is a direct application of a procedure common in medical research called multicenter clinical trial (Horwitz, 1987). In such studies small-scale experiments located in different sites over extended time periods are combined into one large-scale experiment. For example, patients entering any of several hospitals with a given disease might be given an experimental drug or a placebo at random. If the disease is relatively rare, no one hospital's experiment would have an adequate sample size to assess the drug's effects, but combining results over many hospitals over time does provide an adequate sample. In schoolwide reform, the "patient" is an entire grade level in a school, perhaps 100 children. Obtaining an adequate sample of schools at any point in time would require involving thousands of children and hundreds of teachers.

The idea of combining results across experiments is not, of course, foreign to educational research. This is the essence of meta-analysis (Glass, McGaw, & Smith, 1981). However, metaanalyses combine effect sizes (proportions of a standard deviation separating experimental and control groups) across studies with different designs, measures, samples, and other features, leading to such charges as that they may mislead readers by "combining apples and oranges" or by missing unwritten or unpublished studies in which effects were zero or negative (Matt & Cook, 1994; Slavin, 1986).

Combining results across geographically separated experiments into one study is also not unheard-of in educational research. For example, Pinnell, Lyons, DeFord, Bryk, and Seltzer (1994) studied the Reading Recovery tutoring model in ten Ohio districts. Three variations of Reading Recovery were compared to control groups in each district, and results were then aggregated using the cohort of tutored first graders as the unit of analysis. A multi-site replicated experiment only adds to this design the accumulation of experimental-control differences over time.

In addition to applying the multi-site replicated experiment design to data from Success for All schools, this paper also summarizes results of several studies in particular subsets of schools. These include studies of outcomes of the Spanish version of Success for All, *Lee Conmigo;* studies of Success for All with students in English as a Second Language (ESL) programs; studies of special education outcomes of the model; and studies comparing Success for All and Reading Recovery. This paper summarizes the state of research on Success for All in all study sites as of the seventh year of program implementation.

Table 1   Characteristics of Success for All Schools in the Longitudinal Study

District/School	Enrollment	% Free Lunch	Ethnicity	Date Began SFA	Collected	Pre-School?	Full day K?	Comments
Baltimore								
B1	500	83	B-96% W-4%	1987	88-94	yes	yes	First SFA school; had additional funds first 2 years.
B2	500	96	B-100%	1988	89-94	some	yes	Had add'l funds first 4 years.
B3	400	96	B-100%	1988	89-94	some	yes	
B4	500	85	B-100%	1988	89-94	some	yes	
B5	650	96	B-100%	1988	89-94	some	yes	
Philadelphia								
P1	620	96	A-60% W-20% B-20%	1988	89-94	no	yes	Large ESL program for Cambodian children.
P2	600	97	B-100%	1991	92-93	some	yes	
P3	570	96	B-100%	1991	92-93	no	yes	
P4	840	98	B-100%	1991	93	no	yes	
P5	700	98	L-100%	1992	93-94	no	yes	Study only involves students in Spanish bilingual program.
Charleston, SC								
CS1	500	40	B-60% W-40%	1990	91-92	no	no	
Memphis, TN								
MT1	350	90	B-95% W-5%	1990	91-94	yes	no	Program implemented only in grades K-2.
MT2	530	90	B-100%	1993	94	yes	yes	
MT3	290	86	B-100%	1993	94	yes	yes	
MT4	370	90	B-100%	1993	94	yes	yes	
Ft. Wayne, IN								
F1	330	65	B-56% W-44%	1991	92-94	no	yes	SFA schools (& controls) are part of desegregation plan.
F2	250	55	B-55% W-45%	1991	92-94	no	yes	SFA schools (& controls) are part of desegregation plan.
Montgomery, AL								
MA1	450	95	B-100%	1991	93-94	no	yes	
MA2	460	97	B-100%					
Caldwell, ID								
C11	400	20	W-80% L-20%	1991	93-94	no	no	Study compares two SFA schools to Reading Recovery school.
Modesto, CA								
MC1	640	70	W-54% L-25%	1992	94	yes	no	Large ESL program for students speaking 17 languages.
MC2	560	98	A-17% B-4% L-66% W-24% A-10%	1992	94	yes	no	Large Spanish bilingual program.
Riverside, CA								
R1	930	73	L-54% W-33% B-10% A-3%	1992	94	yes	no	Large Spanish bilingual and ESL programs. Year-round school.

Key: B - African-American; L - Latino; A - Asian American; W- White

Evaluation Design

A common evaluation design, with variations due to local circumstances, has been used in all Success for All evaluations. Every Success for All school involved in a formal evaluation is matched with a control school that is similar in poverty level (percent of students qualifying for free lunch), historical achievement level, ethnicity, and other factors. Children in the Success for All schools are then matched on district-administered standardized test scores given in kindergarten or (starting in 1991 in four districts) on Peabody Picture Vocabulary Test (PPVT) scores given by the project in the fall of kindergarten or first grade. In some cases, analyses of covariance rather than individual child matches were used, and at Key School in Philadelphia schools were matched but individual children could not be (because the school serves many limited English proficient students who were not tested by the district in kindergarten).

The measures used in the evaluations were as follows:

1. *Woodcock Reading Mastery Test.* Three Woodcock scales, Word Identification, Word Attack, and Passage Comprehension, were individually administered to students by trained testers. Word Identification assesses recognition of common sight words, Word Attack assesses phonetic synthesis skills, and Passage Comprehension assesses comprehension in context. Students in Spanish bilingual programs were given the Spanish versions of these scales.

2. *Durrell Analysis of Reading Difficulty.* The Durrell Oral Reading scale was also individually administered to students in grades 1-3. It presents a series of graded reading passages which students read aloud, followed by comprehension questions.

3. *Gray Oral Reading Test.* Comprehension and passage scores from the Gray Oral Reading Test were obtained from students in grades 4-5.

Except at Key, analyses of covariance with pretests as covariates were used to compare raw scores in all evaluations, and separate analyses were conducted for students in general and for students in the lowest 25% of their grades. At Key, analyses of variance were used and results were reported separately for Asian (mostly Cambodian) students and for non-Asian students.

The tables and figures presented in this paper summarize student performance in grade equivalents (adjusted for covariates) and effect size (proportion of a standard deviation separating the experimental and control groups), averaging across individual measures. Neither grade equivalents nor averaged scores were used in the analyses, but they are presented here as a useful summary. Outcomes are presented for all students in the relevant grades in Success for All and control schools, and also those for the students in the lowest 25% of their grades, who are most at risk. In most cases the low 25% was determined based on Peabody Picture Vocabulary Test scores given as pretests. In Baltimore and Charleston, South Carolina, however, Peabody pretests were not given and low 25% analyses involve the lowest-performing students at posttest. At Philadelphia's Key School, outcomes are shown separately for Asian and non-Asian students.

Each of the evaluations summarized in this paper follows children who began in Success for All in first grade or earlier, in comparison to children who had attended the control school over the same period. Because Success for All is a prevention and early intervention program, students who start in it after first grade are not considered to have received the full treatment (although they are of course served within the schools). For more details on methods and findings, see Slavin et al. (1992) and the full site reports.

Reading Outcomes

The results of the multi-site replicated experiment evaluating Success for All are summarized in Figure 1 and Tables 2-6 for each grade level, 1-5. Each table shows means in raw scores, grade equivalents, and effect sizes. The analyses compare cohort means for experimental and control schools; for example, the t statistics presented in Table 2 compare 55 experimental to 55 control cohorts, with cohort (50-150 students) as the unit of analysis. The standard deviations show variation among school means, but effect sizes are means of all experimental/control comparisons, which are computed using individual student data Grade equivalents are based on the means, and are only presented for their information value. No analyses were done using grade equivalents.

The results summarized in Tables 2-6 show statistically significantly (p=.05 or better) positive effects of Success for All (compared to controls) on every measure at every grade level, 1-5. For students in general, effect sizes averaged around a half standard deviation at all grade levels. Effects were somewhat higher than this for the Woodcock

Table 2 Cohort Means for Success for All and Control Schools

	Grade 1 (N=55)					
	All Students				*Lowest 25%*	
	SFA		*Control*	*SFA*		*Control*
Durrell Oral Reading						
Mean	6.38		4.75	3.27		2.07
(SD)	(1.83)		(1.49)	(2.61)		(2.03)
GE	1.98		1.73	1.48		1.27
t		6.73***			5.41***	
ES		+0.43			+0.65	
Woodcock Passage Comprehension						
Mean	15.15		12.48	8.76		6.18
(SD)	(2.65)		(3.26)	(3.96)		(3.84)
GE	1.61		1.47	1.29		1.16
t		6.24***			6.32***	
ES		+0.42			+0.86	
Woodcock Word Attack						
Mean	13.60		8.08	8.58		3.59
(SD)	(3.60)		(3.43)	(5.33)		(3.32)
GE	1.82		1.50	1.53		1.16
t		16.32***			9.78***	
ES		+0.79			+1.70	
Woodcock Word Identification						
Mean	29.05		23.23	18.31		13.22
(SD)	(6.78)		(6.22)	(8.18)		(6.40)
GE	1.79		1.60	1.45		1.28
t		9.58***			7.27***	
ES		+0.49			+0.80	
Mean GE	1.80		1.57	1.44		1.22
Mean ES		+0.53			+1.03	

*	$p < .05$
**	$p < .01$
***	$p < .001$

Table 3 Cohort Means for Success for All and Control Schools

		Grade 2 (N=36)		
	All Students		*Lowest 25%*	
	SFA	*Control*	*SFA*	*Control*
Durrell Oral Reading				
Mean	12.39	10.10	7.08	4.47
(SD)	(2.27)	(2.22)	(3.36)	(2.60)
GE	3.00	2.61	2.11	1.68
t	4.97***		7.04***	
ES	+0.38		+1.01	
Woodcock Passage Comprehension				
Mean	25.30	21.52	17.12	12.54
(SD)	(2.48)	(3.71)	(7.22)	(4.90)
GE	2.43	2.05	1.71	1.48
t	4.61***		3.70**	
ES	+0.41		+0.96	
Woodcock Word Attack				
Mean	20.15	14.28	12.46	5.91
(SD)	(3.87)	(3.94)	(6.94)	(4.41)
GE	2.42	1.86	1.74	1.34
t	8.52***		8.68***	
ES	+0.71		+1.75	
Woodcock Word Identification				
Mean	46.29	39.08	33.84	23.92
(SD)	(5.79)	(6.80)	(11.11)	(9.07)
GE	2.52	2.15	1.93	1.63
t	7.27***		6.13***	
ES	+0.48		+0.87	
Mean GE	2.59	2.17	1.87	1.53
Mean ES	+0.50		+1.15	

* $p < .05$
** $p < .01$
*** $p < .001$

Table 4 Cohort Means for Success for All and Control Schools

		Grade 3 (N=33)		
	All Students		*Lowest 25%*	
	SFA	*Control*	*SFA*	*Control*
Durrell Oral Reading				
Mean	17.65	14.85	10.56	7.23
(SD)	(2.50)	(2.56)	(2.97)	(3.31)
GE	3.87	3.41	2.69	(2.14)
t	5.42***		6.76***	
ES	+0.36		+0.96	
Woodcock Passage Comprehension				
Mean	30.98	25.78	21.56	14.20
(SD)	(3.69)	(2.87)	(4.68)	(5.47)
GE	3.03	2.48	2.06	1.56
t	5.63***		6.10***	
ES	+0.51		+1.78	
Woodcock Word Attack				
Mean	**24.06**	19.21	14.35	9.40
(SD)	(4.03)	(4.37)	(7.06)	(6.62)
GE	2.91	2.32	1.87	1.57
t	6.07***		6.43***	
ES	+0.45		+1.18	
Woodcock Word Identification				
Mean	55.38	48.13	41.90	32.05
(SD)	(5.87)	(5.03)	(7.37)	(7.39)
GE	3.24	2.64	2.30	1.88
t	6.87***		7.24***	
ES	+0.39		+0.85	
Mean GE	3.26	2.71	2.23	1.79
Mean ES	+0.43		+1.19	

* $p < .05$
** $p < .01$
*** $p < .001$

Table 5 Cohort Means for Success for All and Control Schools

			Grade 4 (N=13)		
		All Students		*Lowest 25%*	
	SFA	*Control*	*SFA*		*Control*
Gray Comprehension					
Mean	22.38	18.01	14.17		5.42
(SD)	(3.33)	(2.74)	(4.04)		(2.63)
GE	3.78	3.10	2.43		1.44
t		3.67**		4.19**	
ES		+0.44		+2.21	
Gray Passage					
Mean	32.78	24.37	13.20		2.97
(SD)	(7.40)	(4.90)	(5.11)		(2.87)
GE	4.48	3.64	2.44		1.20
t		3.79**		4.68**	
ES		+0.51		+1.64	
Woodcock Passage Comprehension					
Mean	34.31	28.80	24.60		13.18
(SD)	(2.84)	(1.93)	(2.97)		(4.03)
GE	3.46	2.78	2.36		1.54
t		6.11***		15.27***	
ES		+0.62		+1.61	
Woodcock Word Attack					
Mean	26.27	19.51	11.60		4.87
(SD)	(4.97)	(2.28)	(3.93)		(3.21)
GE	3.35	2.35	1.68		1.27
t		4.46***		2.41**	
ES		+0.47		+2.26	
Woodcock Word Identification					
Mean	63.65	55.11	47.89		31.53
(SD)	(4.64)	(4.01)	(5.16)		(6.17)
GE	4.13	3.21	2.62		1.87
t		5.77***		4.67**	
ES		+0.61		+2.87	
Mean GE	3.84	3.02	2.31		1.46
Mean ES		+0.53		+1.68	

* $p < .05$
** $p < .01$
*** $p < .001$

Word Attack scale in first and second grades, but in grades 3-5 effect sizes were more or less equivalent on all aspects of reading Consistently, effect sizes for students in the lowest 25% of their grades were particularly positive, ranging from ES = +1.03 in first grade to ES = +1.68 in fourth grade. Again, cohort-level analyses found statistically significant differences favoring low achievers in Success for All on every measure at every grade level.

Changes in Effect Sizes Over Years of Implementation

One interesting trend in outcomes from comparisons of Success for All and control schools relates to changes in effects sizes according to the number of years a school has been implementing the program. Figure 2, which summarizes these data, was created by pooling effect sizes for all cohorts in their first year of implementation, all in their second year, and so on, regardless of calendar year.

Figure 2 shows that mean reading effect sizes progressively increase with each year of implementation. For example, Success for All first graders score substantially better than control first graders at the end of the first year of implementation (ES = +0.49). The experimental-control difference is even higher for first graders attending schools in the second year of program implementation (ES = +0.53), increasing to an effect

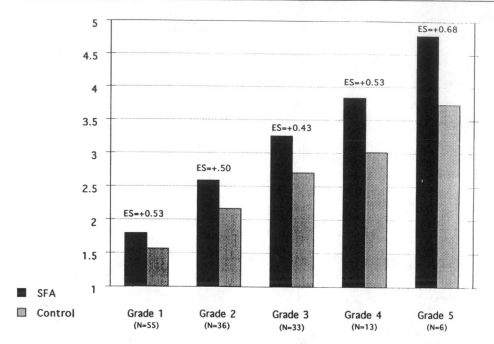

Figure 1   Comparison of Success for All and Control Schools in Mean Reading Grade Equivalents
and Effect Sizes 1988-1994

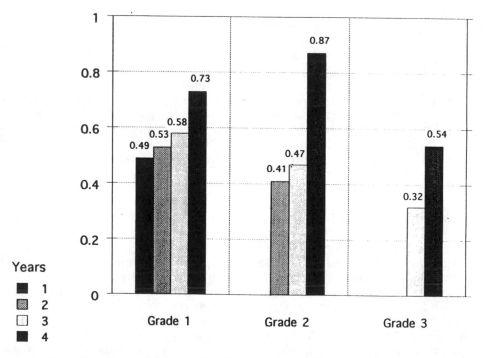

Figure 2   Effect Sizes Comparing Success for All and Control Schools According to Implementation
Year

Table 6 Cohort Means for Success for All and Control Schools

| | Grade 5 (N=6) | | | |
| | All Students | | Lowest 25% | |
	SFA	Control	SFA	Control
Gray Comprehension				
Mean	27.36	21.65	17.79	8.98
(SD)	(3.21)	(3.66)	(4.62)	(2.31)
GE	5.27	3.63	3.06	1.80
t		3.55**		3.66*
ES		+0.59		+ 1.35
Gray Passage				
Mean	43.32	31.37	17.20	8.36
(SD)	(7.72)	(6.17)	(4.80)	(3.52)
GE	5.43	4.37	2.92	1.74
t		6.61***		6.20**
ES		+0.67		+1.36
Woodcock Passage Comprehension				
Mean	37.48	32.60	28.76	22.81
(SD)	(2.43)	(1.42)	(2.45)	(2.99)
GE	4.10	3.23	2.78	2.18
t		7.51***		3.50*
ES		+0.69		+0.79
Woodcock Word Attack				
Mean	29.60	21.73	17.49	8.23
(SD)	(3.25)	(1.07)	(3.00)	(1.37)
GE	4.50	2.61	2.15	1.51
t		6.22**		8.27***
ES		+0.74		+1.83
Woodcock Word Identification				
Mean	69.94	60.35	54.20	40.78
(SD)	(4.63)	(2.76)	(5.25)	(3.72)
GE	4.79	3.74	3.12	2.24
t		5.36**		5.23**
ES		+0.71		+1.15
Mean GE	4.82	3.52	2.81	1.89
Mean ES		+0.68		+1.29

*   *p* <.05
**   *p* <.01
*** *p* <.001

size of +0.73 for schools in their fourth implementation year. A similar pattern is apparent for second and third grade cohorts.

There are two likely explanations for this gain in experimental-control differences. One is that as schools get better at implementing Success for All, outcomes improve. This is a logical outcome, which gives evidence of the degree to which on-going professional development, coaching, and reflection enable school staffs to progressively improve student achievement over time. However, it is also important to note that while first-year first grade cohorts started the program in first grade, second-year cohorts started in kindergarten and most third- and fourth-year cohorts started in prekindergarten. Some or all of the gain in effect sizes could be due to a lasting effect of participation in the Success for All prekindergarten and kindergarten program.

Whatever the explanation, the data summarized in Figure 2 show that while Success for All has an immediate impact on student reading achievement, this impact grows over successive years of implementation. Over time, schools may become increasingly able to provide effective instruction to all of their students, to approach the goal of success for *all*.

Success for All and English Language Learners

The education of English language learners is at a crossroads. For many years, researchers, educators, and policy makers have debated questions of the appropriate language instruction for students who enter elementary school speaking languages other than English. Research on this topic has generally found that students taught to read their home language and then transitioned to English ultimately become better readers in English than do students taught to read only in English (Garcia, 1991; Willig, 1985; Wong-Fillmore & Valdez, 1986). More recently, however, attention has shifted to another question. Given that students are taught to read their home language, how can we ensure that they *succeed* in that language? (See, for example, Garcia, 1994). There is no reason to expect that children failing to read well in Spanish, for example, will later become good readers and successful students in English. On the contrary, research consistently supports the common-sense expectation that the better students in Spanish bilingual programs read Spanish, the better their English reading will be (Garcia, 1991; Hakuta & Garcia, 1989). Clearly, the quality of instruction in home-language reading is a key factor in the ultimate school success of English language learners, and must be a focus of research on the education of these children.

Even if all educators and policy makers accepted the evidence favoring bilingual over English-only instruction, there would still be large numbers of English language learners being taught to read in English. This is true because of practical difficulties of providing instruction in languages other than English or Spanish; teachers fully proficient in Southeast Asian languages, Arabic, and other languages are in short supply, as are materials to teach in these languages. Speakers of languages other than English or Spanish are among the fastest-growing groups in our nation's schools (GAO, 1994). Further, many Spanish-dominant students are taught to read in English, either because of shortages of bilingual teachers, insufficient numbers of Spanish-dominant students in one school, parental desires to have their children taught in English, or other factors. For these reasons, a large percentage of English language learners will always be taught in English only, with instruction in English as a second language (ESL). As with bilingual programs, the quality of reading instruction, ESL instruction, and the integration of the two are essential in determining the success of English language learners being taught in English only.

There is remarkably little research evaluating programs designed to increase the Spanish reading performance of students in bilingual programs. Hertz-Lazarowitz, Ivory, & Calderon (1993) evaluated a bilingual adaptation of Cooperative Integrated Reading and Composition (BCIRC) in El Paso elementary schools starting in second grade. This program, an adaptation of the CIRC program that forms the basis of the upper-elementary reading program used in Success for All, involves having students work in small cooperative groups. Students read to each other, work together to identify characters, settings, problems, and problem solutions in narratives, summarize stories to each other, and work together on writing, reading comprehension, and vocabulary activities. Students in BCIRC classes scored significantly better than control students on the Spanish Texas Assessment of Academic Skills (TAAS) at the end of second grade, and as they transitioned to English in third and fourth grades they performed significantly better than control students on standardized reading tests given in English.

The first application of Success for All to English language learners began in Philadelphia's Francis Scott Key School, which serves a high-poverty neighborhood in which more than 60% of students enter the schools speaking Cambodian or other Southeast Asian languages. An adaptation of Success for All was designed to meet the needs of these children. This adaptation focused on integrating the work of ESL teachers and reading teachers, so that ESL teachers taught a reading class and then helped limited English proficient students with the specific language and reading skills needed to succeed in the school's (English) reading program. In addition, a cross-age

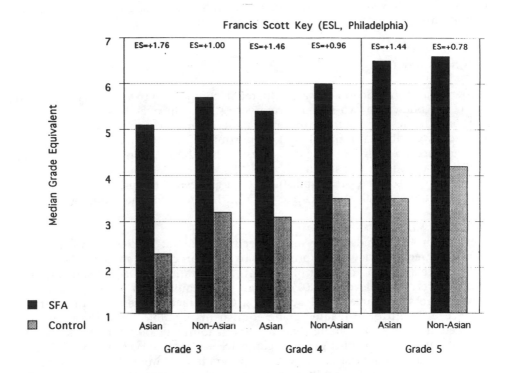

Figure 3 Achievement Medians (Grade Equivalents and Effect Sizes) for Success for All and Control Schools

tutoring program enabled fifth graders, now fully bilingual in English and Cambodian, to help kindergartners succeed in the English program. The performance of students at Francis Scott Key has been compared to that of students in a matched comparison school each year, and the results have consistently favored Success for All for Asian as well as non-Asian students (Slavin & Yampolsky, 1991). The present paper reports the reading performance of the English language learners in grades 3-5 at Key and its comparison school as of spring, 1994, the end of the sixth year of program implementation (see Slavin & Madden, 1995).

In 1992, a Spanish adaptation of the Success for All reading program called *Lee Conmigo* ("Read With Me") was developed for use in Spanish bilingual programs. During the 1992-1993 school year the entire Success for All program (including *Lee Conmigo* for LEP students) was implemented in one Philadelphia school serving a predominately Latino (mostly Puerto Rican) student body. The first year results showed the Spanish bilingual students to be performing substantially better than controls on individually administered tests of Spanish (Slavin & Madden, 1994). This paper reports the results for the second graders who completed their second year in *Lee Conmigo* (see Slavin & Madden, 1995).

A third evaluation of Success for All with English language learners was carried out by Dianda & Flaherty (1995) at the Southwest Regional Laboratory in Southern California. This study involved three schools. Fremont Elementary in Riverside, California, and Orville Wright Elementary in Modesto, are schools with substantial Spanish bilingual programs. The third, El Vista Elementary, also in Modesto, served a highly diverse student body speaking 17 languages using an ESL approach. Students in all three schools were compared to matched students in matched schools. In each case, students were assessed in the language of instruction (English or Spanish).

Francis Scott Key (ESL)

The program at Francis Scott Key was evaluated in comparison to a similar Philadelphia elementary school. The two schools were very similar in overall achievement level and other variables. Thirty-three percent of the comparison school's students were Asian (mostly Cambodian), the highest proportion in the city after Key. The percentage of students receiving free lunch was very high in both schools, though higher at Key (96%) than at the comparison school (84%).

The data reported here are for all students in grades 3-5 in Spring, 1994. With the exception of transfers, all students had been in the program since kindergarten.

Results: Asian Students
The results for Asian students are summarized in Figure 3. Success for All Asian students at all three grade levels performed far better than control students. Differences between Success for All and control students were statistically significant on every measure at every grade level (p<.001). Median grade equivalents and effect sizes were computed across the three Woodcock scales. On average, Success for All Asian students exceeded control in reading grade equivalents by almost three years in third grade (Median ES = +1.76), more than 2 years in fourth grade (Median ES = +1.46), and about three years in fifth grade (Median ES = +1.44). Success for All Asian students were reading more than a full year above grade level in grade 3 and more than a half-year above in fourth and fifth grade, while similar control students were reading more than a year below grade level at all three grade levels.

Results: Non-Asian Students
Outcomes of Success for All for non-Asian students, also summarized in Figure 3, were also very positive in grades 3-5. Experimental-control differences were statistically significant (p<.05 or better) on every measure at every grade level. Effect sizes were somewhat smaller than for Asian students, but were still quite substantial, averaging +1.00 in grade 3, +0.96 in grade 4, and +0.78 in grade 5. Effect sizes were particularly large for the Passage Comprehension measure at all three levels. Success for All students averaged almost two years above grade level in third grade, more than a year above grade level in fourth grade, and about eight months above grade level in fifth grade; at all grade levels, Success for All averaged about 2.5 years higher than control students.

Fairhill (Bilingual)

The bilingual version of Success for All, *Lee Conmigo*, was first implemented at Fairhill Elementary School, a school in inner-city Philadelphia. Fairhill serves a student body of 694 students of whom 78% are Hispanic and 22% are African-American. A matched comparison school was also selected. Nearly all students in both schools qualified for free lunches. Both schools were Chapter 1 schoolwide projects, which means that both had high (and roughly equivalent) allocations of Chapter 1 funds that they could use flexibly to meet student needs.

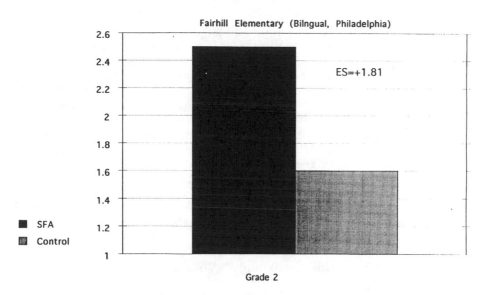

Figure 4 Spanish Reading Achievement Medians (Grade Equivalents and Effect Sizes) For Success for All and Control Schools, Spanish-Dominant Students

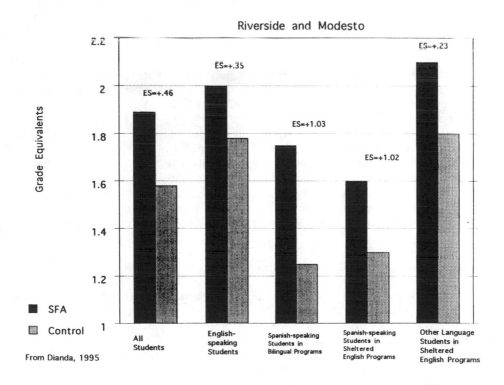

Figure 5 Achievement of Success for All and Control Students by Language Group

Results.

All students defined by district criteria as limited English proficient at Fairhill and its control school were pretested at the beginning of first grade on the Spanish Peabody Picture Vocabulary Test (PPVT). Each following May, these students were tested by native language speakers on three scales of the Spanish Woodcock (Bateria Woodcock de Proficiencia en el Idioma): Letter/Word Identification (Identificacion de Letras y Palabras), Word Attack (Analisis de Palabras), and Passage Comprehension (Comprension de Textos).

ANCOVA's controlling for pretests showed that at the end of grade 2 Success for All students scored substantially higher than controls on every measure (p< .01 or better). Figure 4 summarizes mean grade equivalents and effect sizes. Control second graders scored far below grade level on all three scales. In contrast, Fairhill students averaged near grade level on all measures. Effect sizes on all measures were substantial. Fairhill students exceeded control by 1.8 standard deviations on Letter-Word Identification, 2.2 on Word Attack, and 1.3 on Passage Comprehension.

Fremont (Bilingual), Wright (Bilingual), and El Vista (ESL)

Data from first graders in the three California Success for All schools were analyzed together by Dianda and Flaherty (1995), pooling data across schools in four categories: English-dominant students, Spanish-dominant students taught in Spanish (Lee Conmigo in Success for All schools), Spanish-dominant students taught in English ("sheltered students"), and speakers of languages other than English or Spanish taught in English. The pooled results are summarized in Figure 5 (from Dianda, 1995).

As is clear in Figure 5, all categories of Success for All students scored substantially better than control students. The differences were greatest, however, for Spanish-dominant students taught in bilingual classes (ES = +1.03) and those taught in sheltered English programs (ES = +1.02). The bilingual students scored at grade level, and more than six months ahead of controls. The sheltered students scored about two months below grade level, but were still four months ahead of their controls. Both English-speaking students and speakers of languages other than English or Spanish scored above grade level and about two months ahead of their controls.

The effects of Success for All on the achievement of English language learners are substantially positive. Across three schools implementing *Lee Conmigo,* the Spanish curriculum used in bilingual Success for All schools, the average effect size for first graders on Spanish assessments was +0.88; for second graders (at Philadelphia's Fairhill Elementary) the average effect size was +1.77. For students in sheltered English instruction, effect sizes for all comparisons were also very positive, especially for Cambodian students in Philadelphia and Mexican-American students in California.

Comparing Success for All and Reading Recovery

Reading Recovery is one of the most extensively researched and widely used innovations in elementary education. Like Success for All, Reading Recovery provides one-to-one tutoring to first graders who are struggling in reading. Research on Reading Recovery has found substantial positive effects of the program as of the end of first grade, and longitudinal studies have found that some portion of these effects maintain at least through fourth grade (DeFord, Pinnell, Lyons & Young, 1988; Pinnell, Lyons, DeFord, Bryk, & Seltzer, 1991).

Schools and districts attracted to Success for All are also often attracted to Reading

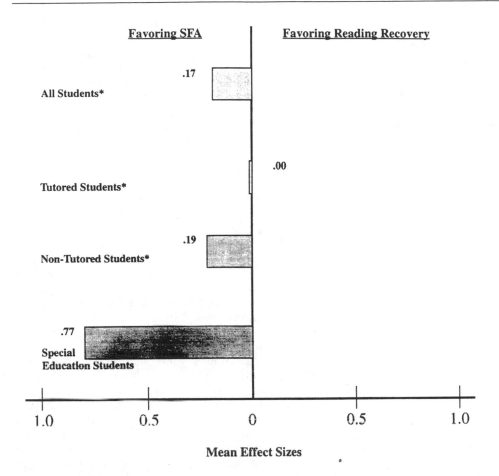

Figure 6 Comparison of Success for All and Reading Recovery Students in Mean Effect Sizes

Note: Adapted from Ross et al., 1995.
* Excludes special education students

Recovery, as the two programs share an emphasis on early intervention and a strong research base. Increasing numbers of districts have both programs in operation in different schools. One of the districts in the Success for All evaluation, Caldwell, Idaho, happened to be one of these. Ross, Smith, Casey, & Slavin (1995) used this opportunity to compare the two programs.

Reading Recovery tutoring is similar to that used in Success for All in that it is done by certified teachers and in that it emphasizes "learning to read by reading" and direct teaching of metacognitive skills. However, it is also different in many important ways. Reading Recovery tutors receive substantially more training than do Success for All tutors. Reading Recovery tutoring sessions are longer than those used in Success for All (30 vs. 20 minutes). Success for All places a great deal of emphasis on a linkage between tutoring and classroom reading instruction; tutors usually use the same books as those used in the reading class and emphasize the same objectives. Tutors in Success for All teach a reading class, so it is easy for them to maintain a consistency of approach. Reading Recovery does not emphasize coordination between tutoring and classroom instruction to this degree largely because the nature and quality of classroom reading instruction is not the central concern of the Reading Recovery program. However, many schools using Reading Recovery do provide classroom reading teachers with professional development to help them create supportive classroom environments that reinforce the strategies used in tutoring. In Caldwell, two schools are using Success for All and one is using Reading Recovery. All three are very similar rural schools with similar ethnic make-ups (10-25% Hispanic, with the remainder Anglo), proportions of students qualifying for free lunch (45-60%), and sizes (411-451). The Success for All schools were somewhat higher than the Reading Recovery school in poverty and percent Hispanic. In 1992-93, one of the Success for All schools was in its second year of implementation and the other was a new school that was in its first year (but had moved a principal and some experienced staff reassigned from the first school). Reading Recovery was in its second year of implementation.

The study compared first graders in the three schools. Figure 6 summarizes the results. As is clear from the figure, students in the Success for All schools performed somewhat better than students in Reading Recovery school overall (ES = +.17). Differences for special education students were substantial, averaging an effect size of +.77. Special education students were not tutored in the Reading Recovery school and were primarily taught in a separate resource room. These students scored near the floor on all tests. In contrast, Success for All special education students were fully mainstreamed and did receive tutoring, and their reading scores, though still low, showed them to be on the way toward success in reading.

Excluding the special education students, there were no differences in reading performance between tutored students in the Success for All and Reading Recovery schools (ES = .00). In light of earlier research, these outcomes suggest that both tutoring programs are highly effective for at-risk first graders.

The comparison of Success for All and Reading Recovery supports a common-sense conclusion. Success for All, which affects all students, has positive effects on all students. Reading Recovery focuses on tutoring and therefore produces its effects on tutored students. These results suggest that Success for All may be most appropriate in schools serving many at-risk students, while Reading Recovery may be the better choice when the number of students at risk of reading failure is small. The results may also justify a merger of the two programs, combining the breadth and comprehensiveness of Success for All with the outstanding professional development for tutors provided by Reading Recovery. Such mergers of Success for All and Reading Recovery are being started in about a half dozen schools located around the U.S.

For more on this study, see Ross et al. (1995).

Success for All and Special Education

Perhaps the most important goal of Success for All is to place a floor under the reading achievement of all children, to ensure that every child performs adequately in this critical skill. This goal has major implications for special education. If the program makes a substantial difference in the reading achievement of the lowest achievers, then it should reduce special education referrals and placements. Further, students who have IEP's indicating learning disabilities or related problems are typically treated the same as other students in Success for All. That is, they receive tutoring if they need it, participate in reading classes appropriate to their reading levels, and spend the rest of the day in age-appropriate, heterogeneous homerooms. Their tutor and/or reading teacher is likely to be a special education teacher, but otherwise they are not treated differently. One-to-one tutoring in reading, plus high-quality reading instruction in the mainstream at the student's appropriate level, should be more effective than the small-group instruction provided in special education classes. For this reason we expect that students who have been identified as being in need of special education services will perform substantially better than similar students in traditional special education programs.

The philosophy behind the treatment of special education issues in Success for All is called "neverstreaming" (Slavin et al., 1991). That is, rather than waiting until students fall far behind, are assigned to special education, and then may be mainstreamed into regular classes, Success for All schools intervene early and intensively with students who are at risk to try to keep them out of the special education system. Once students are far behind special education services are unlikely to catch them up to age-appropriate levels of performance. Students who have already failed in reading are likely to have an overlay of anxiety, poor motivation, poor behavior, low self-esteem, and ineffective learning strategies that are likely to interfere with learning no matter how good special education services may be. Ensuring that all students succeed in the first place is a far better strategy if it can be accomplished. In Success for All, the provision of research-based preschool, kindergarten, and first grade reading, one-to-one tutoring, and family support services are likely to give the most at-risk students a good chance of developing enough reading skills to remain out of special education, or to perform better in special education than would have otherwise been the case.

The data relating to special education outcomes clearly support these expectations. Several studies have focused on questions related to special education. One of the most important outcomes in this area is the consistent funding of particularly large effects of Success for All for students in the lowest 25% of their classes. While effect sizes for students in general have averaged around + 0.50 on individually administered reading measures, effect sizes for the lowest achievers have averaged in the range of +1.00 to +1.50 across the grades. Across five Baltimore schools only 2.2% of third graders averaged two years behind grade level, a usual criterion for special education placement. In contrast, 8.8% of control third graders scored this poorly. Baltimore data have also shown a reduction in special education placements for learning disabilities of about half (Slavin et al., 1992). A recent study of two Success for All schools in Ft. Wayne, Indiana found that over a two year period 3.2% of Success for All students in grades K-1 and 1-2 were referred to special education for learning disabilities or mild mental handicaps. In contrast, 14.3% of control students were referred in these categories (Smith, Ross, & Casey, 1994).

Taken together, these findings support the conclusion that Success for All both reduces the need for special education services (by raising the reading achievement of very low achievers) and reduces special education referrals and placements.

Another important question concerns the effects of the program on students who have already been assigned to special education. Here again, there is evidence from different sources. In the study comparing Reading Recovery and Success for All described above, it so happened that first graders in special education in the Reading Recovery group were not tutored, but instead received traditional special education services in resource rooms. In the Success for All schools, first graders who had been assigned to special education were tutored one-to-one (by their special education teachers) and otherwise participated in the program in the same way as all other students. As noted earlier (recall Figure 6), special education students in Success for All were reading substantially better (ES = +.77) than special education students in the comparison school (Ross et al., in press). In addition, Smith et al. (1994) combined first grade reading data from special education students in Success for All and control schools in four districts: Memphis, Ft. Wayne (IN), Montgomery (AL), and Caldwell (ID). Success for All special education students scored substantially better than controls (mean ES = +.59).

Conclusion

The results of evaluations of 19 Success for All schools in nine districts in eight states clearly show that the program increases student reading performance. In every district, Success for All students learned significantly more than matched control students. Significant effects were not seen on every measure at every grade level, but the consistent direction and magnitude of the effects show unequivocal benefits for Success for All students. This paper also adds evidence showing particularly large impacts on the achievement of limited English proficient students in both bilingual and ESL programs, and on both reducing special education referrals and improving the achievement of students who have been assigned to special education. It compares the outcomes of Success for All with those of another early intervention program, Reading Recovery.

The Success for All evaluations have used reliable and valid measures, individually administered tests that are sensitive to all aspects of reading: comprehension, fluency, word attack, and word identification. Performance of Success for All students has been compared to that of matched students in matched control schools, who provide the best indication of what students without the program would have achieved. Replication of high-quality experiments in such a wide variety of schools and districts is extremely unusual.

An important indicator of the robustness of Success for All is the fact that of the more than 150 schools that have used the program for periods of 1-7 years, only six have dropped out (in all cases because of changes of principals). Many other Success for All schools have survived changes of superintendents, principals, facilitators, and other key staff, major cuts in funding, and other serious threats to program maintenance.

The research summarized here demonstrates that comprehensive, systemic school-by-school change can take place on a broad scale in a way that maintains the integrity and effectiveness of the model. The nineteen schools in nine districts that we are studying in depth are typical of the larger set of schools currently using Success for All in terms of quality of implementation, resources, demographic characteristics, and other factors. Program outcomes are not limited to the original home of the program; in fact, outcomes tend to be somewhat better outside of Baltimore. The widely held idea based on the Rand study of innovation (Berman & McLaughlin, 1978; McLaughlin, 1990) that comprehensive school reform must be invented by school staffs themselves is certainly not supported in research on Success for All. While the program is adapted to meet the needs of each school, and while school staffs must agree to implement the program by a vote of 80% or more, Success for All is an externally developed program with specific materials, manuals, and structures. The observation that this program can be implemented and maintained over considerable time periods and can be effective in each of its replication sites certainly supports the idea that every school staff need not reinvent the wheel.

There is nothing magic about Success for All. None of its components are completely new or unique. Obviously, schools serving disadvantaged students can have great success without a special program if they have an outstanding staff, and other prevention/early intervention models, such as Reading Recovery (Pinnell, 1989) and the School Development Program (Comer, 1988) also have evidence of effectiveness with disadvantaged children. The main importance of the Success for All research is not in validating a particular model or in demonstrating that disadvantaged students can learn. Rather, its greatest importance is in demonstrating that success for disadvantaged students can be routinely ensured in schools that are not exceptional or extraordinary (and were not producing great success before the program was introduced).

We cannot ensure that every school has a charismatic principal or every student has a charismatic teacher. Nevertheless, we can ensure that every child, regardless of family background, has an opportunity to succeed in school.

The demonstration that an effective program can be replicated and can be effective in its replication sites removes one more excuse for the continuing low achievement of disadvantaged children. In order to ensure the success of disadvantaged students we must have the political commitment to do so, with the funds and policies to back up this commitment. Success for All does require a serious commitment to restructure elementary schools and to reconfigure uses of Title 1, special education, and other funds to emphasize prevention and early intervention rather than remediation. These and other systemic changes in assessments, accountability, standards, and legislation can facilitate the implementation of Success for All and other school reform programs. However, we must also have methods known not only to be effective in their original sites, but also to be replicable and effective in other sites. The evaluations presented in this paper provide a practical demonstration of the effectiveness and replicability of one such program.

CHAPTER 5

Roots and Wings

Background and Commentary

In 1992, we received a grant from the New American Schools Development Corporation (NASDC) to develop a "break-the-mold" design for schools for the 21st century. NASDC (now simply called New American Schools) was initially a creation of the Bush administration, but was established as a private foundation which received grants from large corporations and foundations to fund the development and dissemination of school designs tied to new national standards.

The design we developed under NASDC funding is called Roots and Wings (from a quotation from Lao Tzu referring to the two qualities we must develop in children). The "roots" of Roots and Wings refer to building a floor under the achievement of all students. This objective is addressed primarily by program elements drawn directly from Success for All: prekindergarten and kindergarten programs, reading, writing, and language arts programs, tutoring, family support, assessment, and facilitators. To these elements we added "wings," programs in social studies, science, and mathematics designed to build students' problem solving skills, creativity, and knowledge of the world and how it works. A centerpiece of "wings" is WorldLab, an integrated approach to social studies and science in which students engage in elaborate simulations and group investigations. Recall (from Chapter 2) that this was the title of the science program Nancy Madden and I developed as undergraduates twenty-five years ago (our motto: never waste a good idea!). Of course, today's WorldLab is vastly more sophisticated, replicable, and connected to national standards than the original, and it has largely been developed by an extraordinary group of educators working under Nancy Madden's direction.[1] Still, the idea of using simulations to make scientific, economic, and historical ideas relevant and motivating to students working in small groups is precisely what we were trying to do so long ago. Nancy Madden is also working with colleagues Kathy Simons, Honi Bamberger, Barbara Luebbe, and Pat Baltzley to develop MathWings, a constructivist approach to mathematics designed to be usable by a broad range of elementary school teachers.

As of this writing, all elements of Roots and Wings are only in place in our four pilot schools, but more than 30 experienced Success for All schools are phasing in WorldLab and/or MathWings.

The article reprinted in this chapter, from *Educational Leadership*, describes Roots and Wings, but at the time it was written there was only limited evidence of effectiveness, as the program was not fully implemented. In the interim, data on program

---

1   Principally Cecilia Daniels, Stan Bennett, Coleen Furey, Angela Calamari, Susan Dangel, Claire Von Secker, Robin Thorson, Marguerite Collins, and Yael and Shlomo Sharan.

outcomes have been collected, and they show the effects of the program to be very positive (see Slavin, Madden, Dolan, & Wasik, 1996).

Evaluation of Roots and Wings

The Roots and Wings pilot schools are located in rural St. Mary's County, Maryland. Three are in or near the town of Lexington Park, and one is in a completely rural area. An average of 48% of students in the four schools qualify for free or reduced-price lunches. The schools began program implementation in 1992-93, but all program elements were not piloted until the 1994-95 school year.

The assessment of Roots and Wings involves tracking changes over time on the Maryland School Performance Assessment Program (MSPAP), a state-of-the-art performance assessment in which third and fifth graders are asked to design and carry out experiments, write compositions in various genres, read and respond to extended passages, use mathematics to solve complex problems, and so on. Student responses are rated by state contractors against well-validated rubrics on a five-point scale.

Figures 5.1 and 5.2 show gains over three years in MSPAP scores for third and fifth graders, respectively. The figures show substantial gains over that time period in the percentage of students scoring at or above "satisfactory" on all six MSPAP scales. The overall State of Maryland also increased over this time period, but far less than the Roots and Wings schools. Averaging across the six scales, the percentage of Maryland

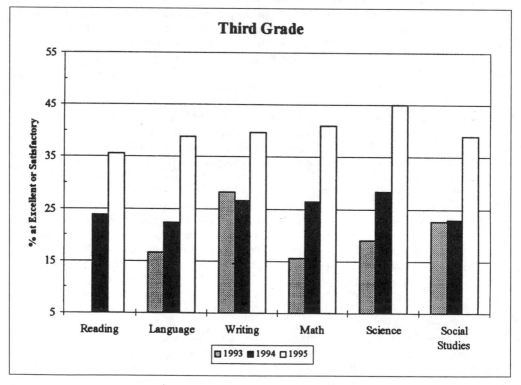

Fig. 5.1 Roots & Wings St. Mary's County Maryland School Performance Assessment Program School Years 1993-1995

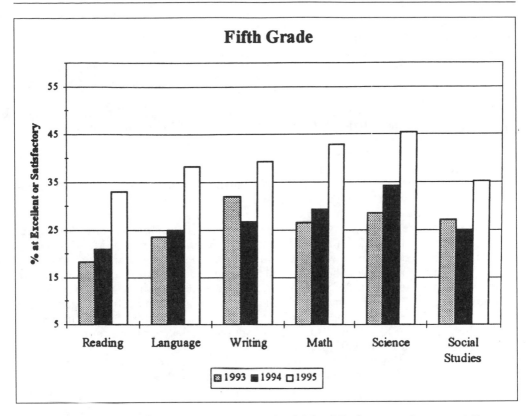

Fig. 5.2 Roots & Wings St. Mary's County Maryland School Performance Assessment Program School Years 1993-1995

third graders scoring satisfactory or better increased in 1993-95 by 8.6 percentage points, in comparison to a gain of 18.9 for Roots and Wings schools. For fifth graders, the state gained an average of 6.4 percentage points, while Roots and Wings schools gained 13.0. This was true despite the fact that the Roots and Wings schools served many more children in poverty, had three times as many Title I students, and had mobility rates twice the state average.

Roots and Wings represents the future of the work we began in Success for All. In its original conception Success for All was never meant to be limited to reading, writing, and language arts, but until our NASDC funding we never had the resources to develop programs in other areas. It is unclear how many schools will ultimately take on all of the components of Roots and Wings, but we now offer elementary schools a full range of curricular options each of which is well worked out, replicable, and known to be highly effective.

At Lexington Park Elementary School in a small town in southern Maryland, 10-year-old Jamal rises to speak. "The chair recognizes the delegate from Ridge School," says the chair, a student from the local high school.

"I'd like to speak in favor of House Bill R130," Jamal begins. "This bill would tell farmers they couldn't use fertilizer on land that is within 200 feet of the Chesapeake Bay. Fertilizer goes into the bay and causes pollution and kills fish. Farmers can still grow a lot of crops even if they don't plant close to water, and we all will have a better life if we can stop pollution in the bay. I yield to questions."

A hand goes up. The chair recognizes a delegate from Carver School. "How does fertilizer harm the bay?" she asks. Jamal explains how the fertilize supplies nutrients to algae, and when too much algae grows it deprives oysters, crabs, clams, and other larger creatures of oxygen.

A delegate from Green Holly School offers another viewpoint: "I'm a farmer," says 11-year-old Maria. "I can hardly pay all my bills as it is, and I've got three kids to feed. I'll go broke if I can't fertilize my whole field!"

The debate on the bill goes on for more than an hour. Student delegates who are playing the role of commercial crabbers and others who have plied the bay for centuries describe how their way of life is disappearing as catches decline due to pollution. Business owners tell how pollution ruins the local economy. Finally, the committee amends the bill to prohibit all but the most impoverished farmers from planting near waterways. The bill passes and goes to the full House of Delegates for a vote.

What is happening at Lexington Park and three other schools in St. Mary's County, Maryland, is a revolution in elementary education. These schools are experimenting with Roots and Wings, a collaborative project of the county school system, the state education department, and The Johns Hopkins University in Baltimore. The pilot project is one of nine "break-the-mold" school restructuring designs being funded by the New American Schools Development Corporation, a private group formed under the Bush administration to solicit business investment in new forms of schools.
Roots and Wings has two main objectives:
1. To guarantee that every child, regardless of family background or disability, will successfully complete elementary school, achieving the highest standards in basic skills such as reading and writing, as well as in mathematics, science, history, and geography (the roots).
2. To engage students in activities that enable them to apply everything they learn so they can see the usefulness and interconnectedness of knowledge (the wings).

The reading, tutoring, and special education approaches are adapted from Success for All, a national program that combines prevention, research-based curriculums and instruction, and family support at elementary schools in areas with high rates of poverty. Success for All is currently being implemented in 57 school districts in 20 states, including Maryland. To this base we are adding science, social studies, and mathematics for a comprehensive alternative to traditional elementary school curriculums.

Integrating Learning in WorldLab

The debate in the "House of Delegates" illustrates one of Roots and Wings' most distinctive and innovative elements: an integrated approach to science, social studies, writing, and other subject areas known as WorldLab. In this 90-minute daily exercise, students play the role of historical figures or people in various occupations, thereby becoming active participants in the scientific discoveries and historical events they study. These simulations not only draw upon the entire content of grades 1-6 science and social studies programs, but also integrate reading, writing, mathematics, and fine arts skills with WorldLab topics.

WorldLab grew out of our conviction that basic skills are essential, but they are not enough for children today. Students must also be able to creatively solve problems, understand their own learning processes, and connect knowledge from different disciplines. To build these higher-order skills, Roots and Wings provides daily opportunities for children to work collaboratively to solve simulated and real-life problems using everything they've learned in class.

Students have been studying the Chesapeake Bay in preparation for a model state legislature, in which they will write, propose, and debate many bills relating to cleaning up the waters of the continent's largest estuary. The bay, as rich a source of science lessons as it is of Atlantic blue crabs, is an ideal subject. The students are learning about causes of pollution, watersheds, tides, the rain cycle, and the life cycle of aquatic plants and animals. The unit also provides an opportunity to integrate lessons on government and politics, economics, and geography.

In other WorldLab units, students play the role of inventors, delegates to the Constitutional Convertion, advisors to the pharaohs of ancient Egypt, 15th century explorers, and so on. In these simulations they work in small, cooperative groups to investigate science and social studies topics. They read books and articles; write broadsides, letters, and proposals; and use fine arts, music, and computer, video, and other technology to prepare newspapers and multimedia reports.

Applying Lessons to Real Life

One problem of traditional, elementary schooling is that the content is not immediately useful to young students. It is entirely possible to be a happy and successful 10-year-old with no knowledge whatsoever of the American Revolution, or the rain cycle, or how to add fractions, or how to write a persuasive letter. Students may work to please their teachers or parents or to get a good grade, or they may be interested in some aspects of the subjects they are studying. But motivation, curiosity, and insight are certain to be much greater when they need information or skills to solve problems that have meaning to them. Simulations provide an ideal opportunity to make information immediately relevant.

In a well-designed simulation, students fully identify with the roles they take on. For example, 11-year-old Maria is an elected representative to the Maryland House of Delegates. She also is a farmer with serious responsibilities: three children, a mortgage, food bills, and taxes. As a real-life kid, Maria cares about the ecology of the Chesapeake Bay. However, as an elected delegate and a farmer, she cares about it from a particular perspective. To participate intelligently in the debates, she needs a basic understanding of government, laws, and the economic impact of the bay. She also needs to know how erosion and eutrophication occur, how sea life is dependent on oxygen and tides, and many other topics. Maria has written impassioned letters to support her views, she has read books on the Chesapeake Bay, and she has used math to figure her expenses. The Bay Unit is not only an interdisciplinary thematic unit, but, because of the simulations, it is also an opportunity to integrate knowledge and skills and make them useful.

Simulations can also give students an emotional investment in the material they are studying. For example, in a two-month unit called Rebellion to Union, principals sent 5th graders notes announcing taxes on certain activities, such as using the pencil sharpeners. Predictably, the 10-year-olds were quick to grasp the unfairness of the levy. They assembled their class governments, wrote notes of protest, and decided to boycott the pencil sharpeners. Some time later they were informed that they would be taxed on their use of desks. They promptly shoved the desks into the hall and sat on the classroom floor. For their part, the principals stuck to the historical plot line: they dissolved the class governments, whereupon the students "seceded" from the schools, justifying their actions with a Declaration of Independence.

Although they knew this was only a simulation, the students were emotionally involved. They wrote letters, picketed their principals' offices, and took great pleasure in defying their authority. When they ultimately read the various drafts of the real *Declaration of Independence,* they were able to identify not only with the framers' words and logic but with their emotions as well. They had, after all, wrestled with similar questions, fears, and uncertainties. Later, they played the parts of delegates to the Constitutional Convention and debated positions appropriate to the states and occupations they represented. These children will never forget the American Revolution or Constitutional Convention. Everything they learned in the unit was relevant and important to them.

Building a Firm Foundation

In addition to basic reading and writing instruction, Roots and Wings intervenes early with supportive classroom and family activities.

MathWings. In our mathematics program, students work in cooperative groups to discover and apply the powerful ideas of mathematics. MathWings, which is keyed to National Council for Teachers of Mathematics standards, balances problem-solving skill and concept development in order to make mathematics come to life. To help the children solve complex problems, teachers make extensive use of calculators, computers, and manipulatives, as well as a host of shared hands-on activities and frequent performance assessments. The goal in math, as in other areas, is to make the subject more than a series of abstract exercises.

Literacy. Integrated reading and writing programs are used in grades 1-5. The students work together in planning, drafting, and revising compositions. The 1st grade program, Reading Roots, is an adaptation of the Success for All beginning reading program, which has been successfully used in more than 100 schools across the country. It integrates phonics and meaning-centered reading to make sure every child is able to read. This program is active and exciting, with teachers supplementing the stories they read with puppet shows, songs, sound and letter games, and opportunities for students to read together.

In the Reading Wings program used in grades 2-5, students are involved in structured peer activities that provide constant opportunities for active learning and immediate feedback. They are placed in cooperative learning groups, where they read to one another and work together to find the main elements of stories. They learn to support one another's reading and to challenge one another to explain and justify their understanding of the material.

Tutoring. For 1st graders who are struggling with beginning reading and are in danger of falling behind, Chapter 1 or special education funds are used to provide one-to-one tutoring. The idea is to employ the most intensive and effective intervention known to bring these children up to expectations quickly.

Early learning. Part of the roots of Roots and Wings is to ensure that all students arrive in 1st grade with good language and pre-reading skills, as well as strong self-esteem, good health, and other prerequisites for success. For children 3 years old and younger, Roots and Wings offers family literacy programs. For preschoolers and kindergartners, we provide various pre-literacy activities, including research-based curriculums focusing on integrated themes, storytelling and retelling, and many opportunities for oral expression.

Neverstreaming. For children at-risk, Roots and Wings stresses preventive services and early intervention – a strategy known as neverstreaming, which is our primary approach to special education and to the federal Chapter 1 funding program for disadvantaged students. Teachers, principals, parents, and community agencies work in a coordinated, comprehensive, and relentless way to see that children receive whatever they need to become competent and confident learners. Our goal is to help most children succeed in mainstream classes and thereby minimize the need for long-term remedial or special education services.

Extended day. To further supplement regular classroom work, an after-school program is offered to all children. Here they may benefit from tutoring, special education, or Chapter 1 services, and a variety of art, music, sports, and computer programs.

Family support and integrated services. A family support team at each Roots and Wings School works to increase parent participation in the school and to improve student attendance and adjustment to classes. The team also coordinates family health, mental health, and social services; and health suites in each school are available for both student and family use. The family support teams also draw upon an extensive partnership with the local Navy base and area businesses. Volunteers from these two sectors work in the schools as tutors, mentors, and activity leaders.

Indicators of Success

There are many early indicators of program outcomes in Roots and Wings. First, the overall reading performance of participating students has improved substantially, and learning disabilities requiring special education have been significantly reduced. This should not be surprising: evaluations of Success for All programs in seven districts around the United States have found similar outcomes (Slavin et al., 1994).

Our Success for All program at The Johns Hopkins University (the parent program of Roots and Wings) also demonstrated that we can help nearly every child succeed in the elementary grades (Slavin et al. 1992, 1994; Madden et al., 1993). Other Maryland schools participating in the program have begun to see improved scores in social studies, science, math, writing, and reading on a statewide performance-based test, the Maryland School Performance Assessment Program. These outcomes are particularly important because the Maryland test is the kind of state-of-the-art performance based assessment that many states are moving toward in their accountability testing programs. And it is the kind of test the federal government is promoting through the recent Goals 2000 legislation and the reauthorization of the Chapter 1/Title 1 accountability program. As both states and the federal government begin emphasizing new forms of performance assessment, school districts will almost certainly be looking for comprehensive, proven models of reform keyed to these changes.

Clearly, in the next stage of school reform, we must have effective, replicable designs for total school restructuring, designs that can be adapted to a wide range of circumstances and needs. Roots and Wings provides one practical vision of what elementary schools can be like if we decide to give every child the academic grounding and the thinking skills, creativity, and broad world view we now expect only of our most gifted children.

SECTION II

Best Evidence Syntheses

CHAPTER 6

Best Evidence Synthesis

Background and Commentary

This section presents a series of reviews of research on issues of school and classroom organization relating to alternative means of accommodating instruction to meet the needs of diverse groups of students. The review method I have used in all of these is called "best-evidence synthesis." This chapter reprints (from the *Educational Researcher*) the first article to give a rationale and description of best-evidence synthesis.

Best-evidence synthesis was created as a reaction to meta-analyses that were appearing in profusion after Glass, McGaw, & Smith (1981) proposed the procedure. My concern was that meta-analysis was developing standard procedures that made bias and systematic error not only possible, but difficult or impossible to detect. This concern was heightened by the publication of meta-analyses in areas I followed closely that did in fact make serious errors. I had earlier written a critique of several meta-analyses (Slavin, 1984). I then wanted to follow up this article with an alternative that preserved the best features of meta-analysis, but altered the review procedures to force reviewers to deal openly with important substantive and methodological issues (especially potential sources of bias) and to discuss in some detail key individual studies that support or contradict the review's conclusion.

The article represented in this chapter was generally well received, and it won the American Educational Research Association's Palmer O. Johnson Award for the best article in an AERA Journal in 1986. However, few researchers in the social sciences have adopted this procedure. I am only aware of less than a half-dozen reviews that explicitly used best-evidence synthesis. Ironically, best-evidence synthesis has been adopted to some extent in epidemiology; in 1995, I published an invited article on the procedure in an epidemiological journal!

If best-evidence synthesis has had an influence on review practices in the social sciences it has been in adding to calls for standards of practice in meta-analysis. Over time, published meta-analyses have increasingly incorporated some of the recommendations I was making, especially regarding requirements that reviewers give information about each study. It is becoming more unusual to see meta-analyses as mysterious in their procedures as those I critiqued. At least in education, however, the number of meta-analyses published has dropped off substantially, perhaps because potential reviewers are daunted by the far more exacting requirements shared by today's best meta-analyses and best-evidence syntheses.

Because of its name, I suppose, some writers have mistakenly asserted that the purpose of best-evidence synthesis is to find a small handful of studies that meet very high standards of quality, discarding all others. As is apparent from the articles reprinted in this volume, this is certainly not the case. All I am trying to do with this procedure is to require that reviewers avoid rigid formats but instead openly and intelligently portray the evidence crucial to forming an informed conclusion on the topic at hand.

In the decade since Glass (1976) introduced the concept of meta-analysis as a means of combining results of different investigations on a related topic, the practice and theory of literature synthesis has been dramatically transformed. Scores of meta-analyses relating to educational practice and policy have appeared, and the number of articles using or discussing meta-analysis in education has approximately doubled each year from 1979 to 1983 (S. Jackson, 1984). Several thoughtful guides to the proper conduct of meta-analyses have been published (see, e.g., Cooper, 1984; Glass, McGaw, & Smith, 1981; Hunter, Schmidt, & Jackson, 1982; Light & Pillemer, 1984; Rosenthal, 1984).

Ever since it was introduced, meta-analysis has been vigorously criticized, and equally vigorously defended. In considering arguments for and against this procedure in the abstract, there is much validity to both sides. Proponents of quantitative synthesis (e.g., Cooper, 1984; Glass et al., 1981; G. Jackson, 1980; Light & Pillemer, 1984) are certainly correct to criticize traditional reviews for using unsystematic and poorly specified criteria for including studies and for using statistical significance as the only criterion of treatment effects. Critics of these procedures (e.g., Cook & Leviton, 1980; Eysenck, 1978; Slavin, 1984; Wilson & Rachman, 1983) are equally justified in objecting to a mechanistic approach to literature synthesis that sacrifices most of the information contributed in the original studies and includes studies of questionable methodological quality and questionable relevance to the issue at hand.

In an earlier article (Slavin, 1984), I evaluated the actual practice of meta-analysis in education by examining eight meta-analyses conducted by six independent sets of investigators, comparing their procedures and conclusions against the studies they included. I found that all of these meta-analyses had made errors serious enough to invalidate or call into question one or more major conclusions. In reviewing several meta-analyses published after my article went to press, I have seen misapplications of the procedure that are at least as serious (Slavin, 1985). Yet the misuses of meta-analysis in education do not in themselves justify a return to traditional review procedures.

In this paper, I propose an alternative to both meta-analytic and traditional reviews that is designed to draw on the strengths of each approach and to avoid the pitfalls characteristic of each. The main idea behind this procedure, which I call "best-evidence synthesis," is to add to the traditional scholarly literature review application of rational, systematic methods of selecting studies to be included and use of effect size (rather than statistical significance alone) as a common metric for treatment effects.

The Principle of Best Evidence

In law, there is a principle that the same evidence that would be essential in one case might be disregarded in another because in the second case there is better evidence available. For example, in a case of disputed authorship, a typed manuscript might be critical evidence if no handwritten copy is available, but if a handwritten copy exists, the typed copy would be inadmissible because it is no longer the best evidence (since the handwritten copy would be conclusive evidence of authorship). I would propose extending the principle of best evidence to the practice of research review. For example, if a literature contains several studies high in internal and external validity, then lower quality studies might be largely excluded from the review. Let's say we have a literature with 10 randomized studies of several months' duration evaluating

Treatment X. In this case, results of correlational studies, small-sample studies, and/or brief experiments might be excluded, or at most briefly mentioned. For example, Ottenbacher and Cooper (1983) located 61 randomized, double-blind studies of effects of medication on hyperactivity, and therefore decided not to include studies of lower methodological rigor. However, if a set of studies high in internal and external validity does not exist, we might cautiously examine the less well designed studies to see if there is adequate unbiased information to come to any conclusion.

The principle of best evidence works in law because there are a priori criteria for adequacy of evidence in certain types of cases. Comparable criteria could not be prescribed for all of educational research, but could be proposed for each subfield as it is reviewed. These criteria might be derived from a reading of previous narrative and meta-analytic reviews and a preliminary search of the literature.

Justification for the "Best Evidence" Principle

The recommendation that reviewers apply consistent, well justified, and clearly stated a priori inclusion criteria is at the heart of the bestevidence synthesis, and differs from the exhaustive inclusion principle suggested by Glass et al. (1981) and others, who recommend including all studies that meet broad standards in terms of independent and dependent variables, avoiding any judgments of study quality. Proponents of meta-analysis suggest that statistical tests be used to empirically test for any effects of design features on study outcomes. The rationale given for including all studies regardless of quality rather than identifying the methodologically adequate ones is primarily that the reviewer's own biases may enter into decisions about which studies are "good" and which are "bad" methodologically. Certainly, studies of interjudge consistency in evaluations of journal articles (e.g., Gottfredson, 1978; Marsh & Ball, 1981; Peters & Ceci, 1982; Scarr & Weber, 1978) show considerable variation from reviewer to reviewer, so global decisions about methodological quality are inappropriate as a priori criteria for inclusion of studies in a research synthesis. It is important to recall that much of the impetus for the development of meta-analysis came from a frequent observation that traditional narrative reviews were unsystematic in their selection of studies, and did a poor job (or no job at all) of justifying their selection of studies, arguably the most important step in the review process (see Cooper, 1984; G. Jackson, 1980; Waxman & Walberg, 1982).

However, while it is difficult to justify a return to haphazard study selection procedures characteristic of many narrative reviews, it is also difficult to accept the meta-analysts' exhaustive inclusion strategy.

The rationale for exhaustive inclusion depends entirely on the proposition that specific methodological features of studies can be statistically compared in terms of their effects on effect size. Cooper (1984) puts the issue this way:

If it is empirically demonstrated that studies using "good" methods produce results different from "bad" studies, the results of the good studies can be believed. When no difference is found it is sensible to retain the "bad" studies because they contain other variations in methods (like different samples and locations) that by their inclusion, will help solve many other questions surrounding the problem area. (pp. 65-66)

In practice, meta-analyses almost always test several methodological and substantive characteristics of studies for correlations with effect size, using a criterion for rejecting the null hypothesis of no differences of .05. However, in order to justify pooling across categories of studies, the meta-analyst must prove the null hypothesis that the categories do not differ. This is logically impossible, and in situations in which the numbers of studies are small and the numbers of categories are large, finding true differences between categories of studies to be statistically significant is unlikely.

One example of this is a recent meta-analysis on adaptive education by Waxman, Wang, Anderson, and Walberg (1985), which coded the critical methodological factor "control method" into eight categories: unspecified, stratification, partial correlation, beta weights in regression, raw or metric weights in regression, factorial analysis of variance, analysis of covariance, or none. In a meta-analysis of only 38 studies, the 8 x 1 ANOVA apparently used to evaluate effects of methodological quality on study outcome had highly unequal and small cell sizes and an extremely high probability of failing to detect any true differences.

The problem of the reviewer's bias entering into inclusion decisions is hardly solved by exhaustive inclusion followed by statistical tests. The reviewer's bias may just as well enter into the coding of studies for statistical analysis (Mintz, 1983; Wilson & Rachman, 1983). Worse, the reader has no easy way to find out how studies were coded. For example, most of the studies coded as "randomly assigned" in a meta-analysis on mainstreaming by Carlberg and Kavale (1980) were in fact randomly selected from nonrandomly assigned groups. To discover this, it was necessary to obtain every article cited and laboriously recode them (Slavin, 1984).

Reviews of social science literature will inevitably involve judgment. No set of procedural or statistical canons can make the review process immune to the reviewer's biases. What we can do, however, is to require that reviewers make their procedures explicit and open, and we can ask that reviewers say enough about the studies they review to give readers a clear idea of what the original evidence is. The greatest problem with exhaustive inclusion is that it often produces such a long list of studies that the reviewer cannot possibly describe each one. I would argue that all other things being equal, far more information is extracted from a large literature by clearly describing the best evidence on a topic than by using limited journal space to describe statistical analyses of the entire methodologically and substantively diverse literature.

Criteria for Including Studies

Obviously, if a priori criteria are to be used to select studies, these criteria must be well thought out and well justified. It is not possible to specify in advance what criteria should be used, as this must depend on the purposes for which the review is intended (see Light & Pillemer, 1984, for more on this point). However, there are a few principles that probably apply generally.

First, the most important principle of inclusion must be germaneness to the issue at hand. For example, a meta-analysis focusing on school achievement as a dependent measure must explicitly describe what is meant by school achievement and must only include studies that measured what is commonly understood as school achievement on individual assessments, not swimming, tennis, block stacking, time-on-task, task completion rate, group productivity, attitudes, or other measures perhaps related to but not identical with student academic achievement (see Slavin, 1984).

Second, methodological adequacy of studies must be evaluated primarily on the basis of the extent to which the study design minimized bias. For example, it would probably be inappropriate to exclude studies because they failed to document the reliability of their measures, as unreliability of measures is unlikely in itself to bias a study's results in favor of the experimental or control group. On the other hand, great caution must be exercised in areas of research in which less-than ideal research designs tend to produce systematic bias. For example, matched or correlational studies of such issues as special education, non-promotion, and gifted programs are likely to be systematically biased in favor of the students placed in regular classes, promoted, or placed in gifted classes, respectively (Madden & Slavin, 1983). In these areas of research, the independent variable is strongly correlated with academic ability, motivation, and many other factors that go into a decision to, for example, promote or retain a student.

Controlling for all these factors is virtually impossible in a correlational study. In research literatures of this kind, random assignment to experimental or control groups is essential. However, in other areas of research, the independent variable is less highly correlated with academic ability or other biasing factors. For example, schools that use tracking may not be systematically different from those that do not. If this is the case, then random assignment, though still desirable, may be less essential; carefully matched or statistically controlled studies may be interpretable.

Third, it is important to note that external validity should be valued at least as highly as internal validity in selecting studies for a best-evidence synthesis. For example, reviews of classroom practices should not generally include extremely brief laboratory studies or other highly artificial experiments. Often, a search for randomized studies turns up such artificial experiments. This was the case with the Glass, Cohen, Smith, and Filby (1982) class size meta-analysis, which found more positive effects of class size in "well controlled" studies than in "less well controlled" studies. Well controlled meant studies using random assignment, but this requirement caused the well controlled study category to include a number of extremely brief artificial experiments, such as a 30-minute study of class size by Moody, Bausell, and Jenkins (1973), as well as a study of effects of class size on tennis "achievement" (Verducci, 1969). Because class size is not strongly correlated with academic ability (see Coleman et al., 1966), this is actually a case in which well designed correlational studies, because of their greater external validity, might be preferred to many of the randomized experimental studies.

One category of studies that may be excluded in some literatures is studies with very small sample sizes. Small samples are generally susceptible to unstable effects. In education, experiments involving small numbers of classes are particularly susceptible to teacher and class effects (see Glass & Stanley, 1970; Page, 1975). For example, if Mr. Jones teaches Class A using Method X and Ms. Smith teaches Class B using Method Y, there is no way to rule out the possibility that any differences between the classes are due to differences in teaching style or ability between Mr. Jones and Ms. Smith (teacher effects) or to effects of students in the different classes on one another (class effects) rather than to any differences between Methods X and Y. To minimize these possibilities, a criterion of a certain number of teachers, classes, and/or students in each treatment group might be established.

In some literatures lacking a body of studies high in internal and external validity, it may be necessary to include (but not pool) germane studies using several methods, each of which has countervailing flaws. For example, if a literature on a particular topic consists largely of randomized experiments low in external validity and correlational studies high in external validity but susceptible to bias, the two types of research might be separately reviewed. If the two groups of studies yield the same result, each buttresses the other. If they yield different results, the reviewer should explain the discrepancy.

Finally, it may be important in some literatures to mention the best designed studies excluded from the review (that is, those that "just missed") to give the reader a more concrete idea of why a study was excluded and what the consequences of that exclusion are. For example, one recent meta-analysis of studies of bilingual education by Willig (1985) devoted considerable attention to describing studies excluded from the review, making the criteria for inclusion clear.

Some arbitrary limitations often placed on inclusion of studies in traditional reviews make little sense, and should be abandoned. Perhaps most common is the elimination of dissertations and unpublished reports (such as government reports or university technical reports). Often, these unpublished reports are better designed than published ones; for example, it may sometimes be easier to get a poorly designed study into a low quality journal than to get it past a dissertation committee. The most important randomized study of special education versus mainstream placement (Goldstein, Moss, & Jordan, 1966) and the Coleman Report (Coleman et al., 1966) are two examples of unpublished government reports essential to their respective literatures.

On the other hand, meta-analyses also exclude one type of study that should not be excluded: studies in which effect sizes cannot be computed. It often happens that studies fail to report standard deviations or other information sufficient to enable computation of effect sizes. While effect sizes can be computed directly from t-scores, $F's$, or p values for two-group comparisons if $N's$ are known (see Glass et al., 1981), there are cases in which important, well designed studies present only p values or $F's$ for complex designs, ANCOVAs, or multiple regression analyses with too little information to allow for computation of effect sizes. Yet there is no good reason to exclude these studies from consideration solely on this basis.

Exhaustive Literature Search

Once criteria for inclusion of studies in a best-evidence synthesis have been established, it is incumbent upon the reviewer to locate every study ever conducted that meets these criteria. Books on meta-analysis (e.g., Cooper, 1984; Light & Pillemer, 1984) give useful suggestions for conducting literature searches using ERIC, Psychological Abstracts, Social Science Citation Index, and bibliographies of other reviews or meta-analyses, among other sources. In some cases, it is necessary to write to authors to request means and standard deviations or other information necessary to understand some aspect of a study. It is particularly important to locate all studies cited by previous reviewers to assure the reader that any differences in conclusions between reviewers are not simply due to differences in the pool of studies located.

Computation of Effect Sizes

In general, effect sizes should be computed as suggested by Glass et al. (1981), with a correction for sample size devised by Hedges (1981; Hedges & Olkin, 1985). The Hedges procedure produces an unbiased estimate of effect size, reducing estimates from studies with total $N's$ (experimental plus control) less than 50.

There are many statistical issues that are important in computing and understanding effect sizes, and many of these have important substantive implications. For example, there are questions of how to interpret gain scores or posttests adjusted for covariates, how to deal with unequal pretest scores in experimental and control groups, and how to deal with aggregated data (e.g., class or school means). Readers interested in statistical issues should refer to the excellent books on the conduct of quantitative syntheses (e.g., Cooper, 1984; Glass et al., 1981; Hedges & Olkin, 1985; Hunter et al., 1982; Rosenthal, 1984).

Averaging effect sizes within studies. Since many studies report a large number of effects, it may be important to compute averages of some effect sizes across particular subsets of comparisons. The amount of averaging to be done depends on the purpose and focus of the best-evidence synthesis. For example, in a general review of the effects of ability grouping on achievement, different measures of reading and language arts might be averaged. However, in a best-evidence synthesis of research on specific reading strategies, we would want to preserve information separately for reading comprehension, reading vocabulary, oral reading, language mechanics, and so on. Similarly, in a review of effects of computer-assisted instruction we might average effects for students of different ethnicities, but in a review of compensatory education, separate effects for different ethnic groups might be preserved. However, when pooling effect sizes across studies, each study (or each experimental-control comparison) must count as one observation with effect sizes from similar measures averaged as appropriate. To count each dependent measure as a separate effect size for pooling purposes, as recommended by Glass et al. (1981), creates serious problems as it gives too much weight to studies with large numbers of measures and comparisons and violates assumptions of independence of data points in any statistical analyses (see Bangert-Drowns, 1986).

Table of Study Characteristics and Effect Sizes

No matter how extensive the literature reviewed, all studies should be listed in a table specifying major design and setting variables and effect sizes for principal studies. This table should include the names of the studies, sample size, duration, research design, subject matter, grade levels, treatments compared, and effect size(s). Other information important in a particular area of research might also be included. For example, the table might indicate which effects were statistically significant in the original research. This table is essential not only in summarizing all pertinent information, but also in making it easier to check the review's procedures and conclusions against the original research on which it was based.

In the table of study characteristics and effects sizes, results from studies for which effect sizes could not be computed may be represented as " + " (statistically significant-positive), "0" (no significant differences), or " − " (statistically significant-negative).

For examples of tables of study characteristics and effect sizes, see Willig (1985), Schlaefli, Rest and Thoma (1985), Kulik and Kulik (1984), and Slavin (1986).

Pooling of Effect Sizes

When there are many studies high in internal and external validity on a well defined topic, pooling (averaging) effect sizes across the various studies may be done. For example, let's say we located a dozen studies of Treatment X in which experimental and control students (or classes) were randomly assigned to treatment groups, the treatment was applied for at least 3 weeks, and fair achievement tests equally responsive to the curriculum taught in the experimental and control groups were used. In this case, we might pool the effect sizes by computing a median across the 12 studies. Medians are preferable to means because they are minimally influenced by anomalous outliers frequently seen in meta-analyses.

In pooling effect sizes, the reviewer must be careful "not to quantitatively combine studies at a broader conceptual level than the readers would find useful" (Cooper, 1984, p. 82). For example, in a quantitative synthesis by Lysakowski and Walberg (1982), it was not useful to pool across studies of cues, participation, and corrective feedback, as these topics together do not form a single well-defined category (see Slavin, 1984).

Pooled effect sizes should be reported as adjuncts to the literature review, not its primary outcome. Pooling and statistical comparisons must be guided by substantive, methodological, and theoretical considerations, not conducted wholesale and interpreted according to statistical criteria alone. For example, many meta-analyses routinely test for differences among effect sizes according to year of publication, a criterion that may be important in some literatures but is meaningless in others, while ignoring more theoretically or methodologically important comparisons (such as plausible interactions among study features).

Pooled effect sizes should never be treated as the final word on a subject. If pooled effects are markedly different from those of two or three especially well designed studies, this discrepancy should be explained. Pooling has value simply in describing the central tendency of several effects that clearly tend in the same direction. When effects are diverse, or the number of methodologically adequate, germane articles is small, pooling should not be done. Hedges and Olkin (1985) have described statistical procedures for testing sets of effect sizes for homogeneity, and these may be useful in determining whether or not pooling is indicated. However, decisions about which studies to include in a particular category should be based primarily on substantive, not statistical criteria.

Literature Review

The selection of studies, computation of effect sizes, and pooling described above are only a preliminary to the main task of a best-evidence synthesis: the literature review itself. It is in the literature review section that best-evidence synthesis least resembles meta-analysis. For example, some quantitative syntheses do use a priori selection, do present tables of study characteristics and effect size, and do follow other procedures recommended for best-evidence synthesis, but it is very unusual for a quantitative synthesis to discuss more than two or three individual studies or to examine a literature with the care typical of the best narrative reviews.

There are no formal guidelines or mechanistic procedures for conducting a literature review in a best-evidence synthesis; it is up to the reviewer to make sense out of the best available evidence.

Formats for Best-Evidence Syntheses

No rigid formula for presenting best-evidence syntheses can be prescribed, as formats must be adapted to the literature being reviewed. However, one suggestion for a general format is presented below. Also, see Slavin (1986) for an example of a best-evidence synthesis.

Introduction. The introduction to a best-evidence synthesis will closely resemble introductions to traditional narrative reviews. The area being studied is introduced, key terms and concepts are defined, and the previous literature, particularly earlier reviews and meta-analyses, is discussed.

Methods. In a best-evidence synthesis, the methods section serves primarily to describe how studies were selected for inclusion in the review. The methods section might consist of the following three subsections.

Best-Evidence Criteria describes and justifies the study selection criteria employed. Clear, quantifiable criteria must be specified, not global ratings of methodological adequacy. Stringent criteria for germaneness should be applied (e.g., studies of individualized instruction in mathematics that took place over periods of at least 8 weeks in elementary schools, using mathematics achievement measures not specifically keyed to the material being studied in the experimental classes). Among germane studies, criteria for methodological adequacy are established, focusing on avoidance of systematic bias (e.g., use of random assignment or matching with evidence of initial equality), sample size (e.g., at least four classes in experimental and control groups), and external validity (e.g., treatment duration of at least eight weeks). The literature search procedure should be described in enough detail that the reader could theoretically regenerate an identical set of articles. A section titled *Studies Selected* might describe the set of studies that will constitute the synthesis, while a section on *Studies Not Selected* characterizes studies not included in the synthesis, in particular describing excluded studies that were included in others' reviews and studies that "just missed" being included.

Literature Synthesis. The real meat of the best-evidence synthesis is in the *Literature Synthesis* section. This is where the research evidence is actually reviewed. This section would first present and discuss the table of study characteristics and effect sizes and discuss any issues related to the table and its contents. If pooling is seen as appropriate, the results of the pooling are described; otherwise, the rationale for not pooling is presented.

In a meta-analysis, the presentation of the "results" is essentially the end point of the review. In a best-evidence synthesis, the table of study characteristics and effect sizes and the results of any pooling are simply a point of departure for an intelligent, critical examination of the literature (see Light & Pillemer, 1984). In the Literature Synthesis section, critical studies should be described and important conceptual and methodological issues should be explored. A best-evidence synthesis should not read like an annotated bibliography, but should use the evidence at hand to answer important questions about effects of various treatments, possible conditioning or mediating variables, and so on. When conclusions are suggested, they must be justified in light of the available evidence, but also the *contrary* evidence should be discussed. Effect size information may be incorporated in the Literature Synthesis, as in the following example:

"Katz and Jammer (19XX) found significantly higher achievement in project classes than in control classes on mathematics computations (ES=.45) and concepts (ES=.31), but not on applications (ES = .02)."

In general, the "best-evidence" studies should be described with particular attention to studies with outstanding features, unusually high or low effect sizes, or important additional data. Studies that meet standards of germaneness and methodological adequacy but do not yield effect size data should be discussed on the same basis as those that do yield effect size data. Studies excluded from the main synthesis may be brought in to illustrate particular points or to provide additional evidence on a secondary issue. Except for the references to effect sizes, the bulk of the Literature Synthesis should look much like the main body of any narrative literature review.

One useful activity in many best-evidence syntheses is to compare review-generated and study-generated evidence (see Cooper, 1984). Review-generated evidence results from comparisons of outcomes in studies falling into different categories, while study-generated evidence relates to comparisons made within the same studies. For example, a reviewer might find an average effect size of 1.0 in methodologically adequate studies of Treatment X, and 0.5 in similar studies of Treatment Y and conclude that Treatment X is more effective than Treatment Y. However, this is not necessarily so, as other factors that are systematically different in studies of the two treatments could account for the apparent difference. This issue could be substantially informed by examination of studies that specifically compared treatments X and Y. If such studies exist and are of good quality, they would constitute the best evidence for the comparison of the treatments. Review-generated evidence can be useful in *suggesting* comparisons to be sought within studies, and may often be the only available evidence on a topic, but is rarely conclusive in itself.

Conclusions. One purpose of any literature review is to summarize the findings from large literatures to give readers some indication of where the weight of the evidence lies. A best-evidence synthesis should produce and defend conclusions based on the best available evidence, or in some cases may conclude that the evidence currently available does not allow for any conclusions.

Summary

The advent of meta-analysis has had an important positive impact on research synthesis in reopening the question of how best to summarize the results of large literatures and providing statistical procedures for computation of effect size, a common metric of treatment effects. It is difficult to justify a return to reviews with arbitrary study selection procedures and reliance on statistical significance as the only criterion for treatment effects. Yet in actual practice (at least in education), meta-analysis has produced serious errors (see Slavin, 1984).

This paper proposes one means, best-evidence synthesis, of combining the strengths of meta-analytic and traditional reviews. Best-evidence synthesis incorporates the quantification and systematic literature search methods of meta-analysis with the detailed analysis of critical issues and study characteristics of the best traditional reviews in an attempt to provide a thorough and unbiased means of synthesizing research and providing clear and useful conclusions. No review procedure can make errors impossible or eliminate any chance that reviewers' biases will affect the conclusions drawn. It may be that applications of the procedures proposed in this paper will still lead to errors as serious as those often found in meta-analytic and traditional reviews. However, applications of best-evidence synthesis should at least make review procedures clear to the reader and should provide the reader with enough information about the primary research on which the review is based to reach independent conclusions.

Chapter 7

Ability Grouping in Elementary Schools

Background and Commentary

The article reprinted in this chapter (from the *Review of Educational Research*) was written as a direct response to meta-analyses by James and Chen-Lin Kulik and their colleagues. I had long been following research on ability grouping, and was frankly amazed to see articles claiming achievement benefits of between-class grouping practices; on further investigation, it was apparent to me that the Kuliks were in error, mostly because they included in their reviews studies of extremely low quality and containing systematic bias and because, in some cases, they combined studies of programs for gifted and talented students with studies of more typical high-average-low ability grouping plans. My review of ability grouping in elementary schools was the first application of best-evidence synthesis, and in one sense this review procedure was designed specifically for the purpose of avoiding the errors I felt were being made by the Kuliks and by other early meta-analysts (including Gene Glass, primary developer of the method).

My best-evidence synthesis of research on elementary ability grouping concluded that there were no positive effects for high, average, or low-achieving students of typical between-class ability grouping plans (e.g., high, middle, and low fourth grades). However, I did find positive effects of the Joplin Plan, cross-grade grouping for reading. This has caused consternation among both proponents and opponents of ability grouping, as opponents are uncomfortable with the Joplin Plan and proponents try to use this finding to justify all forms of ability grouping. To me, however, these seemingly inconsistent conclusions support my most cherished belief: details matter. In my opinion it is foolish to make statements as broad as "ability grouping is good" or "ability grouping is bad." What kind of ability grouping, for what instructional purpose, for what students? We are nowhere near a point of being able to specify precisely how to deal with student diversity in each subject and grade level. It is possible to state that on grounds of promoting democratic, egalitarian values, ability grouping plans should be used only when they have clear evidence of effectiveness and are used in as limited and precise a fashion as possible. However, if some forms of grouping (such as Joplin Plan) do turn out to be effective in enhancing important learning outcomes, then I believe it is appropriate to use them as a part of a broader plan to ensure the success of all children.

Ability grouping is one of the oldest and most controversial issues in education:

Ability grouping is one of the oldest and most controversial issues in education: Hundreds of studies have examined the effects of various forms of between-class ability grouping (e.g., tracking, streaming) and within-class ability grouping (e.g., reading, math groups). By 1930, Miller and Otto had already located 20 experimental studies on ability grouping, and Martin (1927) listed 83 "selected references" on the topic. Scores of reviews of the between-class ability grouping literature have also been written. Almost without exception, reviews from the 1920s to the present have come to the same general conclusion: that between-class ability grouping has few if any benefits for student achievement. However, recent meta-analyses on ability grouping in elementary (C.-L. Kulik & J.A. Kulik, 1984) and in secondary schools (Kulik & Kulik, 1982) have claimed small positive achievement effects of between-class ability grouping, with high achievers gaining the most from the practice.

Despite a half-century of widespread agreement (among most researchers, at least) that between-class ability grouping is of little value in enhancing student achievement, the practice is nearly universal in some form in secondary schools and very common in elementary schools. Recent data are lacking, but over time most teachers at all levels have reported both using and believing in some kind of ability grouping (e.g., National Education Association [NEA], 1968; Wilson & Schmits, 1978). Yet in recent years many districts have begun to reexamine ability grouping, often out of a concern that students low in socioeconomic status, minority students in particular, are discriminated against by being disproportionately placed in low tracks. In fact, ability grouping has become a major issue in many ongoing desegregation cases (e.g., *Hobson v. Hansen,* 1967) where the plaintiffs have argued that ability grouping is used as a means of resegregating black and Hispanic students within ostensibly integrated schools (see McPartland, 1968).

Although many reviews of ability grouping have been written, the most recent comprehensive reviews in this area were written more than 16 years ago (e.g., Borg, 1965; Findley & Bryan, 1971; Heathers, 1969; NEA, 1968). More recent reviews (e.g., Esposito, 1973; Good & Marshall, 1984; Persell, 1977) have referred to the earlier reviews rather than synthesizing the original evidence. The Kuliks' metaanalyses have extracted effect-size data from large numbers of primary studies, but have done little beyond this to explore the substantive and methodological issues underlying these effects (see Slavin, 1984a).

The present paper reviews the literature on ability grouping in elementary schools from the vantage point of the 1980s. It uses a review strategy called "best-evidence synthesis" (Slavin, 1986b), a method that incorporates the best features of both meta-analytic and traditional narrative review (see Slavin, 1987, for an earlier example). The main elements of a best-evidence synthesis are as follows:

• Clearly specified, defensible a priori criteria for inclusion of studies are established.
• All published and unpublished studies that meet these criteria are located and included.
• Where possible, effect sizes for included studies are computed. Effect size is operationalized as the mean of all experimental-control differences on related measures divided by their standard deviations.
• When effect sizes cannot be computed, effects of studies that meet inclusion criteria are characterized as positive, negative, or zero rather than excluded.
• Apart from computation of effect size and use of well-specified inclusion criteria, best-evidence syntheses are identical to traditional narrative reviews. Individual studies and methodological and substantive issues are discussed in the detail typical of the best narrative reviews.

What Is Ability Grouping?

One important problem in discussing "ability grouping" is that the term has many meanings. Several quite different programs or policies go under this heading. In general, ability grouping implies some means of grouping students for instruction by ability or achievement so as to reduce their heterogeneity. However, various grouping plans differ in ways likely to have a considerable impact on the outcomes of grouping. Forms of ability grouping commonly used in elementary schools are described below.

Ability-grouped class assignment. In this plan, students are assigned on the basis of ability or achievement to one self-contained class. A recent survey of 450 Pennsylvania elementary schools found that approximately 25% of first grades, 19% of third grades, and 13% of fifth grades were organized in this way (Coldiron, Braddock, & McPartland, 1987).

Regrouping for reading or mathematics (Ability grouping for selected subjects). Often, students are assigned to heterogeneous homeroom classes for part or most of the day, but are "regrouped" according to achievement level for one or more subjects. In the elementary grades, regrouping is often done for reading (and occasionally mathematics), where all students at a particular grade level have reading scheduled at the same time and are re-sorted from their heterogeneous homerooms into classes that are relatively homogeneous in reading level. When regrouping for reading is done across grade levels, this is called the "Joplin Plan" (see below). In the Coldiron et al. (1987) survey, regrouping for one or two subjects was reported in 24% of first grades, 42% of third grades, and 60% of fifth grades.

Joplin Plan. One special form of regrouping for reading is called the Joplin Plan (Floyd, 1954), in which students are assigned to heterogeneous classes most of the day but are regrouped for reading across grade lines. For example, a reading class at the fifth grade, first semester reading level might include high-achieving fourth graders, average-achieving fifth graders, and low-achieving sixth graders. Reading group assignments are frequently reviewed, so that students may be reassigned to a different reading class if their performance warrants it. One important consequence of cross-grade grouping and flexible assignment is that reading classes contain only one or at most two reading groups, increasing the amount of time available for direct instruction over that typical of reading classes containing three or more reading groups. The Joplin Plan was principally an innovation of the late 1950s and early 1960s, after which time interest in cross-grade grouping turned more toward nongraded plans (see below).

Nongraded plans. The term "nongraded" or "ungraded" refers to a variety of related grouping plans. In its original conception (Goodlad & Anderson, 1963), nongraded programs are ones in which grade-level designations are entirely removed, and students are placed in flexible groups according to their performance level, not their age. Full-scale nongraded plans might use team teaching, individualized instruction, learning centers, and other means of accommodating student differences in all academic subjects. Students in nongraded programs might complete the primary cycle (grades 1-3) in 2 years, or may take 4 years to do so. The curriculum in each subject may be divided into levels (e.g., 9 or 12 levels for the primary grades) through which students progress at their own rates, picking up each year where they left off the previous year.

Some of the nongraded programs evaluated in the 1960s and early '70s did use the flexible, complex grouping arrangements envisioned by Goodlad and Anderson (1963), whereas others did not. For example, several programs described by their authors as "nongraded" or "ungraded" were in fact virtually indistinguishable from the Joplin Plan. That is, students were assigned to heterogeneous classes most of the day but regrouped across grade lines for reading.

Special classes for high achievers. In many elementary and secondary schools, gifted, talented, or otherwise superior students may be assigned to a special class for part or all of their school day, while other students remain in relatively heterogeneous classes.

Special classes for low achievers. One of the most common forms of "ability grouping" is the assignment of students with learning problems to special or remedial classes for part or all of their school day.

Within-class ability grouping. Regardless of the use or non-use of ability grouping of classes, most elementary teachers use some form of within-class ability grouping. The most common form of within-class ability grouping is the use of reading groups, where teachers assign students to one of a small number of groups (usually three) on the basis of reading level. These groups work on different materials at rates unique to their needs and abilities. More than 95% of Pennsylvania first grades, 94% of third grades, and 89% of fifth grades use within-class ability grouping in reading (Coldiron et al., 1987). Similar methods are often used in mathematics, where there may be two or more math groups operating at different levels and rates. In another common form of within-class ability grouping in elementary mathematics, the teacher presents a lesson to the class as a whole, and afterwards, while the students are working problems, the teacher provides enrichment or extension to a high-achieving group, remediation or reexplanation to low achievers, and something in between to average achievers. In the Pennsylvania study, approximately 28% of first grades, 31 % of third grades, and 40% of fifth grades reported some form of within-class ability grouping in math (Coldiron et al.).

Group-paced mastery learning (Block & Anderson, 1975; Bloom, 1976; Slavin, 1987) may be seen as one form of flexible within-class ability grouping, in that students are grouped after each lesson into "masters" and "nonmasters" groups on the basis of a formative test. Nonmasters receive corrective instruction while masters do enrichment activities. Finally, individualized or continuous-progress instruction may be seen as extreme forms of ability grouping, as each student may be in a unique "ability group" of one.

Arguments For and Against Ability Grouping

Very similar lists of advantages and disadvantages of ability grouping have been given by theorists and reviewers for more than 50 years (see, for example, Billett, 1932; Borg, 1965; Esposito, 1973; Findley & Bryan, 1971; Good & Marshall, 1984; Heathers, 1969; Miller & Otto, 1930; NEA, 1968). Ability grouping is supposed to increase student achievement primarily by reducing the heterogeneity of the class or instructional group, making it more possible for the teacher to provide instruction that is neither too easy nor too hard for most students. Ability grouping is assumed to allow the teacher to increase the pace and level of instruction for high achievers and provide more individual attention, repetition, and review for low achievers. It is supposed to provide a spur to high achievers by making them work harder to succeed, and to place success within the grasp of low achievers, who are protected from having to compete with more able agemates (Atkinson & O'Connor, 1963).

Two principal arguments against ability grouping have to do with the fact that this practice must create classes or groups of low achievers. These students are deprived of the example and stimulation provided by high achievers, and the fact of being labeled and assigned to a low group is held to communicate low expectations for students which may be self-fulfilling (see, for example, Good & Marshall, 1984; Persell, 1977). Further, homogeneously low performing reading groups (Allington, 1980; Barr, 1975) and classes (Evertson, 1982; Oakes, 1985) have been observed to experience a slower pace and lower quality of instruction than do students in higher achieving groups. A lack of appropriately behaving models may lead to "behavioral contagion" among homogeneously grouped low achievers (Felmlee & Eder, 1983), so that these groups may spend less time on-task than other groups.

However, perhaps the most compelling argument against ability grouping has little to do with its effects on achievement. This is that ability grouping goes against our democratic ideals by creating academic elites (Persell, 1977; Rosenbaum, 1976; Sorensen, 1970). According to this line of reasoning, all students need opportunities to interact with a wide range of peers. Because ability groupings often parallel social class and ethnic groupings, disproportionately placing low-socioeconomic status (SES), black, and Hispanic students in low tracks (e.g., Haller & Davis, 1980; Heyns, 1978; Rist, 1970), the use of ability grouping may serve to increase divisions along class, race, and ethnic group lines (see Rosenbaum, 1980).

Comprehensive Ability Grouping in the Elementary School

This review focuses on research on comprehensive ability grouping at the elementary level, grouping arrangements that involve all students at particular grade levels except those assigned to special education. This excludes studies of special classes for the gifted and for low achievers. Gifted and special education programs may be conceived of as one form of ability grouping, but they also involve many other changes in curriculum, class size, resources, and goals that make them fundamentally different from comprehensive ability grouping plans. Further, nonrandomized evaluations of gifted and special education/mainstreaming studies suffer from serious problems of selection bias that are less problematic in similar studies of comprehensive ability grouping plans (Madden & Slavin, 1983; Slavin, 1984a). For reviews of research on gifted and accelerated programs, see J. A. Kulik and C.-L. Kulik, 1984, or Passow, 1979; for special education/mainstreaming see Leinhardt and Pallay, 1982, or Madden and Slavin, 1983.

Also excluded from this review are studies of nongraded plans, on the basis that nongraded programs incorporate many elements other than changes in ability groupings (for reviews of research on nongrading, see Pavan, 1973; Slavin, 1986a). For example, nongraded plans often include the use of individualized materials or learning centers, independent study projects, team teaching, and open-space construction (Goodlad & Anderson, 1963). However, several studies that labeled their experimental treatments "nongraded" in fact used the Joplin Plan, which is principally an ability grouping plan. These studies are included along with other evaluations of the Joplin Plan.

Unfortunately, the research comparing alternative ability grouping arrangements in elementary schools has, with few exceptions, failed to specify in detail the changes in instructional practice brought about by the various grouping methods. Clearly, any effects of grouping on achievement are mediated by teacher behaviors. Because of the current state of the evidence, this article focuses primarily on the achievement effects of grouping practices per se, with the recognition that a complete conceptual understanding of these effects is not possible until the relationships between alternative grouping plans, teacher and student behaviors and perceptions, and student achievement are better understood.

The following sections discuss the research on comprehensive ability grouping in elementary schools according to the four principal categories discussed above. Each section contains a table summarizing the principal studies on the ability grouping strategy being presented and a discussion of the studies and the methodological and substantive issues they raise.

Methods

Criteria for Study Inclusion

The studies on which this review is based had to meet a set of a priori criteria with respect to germaneness and methodological adequacy. As stated earlier, all studies had to involve comprehensive studies of ability grouping in elementary schools (grades 1-6). European studies of 11- and 12-year-olds who were in secondary schools (e.g., Douglas, 1973) are excluded, even though studies of students of the same age in elementary schools were included. Studies of within-class ability grouping were included, but other programs related to within-class grouping were excluded. Examples of such excluded programs are mastery learning, individualized and continuous-progress instruction, cooperative learning, multi-age grouping not done for the purpose of reducing student heterogeneity, nongraded plans, open classrooms, and team teaching. No restrictions were placed on year of publication, and every effort was made to locate dissertations and other unpublished documents relating to ability grouping.

Methodological requirements for inclusion. One key element of best-evidence synthesis (Slavin, 1986b) is the a priori establishment of inclusion criteria based on substantive and methodological adequacy. In the present case, criteria were established as follows:

1. Ability-grouped classes were compared to heterogeneously grouped control classes. This requirement excluded several studies that compared achievement gains in experimental classes to "predicted" gains (e.g., Ramsey, 1962) and studies that correlated "degree of heterogeneity" with achievement gains without identifying classes as ability-grouped or heterogeneous (e.g., Leiter, 1983).

2. Achievement data from standardized achievement tests were presented. This excluded scores of anecdotal accounts and several studies of student or teacher attitudes toward ability grouping.

3. Initial comparability of samples was established by use of random assignment, matching of classes, or matching of students within equivalent classes. In cases of matching of classes or students, evidence had to be presented that established that the classes were in fact initially equivalent in IQ or achievement level (within 20% of a standard deviation). Studies in which experimental and control classes were not initially equivalent but gain scores or analyses of covariance were used to adjust scores for these differences (e.g., Moorhouse, 1964) are listed in tables in a separate category, and results of these studies should be interpreted cautiously, as adjustments for pretest differences cannot be assumed to completely control for their influence on posttests (see Reichart, 1979). Several cross-sectional studies providing little evidence of initial equality were excluded. For example, Powell (1964) compared achievement scores of one school using the Joplin Plan to another using a self-contained model, with no evidence that the two schools were initially comparable. Some studies (e.g., Hart, 1959) compared achievement under ability grouping to that under heterogeneous grouping in earlier years in the same schools. Such studies were included if there was evidence that the samples in the earlier years were equivalent in ability or achievement. However, comparisons with previous classes were limited to 2 years, on the assumption that too many unrelated changes could take place over longer periods. This excluded one study that made a comparison over a 10-year period (Cushenberry, 1966) and restricted attention to the first 2 years of an 8-year study by Tobin (1966).

4. Ability grouping was in place for at least a semester. This requirement excluded only one very brief study (Piland & Lemke, 1971).

5. At least three experimental and three control teachers were involved in all included studies. The purpose of this requirement was to minimize the influence of teacher and class effects in small studies (see Slavin, 1984b) on study outcomes. This caused a few very small studies to be excluded (e.g., Johnston, 1973; Putbrese, 1972).

Literature Search Procedures

The studies reviewed here were located in an extensive search. Principal sources included the Education Resources Information Center (ERIC), Psychological Abstracts, Education Index, and Dissertation Abstracts. In these sources, the keywords "ability grouping," "classroom organization," "Joplin Plan," and related descriptions produced hundreds of citations. In addition, all citations in other reviews and meta-analyses were located, and citations made in primary sources were followed up. Every attempt was made to obtain a complete set of published and unpublished studies that met the substantive and methodological criteria outlined above. Further, in a few cases where clarifications were needed about important studies, authors were contacted directly for additional information.

Computation of Effect Sizes

Throughout this review, effects of various ability grouping strategies are referred to in terms of effect size. Effect sizes were generally computed as the difference between the experimental and control means divided by the control standard deviation (Glass, McGaw, & Smith, 1981). The control group was always the heterogeneous grouping plan unless otherwise noted, so that a positive effect size implies greater learning in an ability-grouped plan and a negative value indicates an advantage for heterogeneous grouping. The control group standard deviation (rather than a pooled standard deviation) was used whenever possible because of the possibility that experimental treatments might alter the experimental group's standard deviation. The individual-level control group standard deviation or approximations to it were thus used as the denominator for all effect sizes, as a means of putting all effect sizes in the same metric. However, when means or standard deviations were omitted in studies that met inclusion criteria, effect sizes were estimated when possible from ts, Fs, or exact p values (see Glass et al.). The standard deviations used were always from individual-level scores, not class or school means, because use of aggregated data can spuriously inflate effect size estimates.

Many of the studies in this review presented data on gain scores without presenting pre- or posttest data. Effect sizes from achievement gain scores are typically inflated, as standard deviations of gain scores are less than those of pre- or posttest scores to the degree that pre-post correlations exceed 0.5. If pre-post correlations are known, effect sizes from gain scores can be transformed to the scale of posttest values using the following multiplier:

$$ES = (ES_{gain})(\sqrt{2} \ (1 - r_{pre-post})$$

However, because few studies presenting gain scores also provide pre-post correlations, a pre-post correlation of +0.8 was assumed. This figure is a characteristic correlation between fall and spring scores on alternate forms of the California Achievement Test in the upper elementary grades (CTB/McGraw-Hill, 1979). Substituting 0.8 in the formula, a multiplier of 0.632 is derived, which was used to deflate effect size estimates from gain score data. The intention of this (and other) procedures was to put all effect sizes in the same metric, the unadjusted control group standard deviation. However, because the multiplier is only a rough approximation, effect sizes from gain score data should be interpreted with even more caution than is warranted for effect sizes in general.

In studies in which pretest data were provided, effect sizes were computed as the difference between experimental and control *gain* scores divided by the control group's posttest standard deviation. For example, if an experimental group gained from 10 to

20 (gain = 10) and a control group gained from 12 to 20 (gain = 8), the numerator for computing effect size would be 10 – 8 = 2. If posttests were used, as suggested by Glass et al. (1981), the effect size in this case would have been zero. Similarly, in studies that statistically adjusted posttest scores for one or more covariates, the difference in *adjusted* scores was divided by the unadjusted control group standard deviation. Since all studies that met inclusion criteria presented either gain scores, pre- and posttest scores, or adjusted scores, or else matched on pretests, all effect sizes were adjusted for initial starting points.

If studies did not present enough data to allow for computation of effect size but otherwise met criteria for inclusion, they were included in tables with an indication of the direction and consistency of any achievement differences. In some cases only grade equivalent differences were given, and these are presented in the table. Since the standard deviations of grade equivalents are around 1.0 in upper elementary school, grade equivalent differences may be considered very rough approximations of effect size.

In general one overall effect size is presented for each study, unless two or more different ability grouping plans were compared to heterogeneous control groups in the same study (e.g., Cartwright & McIntosh, 1972) or two distinct samples were studied (e.g., Borg, 1965). Multiple effects within a study were averaged to obtain the overall effect size estimate (see Bangert-Drowns, 1986). If studies presented adequate data, overall effect sizes were also broken down by subject (e.g., reading, mathematics) or type of achievement measure and by achievement or ability level. Effect sizes by ability level should be interpreted with particular caution, as they are often inflated because standard deviations within subgroup categories are restricted in range.

In this best-evidence synthesis, every effort was made to make each effect size a meaningful representation of the effect of ability grouping on student posttest achievement, holding the posttest standard deviation as the common metric. In all tables, randomized studies are listed first, followed by matched studies presenting evidence of initial equality between experimental and control groups, and then matched studies lacking evidence of initial equality. Within categories, studies with the largest sample sizes are listed first. These procedures mean that effect sizes from studies listed earlier in each table should generally be given more weight than those listed later. However, it is important to remember that any effect size is only a rough indicator of the effect of a treatment. Many factors may influence effect size such as differences in subjects, measures, experimental procedures, and study durations.

In several cases, median effect sizes were calculated for substantively and theoretically homogeneous subsets of studies. Medians rather than means were used to avoid giving too much weight to outliers, as effect size data are often nonnormally distributed and cannot be considered a representative sample of all possible effect sizes from studies of a particular issue. When medians were computed, studies coded as + or – (i.e., for which effect sizes could not be computed) were given a numerical value equal to the median effect size for studies in that category with positive or negative values, respectively. For example, if effect sizes for a given table were +.60, +.40, +.20, .00, -.20, a study coded as "+" would be assigned a numerical value of +.40 (the median positive value) for purposes of computing a median. Studies coded as 0 were given the numerical value 0.00. Medians are presented merely to simplify discussion of the central tendency of homogeneous data, not as reliable findings. As is the case for the effect size themselves, medians should be interpreted in light of the quality and consistency of the studies from which they are derived.

Research on Comprehensive Ability Grouping in Elementary Schools

Ability-Grouped Class Assignment

A total of 14 studies of comprehensive ability-grouped class assignment plans were located.

The major characteristics and findings of the 14 studies are summarized in Table 1. The randomized study is listed first, followed by matched studies in descending order of sample size.

Inspection of Table 1 clearly indicates that the effects of comprehensive ability-grouped class assignment on student achievement are zero. The median total effect size across the 17 comparisons in 14 studies is exactly zero, and the effect sizes cluster closely around this value; of 13 comparisons that yielded effect size data, 8 fell in the range -.10 to +.10 and 11 in the range of -.15 to +.15. Effect sizes for reading and mathematics did not exhibit any pattern different from that for overall effects. Further, little support appears in Table 1 for the assertion that high achievers benefit from ability grouping, whereas low achievers suffer. Three studies (Borg, 1965; Flair, 1964; Tobin, 1966) found such a pattern, but three others (Bremer 1958; Hartill, 1936; Morgenstern, 1963) found just the opposite, and Barker-Lunn (1970), Goldberg, Passow, and Justman (1966), Loomer (1962), and Rankin, Anderson, and Bergman (1936) found no differences according to achievement level

The only randomized study of ability-grouped class assignment is one by Cartwright and McIntosh (1972), who compared three grouping methods in a school in Honolulu attended by disadvantaged students from a housing project. The students were ethnically diverse, and most came to school speaking Pidgin English and had to learn standard English as a second language. Students in grades 1-2 were randomly assigned to one of three treatments: self-contained heterogeneous grouping, self-contained ability grouping, and flexible. The ability-grouped students were assigned to relatively homogeneous classes according to intellectual ability and reading achievement without regard for grade level, so that the individual classes were somewhat heterogeneous in chronological age. The flexible classes were grouped for various subjects according to their performance level in those subjects, again without regard for grade level, and were frequently regrouped as their progress during the year warranted. All three treatments were begun when students entered the first and second grades and were continued for 2 years.

The dependent measures were scores on the Metropolitan Achievement Test. As shown in Table 1, the heterogeneous classes had higher scores in reading (*ES* = -.17) and in mathematics (*ES* = -.52) than the ability-grouped classes. The heterogeneous classes also achieved more in reading than the flexible classes (*ES* = -.28), but there were no differences in mathematics.

Even though the Cartwright and McIntosh (1972) study used random assignment to treatments, its results cannot be considered conclusive evidence against the use of ability-grouped class assignment. First, there was only one class at each grade level in each treatment, restricting the possibilities for reducing heterogeneity by ability grouping even more than is usually the case in elementary ability grouping studies. Second, the population involved is quite atypical, and generalization to other settings, even other disadvantaged schools, is difficult to justify.

Because there is only one randomized study of ability-grouped class assignment and it has some important limitations, we must look at the best of the nonrandomized studies, those that used matching procedures to equate nonrandomly assigned groups and presented data to indicate that the groups were, in fact, initially equivalent in achievement or ability.

Table 1　Ability-grouped class assignment

Article	Grades	Location	Sample size	Duration	Grouping criteria	Design	Effect sizes By achievement	By subject	Total
Randomized studies									
Cartwright & Mcintosh 1972	1-3	Honolulu (low income)	262 (9 cl.)	2 yrs.	IQ, ach.	Random asgmt. to 3 trts.:: Het, XG, "Flexible"; XG classes grouped across grade lines		XG vs Het: Rdg. -.17 Math -.52 Flex.vs Het: Rdg. -.28 Math .00	-.34 -.14
Matched studies with evidence of initial equality									
Barker-Lunn, 1970	2-5	England, Wales	5,500 (72 sch.)	4 yrs.	IQ, ach.	Matched schools in social class	Hi 0 Av 0 Lo 0	Rdg./Eng.0 Math 0	0
Goldberg, Passow, & Justman, 1966	5-6	New York City (middle class)	2,219 (86 cl., 45 sch.)	2 yrs.	IQ	Compared classes w/specified IQ ranges; students kept in same classes for 2 yrs.	V. Hi 0 Hi (-) Hi Av (-) Av (-) Lo (-)	Rdg. (-) Math (-)	(-)
Borg, 1965	4-7 6	Utah	4-667 (22 cl.) 6-875 (28 cl.)	4 yrs. I yr.	Gen. ach. Av 0	Compared 2 districts, Het vs. AG Math 0	Gr. 4: Hi (+) Lo (-) Gr. 6: Hi + Av (+) Lo 0	Rdg. 0	0 (+)
Hartill,1936	5-6	New York City	1,374 (15 sch.)	1 sem.(see design)	(see Gen.ach.)	Matched groups-each group Het 1 sem, AG I sem.; scores are gains	Hi -12 Av .00 Lo +.18	Rdg. +0.5 Math +.01	.00
Barthelmess & Boyer, 1932	4-5	Philadelphia	1,130 (10 sch.)	1 yr.	IQ	Schools matched, then students matched; scores are gains	Hi +.18 Av +.22 Lo +15	Composite ach.	+.21
Tobin, 1966	2-6	Marshfield, MA (rural)	989 (10 cl.)	2 yrs.	Rdg. ach.	Compared AG to year before AG introduced; only first 2 yrs. data	Hi +13 Av +.03 Lo -.46	Gen. ach. +.05 Rdg. +.13	+.05

Study	Grades	Location	N	Duration	Matching var.	Design	Results	Achievement	Composite ach.
Breidenstine, 1936	2-6	Soudersburg, PA (rural)	714 (11 sch.)	see design	—	Students in AG or Het classes since 1st grade matched on IQ		Composite ach.	−.08
Rankin, Andersen, & Bergman, 1936	3-6	Detroit	≈600 (7 sch.)	2 yrs.	IQ	Schools matched on student ach.; scores are gains; Het used no rdg. groups; AG used mastery learning in math	AG vs Het Hi +.10 Av +.03 Lo +.12 XG vs Het Hi +.10 Av +.05 Lo +.13	Rdg +.03 Math +.07 Rdg. +.11 Math +.03	+.05 +.07
Daniels, 1961	2-5	England	521 (4 sch.)	3.5 yrs.	Gen. ach.	Schools matched, then students matched on IQ		Rdg./LA −.25 Math −.27	−.26
Bremer, 1958	1	Amarillo, TX (Anglo)	510	1 yr.	Rdg. readiness	Compared AG to yr. before AG introduced	Hi −.24 Av .00 Lo −.06	Rdg. ach.	−.10
Loomer, 1962	4-6	Iowa (rural)	490 (23 cl.)	1 yr.	IQ, gen. ach., judgment	Schools matched on gen. ach.	Hi −.02 Av +.04 Lo −.06	Composite ach.	−.04
Flair, 1964	1	Skokie, IL (suburban)	441 (17 cl.)	1 yr.	Rdg. tchr. prognosis	Schools matched on gen. ach.	V. Hi +.54 Hi −.21 Av −.11	Rdg. +.03 Math −.14	−.06
Morgenstern, 1963	4-6	Uniondale, NY (suburban)	119 (7 cl.)	3 yrs.	—	Schools matched on IQ	Hi −.22 Av +.15 Lo +.64	Rdg./LA +.17 Math +0.6	+.15

Key: AG = Ability-grouped class assignment
  Het = Heterogeneous class assignment
  XG = Ability grouping across grade lines
  + = Results clearly favor ability-grouped classes
  (+) = Results generally favor ability-grouped classes
  O = No trend in results
  (−) = Results generally favor heterogeneous classes
  − = Results clearly favor heterogeneous classes

Three large, longitudinal studies done in the 1960s stand out in the study of ability-grouped class assignment: Barker-Lunn's (1970) study of streaming in English and Welsh junior schools, the Goldberg et al. (1966) study of different grouping patterns in New York City schools, and Borg's (1965) study in two Utah school districts.

Of these three studies, the Goldberg et al. (1966) study is perhaps the most remarkable. This study involved 86 grade 5 classes in 45 New York City elementary schools. Principals of all New York schools submitted Otis IQ distributions of their fourth grades. Only those schools with at least four students with IQs of at least 130 were included in the study sample, which had the effect (according to the authors) of restricting the sample to more middle-class parts of the city. The principals of the selected schools were asked to assign students to classes for the fifth grade to conform to any of 15 grouping patterns, ranging from extremely narrow to extremely broad. "Extremely narrow" classes included students falling within 1 IQ decile (e.g., 120-130), or those restricted to IQ 130 and up or 99 and below. "Extremely broad" classes included a full range of students from below 99 to above 130. Between these extremes were various moderately narrow and moderately broad patterns, classes containing students in 2 to 4 contiguous IQ deciles.

The principals were asked to keep students in the designated grouping patterns for 2 years, throughout grades 5 and 6, and only those students who were in the same schools for the entire 2-year period (79% of the original sample) were included in the data analyses.

With classes in 15 grouping patterns, Goldberg et al. (1966) were able to simulate many alternative grouping arrangements. For example, they could compare very homogeneous to very heterogeneous grouping plans by comparing the achievement gains of all students in 1-decile classes to those in 5-decile classes. They could simulate provision of special classes for the gifted by comparing 5-decile (heterogeneous) classes to combinations of 1-decile (130+) and 4-decile (less than 99 to 130) classes, and so on.

Unfortunately, Goldberg et al. do not present their actual achievement data, but only describe significant differences and patterns of findings. However, the patterns they present consistently favor broad, heterogeneous grouping plans for all students except for the most gifted (130+), who did equally well in broad- or narrow-range classes. Presence of gifted students was beneficial for the achievement of most students in most subjects, whereas the presence of low achievers was neither beneficial nor detrimental overall.

The Goldberg et al. (1966) study is arguably the best evidence in existence against the possibility that reductions in IQ heterogeneity can enhance student achievement in the upper elementary grades. The size and rigor of the experiment make it highly unlikely that any nontrivial positive effect of ability grouping could have been missed. While most achievement comparisons in the Goldberg et al. study were nonsignificant, the patterns of mean differences and of those differences that were statistically significant support heterogeneous rather than ability-grouped class assignments.

The Barker-Lunn (1970) study in England and Wales similarly provides little support for ability-grouped class assignment. This study compared the achievement gains of students in 36 streamed junior schools (serving students aged 7 to 11) and 36 unstreamed schools, matched on social class. The streamed schools had a slight advantage in achievement after 1 year, so the initial comparability of the samples was questioned by the author, the year-to-year scores and 4-year longitudinal comparisons used first-year scores as covariates to control for this initial difference.

As in the case of the Goldberg et al. (1966) study, Barker-Lunn (1970) presents only statistically significant differences and trends. Overall, there were no meaningful trends favoring streamed or nonstreamed schools. All but a few comparisons were nonsignificant, and those that were significant were equally likely to favor streamed or unstreamed schools. Again, if there were any consistent effect of ability grouped class assignment on student achievement, a study the size and quality of Barker-Lunn's would be very likely to find it.

The third large, longitudinal study is one by Borg (1965), who compared achievement gains in two adjacent districts in Utah, one of which used homogeneous and one heterogeneous grouping. Two elementary cohorts were studied. One began in the fourth grade and was followed through grade 7. Another began in the sixth grade and was followed through grade 9. The results for the sixth grade sample presented in Table 1 are only for the first year, as these students went on to junior high school beginning in grade 7. Even though this study used as large a sample as Goldberg et al. (1966) and Barker-Lunn (1970) and was also carefully controlled, the nature of the samples involved make the results of this study less conclusive. First, only two districts were involved, and any differences between the districts other than the use of ability grouping are completely confounded with grouping practice. One district served a small city, whereas the other served its outlying area, so unmeasured population differences may have been operating. Second, whereas the two districts' mean pretest scores were equal within each IQ category, the proportion of high-IQ students was higher in the district using heterogeneous grouping, particularly in the sixth grade sample. Third, the districts involved had been using their respective grouping methods for many years, so that the students being studied had been in ability-grouped or heterogeneous classes for 3 years (the grade 4 sample) or 5 years (the grade 6 sample). Any effects of grouping may have already been registered before the study began.

The results of the Borg (1965) study were inconsistent, but in general the longitudinal data following fourth graders indicated that ability grouping was beneficial for the achievement of high-IQ students, detrimental for that of low-IQ students, and neutral for average-IQ students. After 1 year, high- and average-IQ students scored higher in ability-grouped than in heterogeneous classes, but by the seventh grade this difference had disappeared. High- and average-achieving sixth graders gained more in ability-grouped than in heterogeneous classes, but these differences also dissipated in junior high school.

One well-designed (but 50-year-old) study by Hartill (1936) compared ability-grouped to heterogeneous class assignment in 15 New York City schools. Students in grades 5 and 6 were assigned to ability-grouped or heterogeneous classes for one semester, then reassigned to classes according to the opposite grouping pattern for a semester. Not only were the students their own controls (since they experienced both grouping plans) but they were individually matched with one another, so that the group that experienced ability grouping first was identical in IQ to that which experienced heterogeneous grouping first. The results indicated that low-IQ students achieved slightly better in ability-grouped classes ($ES = +.18$), high-IQ students achieved slightly better in heterogeneous classes ($ES = -.12$), and average-IQ students achieved equally well in the two grouping plans. Overall, achievement gains were identical in the ability-grouped and heterogeneous classes.

Another important early study was one by Rankin et al. (1936), who compared students matched on achievement level in three programs. One was traditional ability-grouped class assignment done within grade levels, except that in mathematics these classes used a program essentially identical to modern group-paced mastery learning (see Block & Anderson, 1975). Another, called "vertical grouping," assigned students to classes according to their level of achievement without regard for their grade level. This procedure produced such homogeneous classes that reading groups within the classes were considered unnecessary. The third plan involved heterogeneous grouping of classes, with the additional requirement that within-class ability grouping (including use of reading groups) was not allowed. Teacher and administration attitudes toward this heterogeneous plan were quite negative, as the degree of heterogeneity in these classes was great and teachers were unable to use any form of grouping to accommodate student differences. However, achievement differences between the two ability-grouped plans and the heterogeneous classes were small for the ability-grouped plan (*ES* = +.05) and for the "vertical" plan (*ES* = +.07).

A 4-year study of streamed and unstreamed junior schools in England by Daniels (1961) compared two pairs of schools. The schools themselves were selected from the same districts and were found to have nearly identical mean IQs. Students were individually matched on IQ within the schools. After 3H years, the students in the unstreamed schools were achieving at a significantly higher level than those in the streamed schools (*ES* = -.26). However, other multiyear studies (Breidenstine, 1936; Tobin, 1966) found effects near zero, and Morgenstern (1963), who followed students from the fourth to the sixth grade, found a small benefit of ability grouping (ES= +.15). Some authors (e.g., Good & Brophy, 1984) have suggested the use of a modified ability grouping plan in which high and average achievers are mixed and average and low achievers are mixed. However, a study by Loomer (1962) found no achievement benefits of such a plan (*ES* = -.04).

In addition to the studies listed in Table 1, a few studies have correlated the degree of heterogeneity in classes with student achievement. Justman (1968) found that the reading achievement of third graders increased slightly more in heterogeneous than homogeneous classes, with average- and low-achieving classes gaining the most from heterogeneity. Leiter (1983) found no correlation between class homogeneity and third grade reading and mathematics achievement, controlling for the previous year's scores, although there was a nonsignificant trend toward higher reading achievement and lower mathematics achievement in more homogeneous classes. Edminston and Benfer (1949) divided 16 classes of fifth and sixth graders into classes with wide and narrow IQ ranges. Over 6 months, students in the wide-range classes gained significantly more in composite achievement than did students in narrow-range classes.

Summary and discussion: Ability-grouped class assignment. Given the persistence of the practice over time and the belief teachers typically place in its effectiveness it is surprising to see how unequivocally the research evidence refutes the assertion that ability-grouped class assignment can increase student achievement in elementary schools. There is a considerable quantity of good quality research on this topic such that any impact of grouping on achievement would surely have been detected.

Several earlier reviews have made the claim that ability grouping is beneficial for high-ability students and detrimental for low-ability students (e.g., Begle, 1975 Eash, 1961; Esposito, 1973). This claim is not clearly supported by the present review. It is possible that a clearer pattern emerges in secondary studies, but it is more likely that confusion arises when studies of special programs for the gifted and for low achievers are included in ability grouping reviews. Studies of special programs for the gifted tend to find achievement benefits for the gifted students (J.A. Kulik & C.-L. Kulik, 1984; Passow, 1979), whereas studies of mainstreaming versus special education for students with learning problems tend to favor regular class placement (Madden & Slavin, 1983). For this reason, including studies of special programs for the gifted and learning disabled in reviews of ability grouping as was done by Begle (1975), Borg (1965), Findley and Bryan (1971), and others, would give the impression that ability grouping is beneficial for high achievers and detrimental for low achievers. However, it is likely that characteristics of special accelerated programs for the gifted account for the effects of gifted programs, not the fact of separate grouping per se (see Fox, 1979). Also, problems of selection bias in nonrandomized studies of programs for the gifted and for students with learning problems bias the results of these studies toward the higher placement, spuriously favoring separate programs for the gifted and mainstream placement for low achievers (see Borg, 1965; Slavin, 1984a).

It is interesting to compare the results of the C.-L. Kulik and J.A. Kulik (1984) meta-analysis of research on ability grouping in elementary schools with those summarized in Table 1. The Kuliks report a mean effect size in favor of ability grouping of +.19. However, this combines studies of special programs for the gifted, for which they report a mean of +.49, and of "more representative populations," for which the effect size is reported as +.07. This latter figure is trivially different from zero and depends in large part on one very high effect size (+.61) reported by Atkinson and O'Connor (1963). This study did not meet the criteria for inclusion in the present review on two grounds. First, the form of ability grouping evaluated was really a special program for the gifted, not ability-grouped class assignment. In each of two schools with three sixth-grade classes, one class was designated as an "advanced" section and the remaining two were designated "regular." Students were selected into the advanced sections on the basis of IQs (in excess of 125), achievement test scores, and teacher judgment.

Second, the experimental and control classes were substantially different in IQ and achievement at pretest. The authors claim that because the control group started higher and ended equal to the experimental group, the experimental group gained more. However, the authors also present data to indicate a ceiling effect on the dependent measure. Their use of nonparametric statistics and presentation of data in terms of "percentage above median" rather than means or actual medians makes interpretation of study results problematic.

Further, the Kuliks do not separately examine studies of ability-grouped class assignment from those investigating regrouping for reading and/or math only. The median effect size for studies included by the Kuliks which in fact evaluated abilitygrouped class assignment is zero, identical to that found in the present synthesis.

Regrouping for Reading and Mathematics

In many elementary schools, reading and/or mathematics is scheduled at the same time for all students in a particular grade. At that time, students leave their heterogeneous homeroom classes to receive reading or mathematics instruction in a class that is more homogeneous in the skills in question.

Previous reviews and meta-analyses (e.g., Borg, 1965; Findley & Bryan, 1970; C.-L. Kulik & J.A. Kulik, 1984) have not made a clear distinction between regrouping and ability-grouped class assignment. Yet there are several important theoretical reasons to do so. First, regrouping minimizes the presumed negative psychological effects of ability grouping. Students spend most of the day in heterogeneous classes, with which they almost certainly identify. Second, regrouping is always done on the basis of actual performance in reading or mathematics, not on IQ. Third, assignments to regrouped classes can be relatively flexible. Any errors in assignment or changes in student level of achievement can be accommodated by moving students to different sections rather than by changing their basic homeroom assignments, as would be required in between-class ability grouping. For these reasons, it is likely that regrouping can produce much more homogeneity in the skills being taught than can ability-grouped class assignment, which is usually based on IQ or general achievement and is relatively inflexible.

Unfortunately, there is neither the number nor the quality of studies of regrouping to make possible definitive conclusions concerning the effectiveness of such plans. Only three studies used matching and presented evidence of initial equality. Four additional studies lacked evidence of initial equality but did adjust posttests for pretests and other variables. Overall, five of the seven studies found that students learned more in regrouped than in heterogeneous classes, whereas two found the opposite trend (see Table 2).

Two of the studies investigated the practice of regrouping for reading only. One of these was a large study by Moses (1966) involving 54 classes in rural Louisiana. This carefully controlled study held constant time and instructional materials in matched experimental and control classes. No consistent differences were found in reading achievement. A study by Berkun, Swanson, and Sawyer (1966) did find significantly greater gains for regrouped than for self-contained reading classes *(ES = +.32)*. However, this article provides few details of the treatment procedures and may suffer from pretest differences between the experimental and control groups (all data presented are posttests adjusted for pretests).

A study by Provus (1960) of regrouping for mathematics provides the best evidence in favor of this practice. Experimental students in 11 classes in a suburb of Chicago were regrouped from their heterogeneous homerooms into relatively homogeneous mathematics classes at the same grade level. Achievement gains for students in these classes were compared to those of students matched on IQ who remained in heterogeneous classes all day. One effect of the regrouping was to allow high achievers to be exposed to material far above their grade level; there were cases of fourth graders finishing the year working on eighth grade material. Perhaps for this reason, achievement gains for high-ability students in the regrouping program were much greater than those of comparable control students *(ES = +.79)*, but the program was less spectacularly beneficial for average-ability *(ES = +.22)* and low-ability students *(ES = +.15)*. In contrast, Davis and Tracy (1963) found that regrouping for mathematics was detrimental to the achievement of students in a rural North Carolina town. However, this study compared only two schools, and there were substantial achievement differences at pretest. Also, it is important to note that no attempt was made to provide differentiated materials to the regrouped classes; all classes used grade level-appropriate texts.

Three studies investigated the effects of regrouping in multiple subjects. In a study by Koontz (1961) in Norfolk, Virginia, experimental students were separately grouped according to their achievement in reading, mathematics, and language. At other times students remained in "intact classes," but it is unclear whether these were ability grouped or heterogeneous. This method approaches a departmentalized arrangement, as students changed classes three or four times each day. Its effects on all three subjects involved turned out to be negative, particularly for reading, where the heterogeneous, self-contained classes gained .42 grade equivalents more than the regrouped students. A study by Balow and Ruddell (1963) evaluated regrouping for reading and math, and found positive effects in both subjects for average and low achievers. However, pretest differences favoring the experimental (regrouped) classes throw some doubt on these findings.

Finally, Morris (1969) studied a program in which regrouping was done for reading and math. The program was called a "nongraded primary plan" by the author, but because regrouping was done within grade levels, it was categorized in this review as a regrouping program. Overall student achievement at the end of 3 years was higher in the regrouped classes than in heterogeneous control groups, controlling for IQ *(ES = +.43)*. After 2 more years during which all students, experimental as well as control, were in a regrouping plan, the former experimental students had greatly increased their advantage over the control group *(ES = + 1.20)*.

Summary and discussion: Regrouping for reading and mathematics. Overall, the results of studies of regrouping for reading and mathematics are inconclusive. None of the grouping patterns evaluated were consistently successful, although one study (Provus,

Table 2 Ability grouping for selected subjects

Article	Grades	Location	Sample size	Subjects	Grouping criteria	Duration	Design	By achievement	Effect sizes By subject	Total
Matched studies with evidence of initial equality										
Moses, 1966	4-6	Calcasieu Parish, LA (rural)	900 (54 cl.)	Rdg.	Rdg. ach.	1 sem.	Students matched on grade, sex, age, rdg. ach.	Hi +.10 Av +.04 Lo -.01	Rdg.	+.05
Provus, 1960	4-6	Homewood, IL (suburban, high-ach.)	240 (19 cl.)	Math	Math ach.	1 yr.	Students matched on IQ; scores are gains	Hi +.79 Av +.22 Lo +.15	Math	+.39
Koontz, 1961	4	Norfolk, VA	108 (10 cl.)	Rdg, Math, Lang.	Rdg. ach. Math ach. Lang. ach.	1 yr.	Students matched on gen. ach.; scores are grade equivalents	V. Hi -.12 GE Hi -.25 GE Av -.33 GE Lo -.41 GE	Rdg. -.42 GE Math -.29 GE Lang.-.12 GE	-.28 GE
Matched studies lacking evidence of initial equality										
Berkun, Swanson, & Sawyer, 1966	3-5	Monterey, CA	1,098 (10 sch., 45 cl.)	Rdg.	Rdg. ach.	1 yr.	Compared schools using AG or Het rdg. classes; no evidence of initial equality; posttests adj. for pre	Hi +.44 Lo +.32	Rdg.	+.32
Davish Tracy, 1963	4-6	North Carolina	393 (2 sch.)	Math	Math ach.	1 yr.	Compared schools using AG or Het math classes; pretest diffs. favored AG; post adj. for pre, IQ, other vars.		Math	(-)
Balow & Ruddell, 1963	6	Southern California (suburban)	197 (8 cl.)	Rdg. Math	Rdg. ach. Math ach.	1 yr.	Compared AG to Het schools; pretest diffs. favored AG; scores are gains	Hi O Av + Lo +	Rdg. (+) Math (+)	(+)
Morris, 1969	1-3	UpperMoreland, PA (suburban)	117 (8 cl.)	Rdg. Math	Rdg. ach. Math ach.	3 yrs.	Compared AG to previous year (Het); IQ diffs. favored Het; scores are grade equivalents adjusted for IQ		Composite ach.	+.43 GE

Key: AG = Ability grouping used for selected subjects
Het = Heterogeneous class assignment
+ = Results clearly favor ability-grouped classes
(+) = Results generally favor ability-grouped classes
0 = No trend in results
(-) = Results generally favor heterogeneous classes
- = Results clearly favor heterogeneous classes

1960) gave strong evidence favoring the use of regrouping in mathematics if students are given materials appropriate to their levels of performance. Another study (Morris, 1969) found strong positive effects of regrouping for reading and mathematics. The regrouping plan evaluated in this study also emphasized adaptation of the level of instruction to accommodate student differences. These studies suggest that regrouping for reading and/or mathematics can be effective if instructional pace and materials are adapted to students' needs, whereas simply regrouping without extensively adapting materials or regrouping in all academic subjects is ineffective. Evidence from studies of the Joplin Plan, summarized in the following section, provides further support for this possibility. However, more research is needed to establish the achievement effects of regrouping within grade levels.

Joplin Plan

The Joplin Plan (Floyd, 1954) is in its simplest form an extension of regrouping for reading to allow for grouping by reading level across grade level lines. This practice typically creates reading classes in which all students are working at the same or at most two reading levels, so that within-class ability grouping may be reduced or eliminated. The trade-off between within-class (reading groups) and between-class (Joplin) ability grouping is a pivotal issue in studies of the Joplin Plan, which may be conceptualized not as ability grouping versus heterogeneous grouping but as between- versus within-class grouping (see, for example, Newport, 1967). In contrast, studies of regrouping for reading within grades maintain reading groups within the class, although there may be some reduction in the number of reading groups used.

Nongraded plans share with the Joplin Plan the idea of grouping students according to performance level in a specific skill, ignoring grade level or age. Some forms of nongraded grouping are very similar to the Joplin Plan, except that they are applied in the primary rather than intermediate grades and have been utilized in subjects other than reading. Unlike the complex individualized arrangements described by Goodlad and Anderson (1963), these programs involve cross-grade grouping in only one subject, usually reading. Studies of such programs are included as "Joplin Plan" studies in this review.

Table 3 summarizes the research on the Joplin Plan and Joplin-like nongraded plans. Overall, the evidence in Table 3 consistently supports the use of the Joplin Plan. Joplin classes achieved more than control classes in 11 of 14 comparisons with the remaining three studies finding no differences. The median effect size for these studies is about +.45. Further, the quality of these studies is high, with 2 using random assignment and 10 using matching with good evidence of initial equality.

Morgan and Stucker (1960s conducted a randomized study of the Joplin Plan in rural Michigan. Fifth and sixth grade students were matched on reading achievement and then randomly assigned to four Joplin and four control classes. Teachers were also randomly assigned to treatments. Because there were only two Joplin classes at each grade level, the amount of cross-grade grouping that could be done was limited, and control groups were ability grouped (within grade), yet the authors still document a considerable reduction in class heterogeneity as a result of crossgrade assignment.

Results indicated significantly higher achievement in the Joplin Plan for high and low achievers in fifth grades and low achievers in the sixth grades. The authors explain the failure to find experimental-control differences for high-achieving sixth graders by noting that because of the small number of classes involved in the study, high-achieving sixth graders could not be accelerated as much as would have been possible with larger numbers of classes. Whatever the explanation, larger experimental-control differences for low achievers *(ES = +.94)* than for high achievers *(ES =*

+.32) are entirely due to the lack of differential gains for high-achieving sixth graders. Hillson, Jones, Moore, and Van Devender (1964) studied a nongraded program similar to the Joplin Plan. They randomly assigned students and teachers to nongraded or traditional classes. Students in the nongraded classes were assigned to heterogeneous classes but regrouped across grade levels for reading. They proceeded through nine reading levels and were continually regrouped on the basis of their reading performance. Within each reading class teachers had multiple reading groups and used traditional basal readers and instructional methods (J.W. Moore, personal communication, January 23, 1986).

The results of this study supported the efficacy of the nongraded program. After 3 semesters, reading scores for experimental students on three standardized scales were considerably higher than for control students (*ES* = +.72, or about .41 grade equivalents). After 3 years in the program, experimental-control differences had diminished but were still moderately positive (*ES* = +.33) (Jones, Moore, & Van Devender, 1967). Ten studies compared Joplin or Joplin-like nongraded classes to matched control classes and presented evidence of initial comparability. The largest of these (Russell, 1946) was done before Floyd (1954) first described the Joplin Plan, but evaluated a very similar intervention. Students in grades 4-6 were regrouped for reading without regard to grade level. This created relatively homogeneous groups, but homogeneity was increased still further by the use of reading groups within the classes (usually two) and by reviewing and modifying group assignments four times per year. Students in this plan, called "circling," were matched with students in other schools that did not regroup for reading and were followed for 2 years, from the beginning of grade 4 to the beginning of grade 6. Results indicated no differences between the two types of grouping plans (*ES* = .00).

It is interesting to note that the only other matched equivalent study to find no advantage for the Joplin Plan also used reading groups within the regrouped reading classes. This was a study by Carson and Thompson (1964), in which students in grades 4-6 were regrouped across grade lines for reading but were still assigned to reading groups within their reading classes. These students' gains in reading achievement were compared to those of students assigned by ability (within grade) to self-contained classes.

The eight remaining matched equivalent studies all found positive effects of Joplin or Joplin-like nongraded plans on student achievement. For example, Green and Riley (1963) compared the Joplin Plan to the traditional methods in use in the same schools during the previous year. Students in the Joplin classes gained significantly more in reading achievement than did students in the earlier years (*ES*= +.36).

In a study by Hart (1959), grade 4-5 students were regrouped into nine reading classes. Seven of these had only one reading level, one had two, and one had four very low-achieving groups (but may have used fewer than four reading groups in the classroom). The top class was reading at the seventh grade level, and the bottom class contained students ranging from primer to second grade, second semester. Students' scores were compared to those of students taught by the same teachers the previous year. Gains on the California Achievement Test strongly favored the Joplin approach (*ES* = +.89). In both the fourth and fifth grades, Joplin groups gained about a full grade equivalent more than did heterogeneously grouped classes in earlier years.

Rothrock (1961) also compared Joplin Plan classes to heterogeneous classes that used within-class ability grouping, and found significantly positive effects on student reading achievement and work-study skills, averaging .44 grade equivalents more than in heterogeneous classes. An individualized reading program fell between the Joplin and heterogeneous programs in achievement effects. Green and Riley (1963) found consistently greater reading achievement gains in Joplin Plan classes than in matched

Table 3 Joplin Plan

Article	Grades	Location	Sample size	Subject	Duration	Design	Effect sizes		
							By achievement	By subject	Total
Randomized studies									
Morgan & Stucker, 1960	5-6	Dundee, MI	360 (8 cl.)	Rdg.	1 yr.	Students & teachers randomly assigned to Joplin or AG; authors note limited opportunity for high-ach. 6th graders to work above grade level	Hi +.32 Lo +.94	Rdg.	+.30
Hillson, Jones, Moore, & Van Devender, 1964; Jones, Moore, & Van Devender, 1967	1-3	Shamokin, PA	52 (6 cl.)	Rdg.	1.5 yrs. 3 yrs. (follow-up)	Students & teachers randomly assigned to NG/Joplin or Het classes for reading only		Rdg. After 3 yrs.: +.33	+.72
Matched studies with evidence of initial equality									
Russell, 1946	4-5	San Francisco	526 (6 sch.)	Rdg.	2 yrs.	Matched students in Joplin & Het schools on IQ; study predated use of term "Joplin Plan"; two reading groups used within regrouped classes		Rdg.	.00
Green & Riley, 1963	4-6	Atlanta (whites)	4 sch.	Rdg.	1 yr.	Compared Joplin to previous yr. (Het)		Rdg.	+.36
Ingram, 1960	1-3	Flint, MI	68 exp. 377 cont.	Rdg.	3 yrs.	Compared NG/Joplin in reading only to previous yr. (Het)		Rdg.	+.55
Halliwell, 1963	1-3	New Jersey	295	Rdg.	1 yr.	Compared NG/Joplin in rdg. & lang. to previous yr. (Het); exp. also sig. higher in math, which was not in program		Rdg. +.53 (gr. 1-3) Spel. +.58 (gr. 2-3) Lang. +.26 (gr. 3)	+.59
Carson & Thompson, 1964	4-6	Sebastopol, CA	250	Rdg.	1 yr.	Compared Joplin to AG classassignment matching on IQ: Rdg. grps. were used within Joplin classes		Rdg.	0
Anastasiow, 1968	4-6	North Carolina	217 (2 sch.)	Rdg.	1 yr.	Matched Joplin & Het schools on IQ, SES; scores are gains		Rdg.	+.15
Hart, 1959	4-5	Hillsboro, OR	190	Rdg.	1 yr.	Compared Joplin to previous yr. (Het); scores are gains		Rdg.	+.89
Rothrock, 1961	4-5	McPherson, KS	124 (8 cl.)	Rdg.	1 yr.	Matched classes, Joplin vs. Het		Rdg.	+.44 GE

Study	Grades	Location	N	Subject	Duration	Notes	Results		Effect
Skapski, 1960	3	Burlington, VT	110 (3 sch.)	Rdg.	3 yrs	Compared schools NG/Joplin in rdg. only to Het, matched on IQ; no exp.-cont. diffs. in math, which was not in program	V. Hi +.91 Hi +.48 Av +.52	Rdg.	+.57
Hart, 1962	1-3	Hillsboro, OR	100 (2 sch.)	Math	3 yrs.	Compared NG/Joplin in math only to Het, students matched on IQ, sex, SES; same school studied in Hart (1959) Joplin study		Math	+.46
Matched studies lacking evidence of initial equality									
Moorhouse, 1964	4-6	Laramie, WY	169 (4 cl.)	Rdg.	1 sem. 5 sem. (fol- low-up)	Matched 2 schools, Joplin vs. Het; 4th and 6th grade Het started with higher pretests; scores are grade equivalent gains	Hi +.61 GE Av +.74 GE Lo +.34 GE By end of gr. 6: Gr. 4 (5 sem.) +.50 GE Gr. 5 (3 sem.) +.40 GE		+.63 GE
Kierstead, 1963	3-8	Rutland, VT	277 (2 sch.)	Rdg.	1 yr	Matched 2 schools, Joplin vs. AG class assignment; Joplin higher at pretest; scores are grade equivalent gains	Hi -.01 GE Av +.08 GE Lo -.14 GE		-.02 GE

Key. Joplin = Joplin Plan
NG/Joplin = Program identified as nongraded or ungraded which used Joplin organization (cross-grade grouping for only one subject)
AG = Ability-grouped class assignment
Het = Heterogeneous class assignment
+ = Results clearly favor Joplin classes
(+) = Results generally favor Joplin classes
0 = No trend in results
(−) = Results generally favor control classes
− = Results clearly favor control classes

heterogeneously grouped classes in different schools (*ES* = +.36). Every experimental class gained significantly more than its corresponding control class, and the average experimental-control difference in grade equivalents was .54. Anastasiow (1968) found no significant differences between a Joplin-type regrouping plan and heterogeneous grouping, but the trends favored the Joplin Plan groups (*ES*= +.15).

Moorhouse (1964) compared a school using the Joplin Plan to a heterogeneously grouped control school. The students in grades 4-6 were grouped in seven reading classes. Three classes contained one reading level, three contained two, and one (the lowest) contained students from five levels, with a range from first to third grade levels. The top class had students working at the seventh and eighth grade levels. The authors note that three quarters of all students in the Joplin classes were working at a level different from the ones usually used at their grade level. Unfortunately, the results of the Moorhouse study are marred by pretest differences favoring the control groups in grades 4 and 6. However, at all three grade levels (including grade 5, where there were no pretest differences), Joplin classes gained considerably more in reading achievement than heterogeneous control classes, averaging gains of 1.24 grade equivalents (*GE*) in one semester, more than twice the gains seen in control classes (.61 *GE*). Experimental and control classes were followed for a total of five semesters. By the end of sixth grade, fourth graders had gained a total of .50 grade equivalents more than control. Fifth graders had gained about .40 grade equivalents by the end of sixth grade, but lost this advantage by the eighth grade. Sixth graders, who made the greatest gains initially, maintained most of that gain through the eighth grade. Overall, the patterns of results indicate that achievement gains due to the Joplin Plan were primarily seen early in the program implementation and then diminished as students entered the junior high school (which was not using the Joplin Plan).

Ingram (1960) evaluated a nongraded program very similar to that studied by Hillson et al. (1964). Students in grades 1-3 were assigned to one of nine reading levels without regard to age. As students moved from grade to grade, they picked up where they had left off in the previous year. Teachers generally had more than one reading group within their reading classes. As in the Hillson et al. study, the results supported the nongraded approach By the end of 3 years, students in the nongraded program were achieving approximately .7 grade equivalents ahead of similar students in earlier years before nongrading (*ES* = +.55) on standardized reading, spelling, and language tests.

Halliwell (1963) evaluated a nongraded primary program that was virtually identical to the Joplin Plan. Students in grades 1-3 were regrouped for reading only, and remained in heterogeneous classes the rest of the day. Spelling was also included in the regrouped classes for second and third graders. The article is unclear as to whether within-class grouping was used in regrouped reading classes, but there is some indication that reading groups were not used. Results indicated considerably higher reading achievement in nongraded classes than in the same school the year before nongrading was introduced (*ES* = +.59). Scores were higher for nongraded students at every grade level, but by far the largest differences were for first graders, who exceeded earlier first grade classes by .94 grade equivalents (*ES* = + 1.22).It is important to note that mathematics achievement, measured at the second and third grade levels, also increased significantly more in the nongraded classes than in previous years (*ES* = +.51). Because mathematics was not part of the nongraded program, this finding suggests the possibility that factors other than the nongraded program might account for the increases in student achievement. However, the author notes that teachers claimed to have been able to devote more time to mathematics because the nongraded program required less time for reading, spelling, and language instruction than they had spent on these subjects in previous years.

A study by Skapski (1960) also evaluated the use of a Joplin-like nongraded organization for reading only. The details of the nongraded program were not clearly described, but it appears that reading groups were not used in regrouped classes and that curricula and teaching methods were traditional. Two comparisons were made. First, the reading scores of students in the nongraded program were compared to the same students' arithmetic scores, on the assumption that because arithmetic was not involved in the nongraded plan any differences would reflect an effect of nongrading. Results of this comparison indicated that second and third grade-aged students achieved an average of 1.1 grade equivalents higher in reading than in arithmetic. Further, scores of third graders who had spent 3 years in the nongraded program were compared to those of students in two control schools matched on IQ. Results indicated that the nongraded students achieved at a much higher level in reading than did control students (*ES* = +.57), but there were no differences in arithmetic (which was not involved in the nongraded program). Differences were particularly large for students with IQs of 125 or higher (*ES* = +.97), but were still quite substantial for students with IQs in the range 88-112 (*ES*= +.52).

Only one study evaluated the use of a Joplin-like program in mathematics. This is a study by Hart (1962), which took place in the same school that evaluated the Joplin Plan in reading in its intermediate grades (Hart, 1959). Experimental students were regrouped for arithmetic instruction across grade lines and were taught as a whole class. Students were frequently assessed on arithmetic skills and reassigned to different classes if their performance indicated that a different level of instruction was needed. Experimental students who had spent 3 years in the nongraded arithmetic program were matched on IQ, age, and SES with students in similar schools using traditional methods. It is not stated whether control classes used within-class ability grouping for arithmetic instruction. Results indicated an advantage of about one half of a grade equivalent for the experimental group (*ES* = +.46).

Summary and discussion: Joplin Plan. Considered together, the results of the Joplin and Joplin-like nongraded plans are remarkably consistent. Both randomized studies found positive effects on student achievement, as did all but 2 of the 10 matched equivalent studies.

Only four studies presented results according to student ability levels. Morgan and Stucker (1960) found stronger positive effects of the Joplin Plan for low than for high achievers, whereas Moorhouse (1964) found the largest gains for high and average achievers, and Kierstead (1963) found no experimental-control differences at any level of ability. Skapski's (1960) results indicated that very able students benefited the most from a Joplin-like nongraded reading program. In no case did one subgroup gain at the expense of another; either all ability levels gained more than their control counterparts or (in the case of the Kierstead study) none did.

It is possible that the positive effects of the Joplin Plan are due to the novelty of the approach rather than to anything inherent in the program itself. This is always a concern in experimental studies. However, in this case the possibility of novelty effects accounting for all of the effects of the Joplin Plan seems remote. Some of the studies with the largest effects involved durations of 3 years (Hart, 1962; Ingram, 1960; Skapski, 1960). Even 1 year, the minimum duration for all Joplin studies, is a long time for a novelty effect to be detectable on a standardized achievement measure.

Within-Class Ability Grouping

Research on within-class ability grouping differs from that on between-class ability grouping in many important ways. First, this research is considerably more likely to use random assignment than is research on between-class ability grouping. Five randomized studies met the criteria for inclusion applied in this review (one additional study, by Putbrese (1972), also used random assignment but was omitted because it had only one experimental and one control class). Second, the duration of within-class ability grouping studies is shorter, with most studies lasting about 1 semester.

Research on within-class ability grouping is summarized in Table 4. Every study that met the inclusion criteria involved the use of math groups, although Jones (1948) also studied grouping in reading and spelling. The lack of studies of grouping in reading is surprising. It may be that this practice is so widespread in elementary schools that formation of ungrouped control groups is difficult to arrange, even on an experimental basis. One recent correlational study did compare grouped and ungrouped reading classes, but presented no data to indicate that these classes were initially equivalent (Sorensen & Hallinan, 1986).

Every study of within-class ability grouping in mathematics favored the practice, though not always significantly. The median effect size for the five randomized studies is +.32; including matched studies makes the median only slightly higher. Five studies broke down achievement effects by student ability levels, but there is no clear pattern of results favoring ability grouping for one or another subgroup. Every subgroup gained more in classes using within-class ability grouping than in control (ungrouped) treatments. However, it is interesting to note that the median effect size for low achievers (*ES* = +.65) was higher than that for average (*ES* = +.27) or high achievers (*ES* = +.41).

Slavin and Karweit (1985) conducted two large randomized studies of within-class ability grouping, one in highly heterogeneous, racially mixed schools in Wilmington. Delaware (Experiment 1), and one in relatively homogeneous, predominately white schools in and around Hagerstown, Maryland (Experiment 2). In Experiment 1, grade 4-6 classes were randomly assigned to one of three treatments. One, an individualized model, is not considered here. A second was a whole-class instructional model called the Missouri Mathematics Program (MMP), which had been found in earlier research (Good & Grouws, 1979) to be more effective than traditional whole-class instruction. The MMP, based on the findings of studies of the practices of outstanding elementary mathematics teachers (e.g., Good & Grouws, 1977), uses a regular sequence of teaching, controlled practice, independent seatwork, and homework, with an emphasis on a high ratio of active teaching to seatwork, teaching mathematics in the context of meaning, and management strategies intended to increase student time on-task (Good, Grouws, & Ebmeier, 1983). The third treatment was a form of within-class ability grouping based on the MMP, but using two ability groups. Specific management strategies were incorporated in this model to help teachers deal with the problem of having one group working independently while the other was receiving instruction. Details of this treatment are presented in Slavin and Karweit (1983).

Results of the semester-long study indicated significantly higher achievement in ability-grouped than in whole-class instruction (*ES* = +.32). Effects were large for mathematics computations (*ES* = +.64), but there were no differences in concepts and applications (*ES* = .00).

Experiment 2 involved the same comparisons, except that an untreated control group was also added. The results very closely paralleled those of Experiment 1; students in the ability-grouped model achieved significantly more than those in the two whole-class methods (*ES* = +.27), with differences larger in mathematics computations (*ES* = +.37) than concepts and applications (*ES* = +.17).

Dewar (1964) randomly assigned sixth grade mathematics classes and their teachers in a suburb of Kansas City, Kansas, to use within-class ability grouping or whole-class instruction for a full school year. Three math groups were used in the grouped classes. Results strongly favored the grouped classes for students who had

Table 4 Within-class ability grouping

Article	Grades	Location	Sample size	No. of groups	Duration (months)	Design	Effect sizes By achieve-	By subject	Total
Randomized studies									
Slavin & Karweit, 1985 (Experiment 2)	3-5	Hagerstown, MD	366 (17 cl.)	2	4	Classes randomly assigned to WCAG, MMP, control	Hi +.41 Av +.21 Lo +.29	Math	+.27
Slavin & Karweit, 1985 (Experiment 1)	4-6	Wilmington, DE	231 (11cl.)	2	5	Classes randomly assigned to WCAG, control	Hi +.13 Av +.30 Lo +.65	Math	+.32
Dewar, 1964	6	Johnson Co., KS (middle class)	199 (8 cl.)	3	8	Classes randomly assigned to WCAG, control	Hi +.55 Av +43 Lo +.67	Math	+.55
Smith, 1960	2-5	Lake Charles, LA	180 (8 cl.)	3	5	Classes randomly assigned o WCAG, control then students matched on sex, age, IQ, math ach.	Hi +.28 Av +.25 Lo +.69	Math	+.41
Wallen & Vowles, 1960	6	Salt Lake City, UT	112 (4 cl.)	4	4	Classes randomly assigned to WCAG, control, for 1 sem.; then classes switched treatments for 2nd sem.		Math	+.07
Matched studies with evidence of initial equality									
Spence, 1958	4-6	Bethel, PA	400 (25 cl.)	3	8	Students in WCAG, control matched on IQ, math ach.; scores are gains		Math	+.44
Jones, 1948	4	Richmond, IN	250 (31cl.)	–	8	Students in WCAG, control matched on IQ; scores are gains	Hi +.48 GE Av +.27 GE Lo +.37 GE	Rdg. +.23 Spel. +.43 Math +.16	+.26
Matched studies lacking evidence of initial equality									
Stern, 1972	3-4	Covina, CA (low achievers)	217 exp. 91 cont. (67 cl.)	–	4	Compared WCAG, control matched on general ach. level, low achievers only; control began higher at pretest		Math	+.36

Key: WCAG = Within-class ability grouping.
MMP = Missouri Mathematics Program.

been in the top, middle, and low groups in comparison to their counterparts in the control group (*ES* = +.55). In a similar study, Smith (1960) randomly assigned grade 2-5 classes to grouped or control conditions for 5 months. Within each grade level, students in experimental and control groups were matched on sex, age, achievement level, and IQ. Results favored within-class ability grouping (*ES* = +.41), especially for students assigned to the lowest group (*ES* = +.69).

The smallest positive effect of any study of within-class ability grouping was reported in a study by Wallen and Vowles (1960), who randomly assigned four sixth grade mathematics classes to ability-grouped or whole-class treatments for 1 semester and then had all classes experience the opposite treatment for 1 semester. After 1 semester, the scores of the ability-grouped students were higher than those of the control students (*ES* = +.30), but second semester results nearly wiped out this difference (*ES* = +.07). It is important to note that this was the only study to use four math groups rather than two or three.

Three nonrandomized studies generally supported the results of the randomized ones. Spence (1958) compared the achievement gains of students in mathematics classes using within-class ability grouping to that of students in control classes matched on IQ and arithmetic achievement. Results indicated significantly greater gains for the grouped students. Stern (1972) compared the achievement of the lowest 20% of students in heterogeneous classes using math groups to that of comparable students in heterogeneous ungrouped classes. Despite matching, achievement pretest differences favored students in the control conditions, but gain scores clearly favored the grouped classes (*ES* = +.36).

A study by Jones (1948) also compared matched students in different grouping arrangements. The experimental treatment used in this study is not well specified, but did involved within-class grouping of some kind for reading, spelling, and mathematics. Results favored grouped classes for all three subjects and for three levels of ability (overall *ES* = +.26). Information adequate for computation of effect sizes was provided for composite achievement only.

Summary and discussion: Within-class ability grouping. Research on the use of math groups consistently supports this practice in the upper elementary grades. Among research on ability grouping in general, this research is of exceptional quality. Five well-controlled studies used random assignment of classes to treatments and sample sizes large enough to minimize the potential impact of teacher effects. However, there is not enough research on within-class ability grouping in reading or in the primary grades to permit any conclusions.

There is no evidence to suggest that achievement gains due to within-class ability grouping in mathematics are achieved at the expense of low achievers; if anything, the evidence indicates the greatest gains for this subgroup.

As in the case of the Joplin Plan, it is possible that the effects of within-class ability grouping in mathematics are due in part to the fact that this procedure is a novelty in this subject. Again, though, this explanation seems unlikely. Within-class ability grouping is hardly seen as an innovative treatment in the elementary school; not only is its use virtually universal in another subject, reading, but grouping in math has been around for decades.

It is not necessarily the case that the effects of within-class ability grouping in math are entirely due to the reductions in heterogeneity it brings about. There is one study that raises some question about this assumption. A dissertation by Eddleman (1971) compared within-class ability grouping in mathematics to a within-class grouping plan in which students were assigned to three *heterogeneous* subgroups. There was some differentiation of instructional level for students in the ability-grouped classes, but in all other respects the teacher's methods in the two groups were identical, with instruction given to one group at a time while the other two groups worked problems at their desks. Classes were randomly assigned to treatments, with the same teachers teaching classes using homogeneous and heterogeneous subgroups. Results of the 9-week study slightly favored the heterogeneous grouping plan (*ES* = -.16). Unfortunately, there was no ungrouped control condition, so it is impossible to determine whether the two forms of subgrouping were equally effective or equally ineffective; the brevity of the study suggests the latter. However, if future research established that within-class ability grouping and within-class heterogeneous grouping were equally effective (and more effective than ungrouped arrangements), we would have to reconceptualize the usual explanations for the effectiveness of within-class ability grouping. For example, it may be that within-class ability grouping increases achievement by reducing the size of instruction groups (say, from 30 to 10) or by structuring the teacher's instructional time more effectively rather than having anything to do with reducing homogeneity (see Slavin & Karweit, 1984). Clearly, research directed at explaining the achievement effects of within-class ability grouping is needed.

Discussion

Many previous reviewers of the ability grouping literature have characterized the evidence as a muddle or a maze (e.g., Borg, 1965; Passow, 1962). However, earlier reviewers have generally combined elementary with secondary research, good quality with hopelessly biased studies, research on comprehensive ability grouping plans with that on special programs for the gifted or learning disabled, and in some cases, research on between-class ability grouping with that on within-class grouping. When the scope of the review is limited to methodologically adequate studies of comprehensive ability grouping at the elementary level and different types of ability grouping are reviewed separately, the results are surprisingly clear cut for most types of grouping. The best evidence from randomized and matched equivalent studies supports the positive achievement effects of the use of within-class ability grouping in mathematics in the upper elementary grades and of the Joplin Plan in reading. In contrast, there is no support for the practice of assigning students to self-contained classes according to general ability or performance level, and there are enough good quality studies of this practice that if there were any effect, it would surely have been detected. Evidence on the effects of regrouping within grade levels for reading and mathematics is unclear, and there is no methodologically adequate evidence concerning the use of reading groups.

The conclusion of the research reviewed here for practice may be quite simple: Use the grouping methods that have been found to be effective (within-class ability grouping in upper-elementary mathematics, Joplin Plan in reading), and avoid those that have not been found to be effective. In particular, there is good reason to avoid ability-grouped class assignment, which seems to have the greatest potential for negative social effects in that it entirely separates students into different streams (see Rosenbaum, 1980). However, there is much more we must understand about how various ability grouping plans have their effects. A theory able to encompass the research findings is needed. In the absence of descriptive studies capable of contrasting teachers' and students' behaviors and perceptions in alternative grouping arrangements, we can only speculate about how effects of various grouping plans may be produced. Such studies are clearly needed. However, the remainder of this paper explores the findings and other evidence that do exist at present in an attempt to extract general principles of grouping for instruction in the elementary school.

Accommodating Instruction to Student Differences

The primary reason educators group students according to ability or performance level is to enable teachers to provide instruction closely suited to the readiness and needs of different students. In a highly diverse class, it is argued, one level and pace of instruction is likely to be too easy for some students and/or too difficult for others. Ability grouping is supposed to reduce student heterogeneity so that an appropriate pace and level of instruction is provided for most students. Having instruction be carefully accommodated to students' level of readiness is probably more important in some subjects than in others. In general, subjects in which skills build upon one another in a hierarchical fashion (e.g., mathematics, reading) should require more accommodation to individual differences in learning rate than subjects in which learning the next skill or concept is less clearly dependent on mastery of earlier material (e.g., social studies, science). The reason for this is that with hierarchically organized subjects there is a risk that, if the teacher proceeds too rapidly, some students will lack the prerequisite skills needed to learn new material, whereas if the teacher takes the time needed to ensure that all students have prerequisite skills, the more able students will waste a great deal of time.

Ability grouping is one logical way out of the dilemma posed by having to choose one instructional pace for a diverse group in a hierarchical subject. Yet if an ability-grouping plan is to actually reduce the heterogeneity of instruction groups in any particular skill, there are (at least in theory) three criteria it should satisfy:

1. The grouping plan must measurably reduce student heterogeneity *in the specific skill being taught;*

2. The plan must be flexible enough to allow teachers to respond to misassignments and changes in student performance level after initial placement; and

3. Teachers must actually vary their pace and level of instruction to correspond to students' levels of readiness and learning rates.

Research on the effect of grouping on class heterogeneity has found that in the situation typical of elementary schools where students are divided into two or three "homogeneous" groups, the actual reduction in heterogeneity brought about may be quite minimal. This is particularly true when students are assigned to classes on the basis of IQ or of a general measure of performance, as imperfect correlations between these measures and actual performance in any particular subject leave a great deal of heterogeneity in the supposedly homogeneous classes (Balow, 1962; Balow & Curtin, 1966; Clarke, 1958; Goodlad & Anderson, 1963).

Thus, ability-grouped class assignment generally fails to meet the first of the three criteria listed above; a one-time assignment by general ability is unlikely to create enough homogeneity on any particular skill to make an instructional difference. Ability-grouped class assignment is also unlikely to fulfill the second criterion, flexibility. Transferring students between self-contained classes is difficult to arrange, so students who are misassigned or whose achievement level markedly changes over time are likely to remain in the self-contained class. In contrast, regrouping and Joplin plans group students based on their performance in a *specific* skill and are inherently more flexible than ability-grouped class assignment, as changing students between regrouped classes involves only one subject, not a change in students' main class identification. Similarly, within-class ability grouping is done based on performance in a particular skill and is the easiest grouping plan to alter based on changes in student performance. To what extent do teachers adapt their level and pace of instruction to the needs of different ability groups? Research comparing alternative grouping arrangements has not examined this question in any depth, but there are some clues. Studies by Barr and Dreeben (1983) found that teachers do adapt their instructional pace to accommodate the aptitudes of reading groups, but they also found considerable variation from school to school and teacher to teacher in pacing for groups of similar aptitudes.

Some indirect evidence suggests the importance of adapting instruction to student differences. One form of grouping often seen in mathematics instruction involves assigning students to three ability groups within the class. The teacher presents one lesson to the class as a whole and, while students are doing seatwork, visits with the low-ability group to provide additional explanation, the high group to provide enrichment, and the middle group to provide some of each. Note that this strategy does not adapt the pace or level of instruction to student needs, as all students still experience the same pace of instruction. Two studies at the secondary level (Bierden, 1969; Mortlock, 1970) and one in a community college (Merritt, 1973) found no significant differences between this type of within-class ability grouping and traditional whole-class instruction, much in contrast to the studies of within-class ability grouping plans in which level and pace of instruction were adapted to student performance levels.

One critical feature of the Joplin Plan is frequent, careful assessment of student performance levels and provision of materials appropriate to these levels regardless of students' grade levels. In this plan, adaptation of instructional pace and level to student needs is considerable. In one study of the Joplin Plan (Moorhouse, 1964) it was noted that "three-quarters of the [grades 4-6 experimental] students were reading material either above or below the grade level they would usually be asked to attempt in the graded system" (pp. 281-282). In contrast, a study of regrouping for reading by Moses (1966) instructed experimental teachers to use only materials appropriate to students' grade levels and to follow the school district's usual course of study. This study found no significant advantages of regrouping (ES = +.05). Similarly, Davis and Tracy (1963), in a study of regrouping for mathematics, held experimental and control teachers to the same grade-level textbooks. In this study, control students gained more in achievement than did students in the within-grade regrouping plan. An otherwise similar study of within-grade regrouping for mathematics by Provus (1960) did allow for differentiation of level and pace of instruction, noting that "it was possible for a fourth grader ... to advance to sixth grade or even eighth grade work by the end of the school year" (p. 392). Control (ungrouped) students were also able to go beyond their designated grade level, but presumably did not do so as often as did experimental students. This study found positive effects of regrouping on mathematics achievement (*ES* = +.39). Of course, the Joplin Plan can be seen as a form of regrouping for reading and/or mathematics that goes to great lengths to adapt the level and pace of instruction to that of the regrouped classes. In fact, it could be argued that it is not the cross-grade aspect of the Joplin Plan that accounts for its effects, but rather the fact that students in this plan are carefully assessed and given instruction appropriate to their needs in the regrouped classes.

Taken together, the evidence points to a conclusion that for ability grouping to be effective at the elementary level, it must create true homogeneity on the specific skill being taught and instruction must be closely tailored to students' levels of performance.

Class Identification

Critics of ability grouping (e.g., Oakes,1985; Rosenbaum,1980; Schafer & Olexa, 1971) have often noted the detrimental psychological effect of being placed in a low-achieving class or track. An interview with a former delinquent about his discovery that he had been assigned to the "basic track" in junior high school illustrates this theme (from Schafer & Olexa):

I felt good when I was with my [elementary] class, but when they went and separated us – that changed us. That changed our ideas, our thinking, the way we thought about each other, and turned us to enemies toward each other – because they said I was dumb and they were smart.

When you first go to junior high school you do feel something inside – it's like a ego. You have been from elementary to junior high, you feel great inside... you get this shirt that says Brown Junior High... and you are proud of that shirt. But then you go up there and the teacher says – "Well, so and so, you're in the basic section, you can't go with the other kids." The devil with the whole thing – you lose – something in you – like it goes out of you. (pp. 62-63)

The anguish expressed by the student who was assigned to the basic track is interesting in light of the high probability that the student had been in low reading groups in elementary school, but he still perceived being "separated" into different classes as a completely new and much more serious affront to his self-esteem. Within-class grouping generally takes place within the context of a more or less heterogeneous class, and a student still identifies with the class as a whole. Ability-grouped class assignment, however, creates a situation in which a low achiever only has a low-achieving group with which to identify (see Richer, 1976).

It may be that flexible skill-based ability grouping has a different impact on students' self perceptions and on teachers' expectations than does ability-grouped class assignment. Students in low-performing classes would seem likely to come to see themselves as a different species of human being, whereas those assigned to low reading or math groups in heterogeneous classes may see this placement as being done to help them, particularly if assignment to groups is flexible and is clearly focused on achievement in a particular subject.

Teachers' expectations and behaviors may also be different in different types of ability grouping. Not surprisingly, teachers prefer to teach higher achieving students (NEA, 1968) and have higher expectations for their achievement. These expectations can have an impact on teachers' behaviors and students' achievement (see Good & Brophy, 1984). For example, in a study of Air Force training, Schrank (1969) had students randomly assigned to classes, but told instructors that the classes were grouped by ability. Classes that had been (falsely) identified as high-achieving in fact achieved more than did classes identified as low-achieving.

The problems of teachers' low expectations for students in low-track classes and their disliking being assigned to these low-achieving classes are largely alleviated in the Joplin Plan, in which reading classes are formed across age lines. This means that a class may contain high-achieving young students and low-achieving older students, so homogeneity for instruction may be achieved without establishing classes that teachers do not want to teach and from whom teachers expect little. Also, as noted by one of the authors of the Hillson et al. (1964) study of a Joplin-like nongraded plan (J.W. Moore, personal communication, January 23, 1986), low-achieving students progressed from reading level to reading level rather than remaining year after year in the low reading group.

Teachers may have low expectations for students in low reading or math groups, but there is some evidence that they try to bring low groups up to the level of the rest of the class. For example, Rowan and Miracle (1983) found that in between-class ability grouping teachers tended to maintain a slow pace of instruction for low-achieving classes, but tended to allocate more time and a more rapid pace of instruction for low reading groups in heterogeneous classes. This and other research (e.g., Alpert, 1974) suggests that in within-class ability grouping, teachers tend to try to equalize the achievement of all students by assigning smaller numbers of students to low groups.

Another issue relating to students' identification with their class is the question of how many times students are regrouped each day. When students are regrouped for reading and/or mathematics, they still typically spend the rest of their school day in heterogeneous homeroom classes, which probably remain their primary reference group. However, as the number of regroupings increases, the situation comes to resemble departmentalization, where students move from teacher to teacher and have no one group with which to identify. Research on departmentalization in upper elementary and junior high schools generally finds this practice to be detrimental to student achievement in comparison to self-contained assignment (Hosley, 1954; Jackson, 1953; Spivak, 1956; Ward, 1970). Departmentalization might reduce students' attachment to school by diffusing their attachments to particular teachers. Indirect evidence of this is a finding by Slavin and Karweit (1982) that student truancy in an urban school district rose from about 8% in the fifth and sixth grades to 26% in the seventh grade, the time of first exposure to a departmentalized (and tracked) school in which no one teacher takes responsibility for any one student. It may be that the Koontz (1961) study, in which students were separately regrouped for reading, mathematics, and spelling, deprived students of an opportunity to identify with a single teacher and a heterogeneous class. In this study, heterogeneous control students gained more in achievement than did those who were regrouped, with low achievers suffering most from regrouping.

The evidence on the importance of having students principally identify with a heterogeneous class is more speculative than conclusive, but there are several indirect indications to support the following conclusion: Students should be assigned to heterogeneous classes for as much of the school day as possible, so that it is the heterogeneous homeroom with which students principally identify. Even if for no other reason than that separating students by ability goes against the grain of our democratic, egalitarian ideals, any ability grouping plan must have clear educational benefits if it is to be justified. Because no achievement benefits of ability-grouped class assignment have been identified, and because more effective grouping methods exist, use of this strategy should be avoided. One instance in which ability-grouped class assignment should particularly be avoided is when this practice creates racially identifiable classes. In fact, in a recent case in South Carolina, an administrative law judge ruled that since ability-grouped class assignment concentrated black students in low-ability classes and had no educational justification, the district was compelled to abandon the practice or lose federal funds (*U.S. Department of Education v. Dillon County School District No. 1*, 1986). However, the judge specifically recommended use of forms of ability grouping that regrouped students only for specific subjects, such as reading and math, leaving students in heterogeneous classes most of the day.

Instructional Time

One issue of considerable importance in relation to within-class ability grouping relates to a trade-off between providing students with instruction appropriate to their needs on the one hand, and providing adequate instructional time on the other (see Slavin, 1984a). When a teacher uses a within-class ability grouping plan with three groups, this means that students must spend at least two thirds of their instructional time working without direct teacher instruction or supervision. Several studies have found that large amounts of unsupervised seatwork are detrimental to student achievement (see Brophy & Good, 1986). Transition times between ability groups further reduce instructional time (Arlin, 1979).

In theory, the amount of instructional time lost because of use of within-class ability grouping would depend directly on the number of groups in use. Division of students into large numbers of ability groups forces the teacher to spend less time with each group and to assign large amounts of independent seatwork, much of which may be of little value beyond keeping students quiet and busy (see Anderson, Brubaker, Alleman-Brooks, & Duffy, 1985). It may not be a coincidence that Wallen and Vowles (1960), the study with the smallest effect size favoring within-class ability grouping in mathematics ($ES = +.07$), is also the only one to use four (rather than two or three) ability groups.

The evidence summarized in Table 4 indicates that regardless of any losses in instructional time associated with within-class grouping, this strategy is instructionally effective in elementary mathematics. Mathematics instruction does require a certain amount of time for students to work problems on their own, so follow-up time (the time during which some students must work by themselves while others are working with the teacher) may be less of a problem in mathematics than in other subjects. However, it still seems apparent that the requirement for large amounts of follow-up time is a drawback in any within-class grouping arrangement.

The problem of follow-up time may be important in explaining the effectiveness of the Joplin Plan for reading. The studies of this program do not typically compare the numbers of ability groups used in experimental and control groups, but it is clear that there are smaller numbers of reading groups in Joplin than in traditional classes, and that in some cases Joplin Plan classes do not use reading groups at all. In fact, some authors (e.g., Newport, 1967) clearly describe studies of the Joplin Plan as comparisons of inter- versus intra-class ability grouping, since control groups in Joplin studies always used reading groups. The cross-grade grouping plans create such homogeneous classes that the need for within-class grouping is diminished or eliminated. The time savings of such plans are therefore considerable, and they may produce their effects in part because students in them receive a greater amount of direct instruction from the teacher and supervision during seatwork than do students in control classes using the more typical three or more reading groups.

This line of reasoning may justify use of smaller numbers of ability groups in heterogeneous reading classes. Unfortunately there is no direct experimental evidence on the optimum number of reading or math groups; the number three is treated as though it were handed down from Mount Sinai. In heterogeneous classes it may be that small numbers of ability groups do not provide adequate homogeneity for effective instruction. However, it should be noted that two studies involving only two ability groups in heterogeneous mathematics classes found significant benefits of ability grouping for student achievement (Slavin & Karweit, 1985).

Alternatives to Ability Grouping

Whatever their achievement effects may be, ability grouping plans in all forms are repugnant to many educators, who feel uncomfortable making decisions about elementary-aged students that could have long-term effects on their self-esteem and life chances. In desegregated schools, the possibility that ability grouping may create racially identifiable groups or classes is of great concern. For these and other reasons, several alternatives to ability grouping have been proposed.

One alternative to ability grouping proposed by Oakes (1985) and Wilkinson (1986), among others, is cooperative learning, instructional methods in which students work in small, mixed-ability learning teams. Research on cooperative learning has found that when the cooperative groups are rewarded based on the learning of all group members, students learn consistently more than do students in traditional methods (Slavin, 1983).

Among the most effective of the cooperative learning models are programs designed to accommodate diverse student needs. These include Team Assisted Individualization-Mathematics (Slavin, in press), and Cooperative Integrated Reading and Composition (Stevens, Madden, Slavin, & Farnish, in press). Thus, cooperative learning may be a plausible alternative to ability grouping that takes student diversity as a resource to be used in the classroom rather than a problem to be solved. However, there is no research at present that specifically compares cooperative learning to ability grouping.

Mastery learning may also be seen as an alternative to ability grouping. In group-based mastery learning (Block & Anderson, 1975), students are tested at the end of a series of lessons, and those who need it receive additional assistance. However, studies that used standardized achievement measures have found few positive effects of group-based mastery learning (Slavin, 1987). On the other hand, continuous-progress mastery learning programs, in which students are constantly regrouped as they proceed at their own rates through a hierarchical curriculum, have been found to be instructionally effective in several studies (Slavin & Madden, 1987), although these methods have not been directly compared to ability grouping. Completely individualized approaches, such as Individually Prescribed Instruction, have not generally been found to be instructionally effective (Horak, 1981).

Research comparing achievement effects of various forms of ability grouping and alternatives to ability grouping is clearly needed. At present, cooperative learning and continuous-progress programs appear to have the greatest potential as alternative means of accommodating student diversity, but the effects of these and other methods relative to those of traditional between- and within-class ability grouping methods are not currently known.

Conclusions

This paper reviewed the best evidence concerning the achievement effects of comprehensive ability grouping plans in elementary schools. Four principal grouping plans were examined: ability-grouped class assignment, regrouping for reading and/or mathematics, the Joplin Plan, and within-class ability grouping. The effects of these grouping methods on student achievement from methodologically adequate studies are summarized below.

Ability-grouped class assignment. Evidence from 17 comparisons in 13 matched equivalent and one randomized study clearly indicates that assigning students to self-contained classes according to general achievement or ability does not enhance student achievement in the elementary school (median $ES = .00$).

Regrouping for reading and mathematics. Research is unclear on the achievement outcomes of grouping plans in which students remain in heterogeneous classes most of the day and are regrouped by ability within grade levels for reading and/or mathematics. There is some evidence that such plans can be instructionally effective if the level and pace of instruction is adapted to the achievement level of the regrouped class and if students are not regrouped for more than one or two different subjects.

Joplin Plan. There is good evidence that regrouping students for reading across grade lines increases reading achievement. The Joplin Plan (Floyd, 1954) and essentially similar forms of nongraded plans (e.g., Hillson et al., 1964) have had relatively consistent positive effects on reading achievement (median $ES = +.44$), and one study (Hart, 1962) found that a similar program could also be effective in mathematics.

Within-class ability grouping. Research on within-class ability grouping is unfortunately limited to mathematics in upper elementary school. However, this research supports the use of within-class grouping (approximate median $ES = +.34$), especially if the number of groups is kept small.

In addition to conclusions about the effects of particular grouping strategies, several general principles of ability grouping were proposed on the basis of the experimental evidence and other research. The following are tentatively advanced as elements of effective ability grouping plans:

1. Students should remain in heterogeneous classes at most times and be regrouped by ability only in subjects (e.g., reading, mathematics) in which reducing heterogeneity is particularly important. Students' primary identification should be with a heterogeneous class.

2. Grouping plans should reduce student heterogeneity in the specific skill being taught (e.g., reading, mathematics), not just in IQ or overall achievement.

3. Grouping plans should frequently reassess student placements and should be flexible enough to allow for easy reassignments after initial placement.

4. Teachers should actually vary their level and pace of instruction to correspond to students' levels of readiness and learning rates in regrouped classes.

5. In within-class ability grouping, numbers of groups should be kept small to allow for adequate direct instruction from the teacher for each group.

Future Directions

Despite more than a half-century of research on the topic, there are many fundamental questions yet to be answered concerning the achievement effects of alternative ability grouping practices. For example, the research on within-class ability grouping is essentially restricted to mathematics in the upper elementary grades. Experimental studies of the use of reading groups are needed, as are studies of optimum numbers of within-class groups for reading and math. Many studies are needed to understand why and under what conditions various grouping plans produce achievement effects. Simple-appearing changes in grouping are likely to have complex effects, any of which may contribute to ultimate effects on student achievement. For example, different between-class grouping plans (e.g., ability-grouped class assignment, Joplin Plan) are likely to have different effects on within-class grouping. Studies are needed to understand the effects of various grouping plans on what actually happens in the classroom, for example, how different plans affect the teacher's pace of instruction and use of class time and the success rate of students close to and far away from the class's mean aptitude. Component analyses are needed to explore the critical features of various grouping plans. For example, which of the elements of the Joplin Plan account for the positive effects seen in the studies of this practice? Is it simply use of flexible, homogeneous grouping across grade lines, is it the reduction in the need for within-class ability grouping, or are other factors involved?

At present, there are many studies examining classroom practices in high- and low-ability classes and ability groups (see, for example, Barr & Dreeben, 1983; Hiebert, 1983; Oakes, 1985; Peterson, Wilkinson, & Hallinan, 1984). Although such studies provide important descriptive information, they cannot answer the question of whether or not the effects of ability grouping per se are positive. Only comparisons of ability-grouped and heterogeneous classes, preferably in experiments in which teachers and students are randomly assigned to treatments, can provide this kind of information. However, experimental studies of ability grouping typically lack adequate descriptive information on what teachers in contrasting grouping arrangements are actually doing. It is unlikely that grouping plans have a *direct* effect on achievement; whatever effects are seen must be due to changes in teachers' instructional practices, yet these changes and their consequences are poorly understood at present. What is needed is a merging of experimental and descriptive research, so that the effects of alternative grouping strategies and the processes by which these effects are brought about can be studied at the same time.

Finally, more research is needed on alternatives to ability grouping, such as cooperative learning, continuous-progress programs, and mastery learning.

There are two particularly important reasons for further investigation of grouping practices in elementary schools. First, every school district, school administrator, and teacher makes decisions about ability grouping at some time, and these decisions should be made in light of reliable evidence. Ironically, the grouping practice with the least support in the research, ability-grouped class assignment, is among the most widely used; schools need effective alternatives to this practice. Second, if educational researchers can identify grouping practices that can accelerate student achievement, this would provide one kind of school reform that would be low in cost, easy to implement, and easy to maintain over time. In a time of increasing demands on education coupled with dwindling resources, research on easily modified school organizational practices seems particularly likely to bear fruit. We have much yet to learn in this area, but this review illustrates that the potential of effective grouping practices for meaningful improvements in the achievement of elementary students is great, and is certainly worthy of further study.

CHAPTER 8

Ability Grouping in Secondary Schools

Background and Commentary

In the course of assembling studies for my review of research on ability grouping in elementary schools, I also collected studies on secondary ability grouping (in fact, I originally intended to review research involving both levels in one article, but found too much research at each level to make this practical). The issues surrounding ability grouping in middle, junior high, and high schools are different in many ways from those in elementary schools, so the topic certainly deserves its own treatment.

At the time I was working on my ability grouping syntheses, research on ability grouping was not a popular focus. Few of the experimental studies I reviewed took place after the early 1970's. Shortly thereafter, perhaps partially in response to my articles and to Jeannie Oakes' writings, research on this topic, especially in secondary schools, has substantially accelerated. However, there have still been very few experimental comparisons of ability grouped and heterogeneous class assignments. Recent research has primarily been either correlational or descriptive. Like earlier studies, correlational studies comparing high, average, and low ability groups in secondary schools continue to find that students in high ability groups perform better than expected while those in low groups do worse (see, for example Brewer, Rees, & Argys, 1995). However, I have doubts about this conclusion, as I continue to believe that these studies under-control for substantial pretest differences in achievement (see Slavin, 1995).

While the academic debate about ability grouping in secondary schools continues, attention is shifting to what I believe to be a more promising direction: description and, even more importantly, development and evaluation of organizational plans for middle and high schools that reduce or eliminate between-class ability grouping. For example, Oakes, Quartz, Gong, Guiton, & Lipton (1993) and Wheelock (1992) have described middle schools undergoing detracking, and my colleagues Douglas Mac Iver and James McPartland are developing and evaluating middle and high school designs that reduce between-class regrouping. At this stage, what holds back the detracking movement is less a commitment among educators to traditional practices than a perceived lack of practical alternatives. The recent and continuing descriptive and experimental work should help to provide alternative visions and replicable models of reformed practice in secondary schools.

For more than 70 years, ability grouping has been one of the most controversial issues in education. Its effects, particularly on student achievement, have been extensively studied over that time period, and many reviews of the literature have been written. In recent years, a comprehensive review of the achievement effects of ability grouping in elementary schools has been published by Slavin (1987), but only brief meta-analyses by Kulik and Kulik (1982, 1987) have reviewed the evidence on ability grouping and heterogeneous placement in secondary schools.

The purpose of this paper is to present a comprehensive review of all research published in English that has evaluated the effects of ability grouping on student achievement in secondary schools. *Secondary schools* are defined here as middle, junior, or senior high schools in the United States, or similarly configured secondary schools in other countries. Secondary schools can include grades as low as five, but they usually begin with sixth or seventh grades. Ability grouping is defined as any school or classroom organization plan that is intended to reduce the heterogeneity of instructional groups; in between-class ability grouping the heterogeneity of each class for a given subject is reduced, and in within-class ability grouping the heterogeneity of groups within the class (e.g., reading groups) is reduced.

Unlike the situation in elementary schools, the type of ability grouping used in secondary schools is overwhelmingly between-class grouping (McPartland, Coldiron, & Braddock, 1987). Several closely related forms of ability grouping are used. Sometimes students are assigned on the basis of some combination of composite achievement, IQ, and teacher judgments to a track, within which all courses are taken. For example, senior high school students are often assigned to academic, general, and vocational tracks; middle/junior high school students are often assigned to advanced, basic, and remedial tracks (in either case, the number of tracks and the names used to describe them vary widely). This type of grouping plan is generally called tracking in the United States or streaming in Europe. It is an example of what Slavin (1987) called "ability-grouped class assignment." In addition to assignment to higher and lower sections of the same courses, tracking in senior high schools usually involves different courses or course requirements. For example, a student in the academic track may be required to take more years of mathematics than a student in the general track, or may take French III rather than metal shop.

A particular form of tracking often seen in middle/junior high schools is block scheduling, where students spend all or most of the day with one homogeneous group of students. Some schools rank-order students from top to bottom and assign 0 them to, say, 7-1, 7-2, 7-3, and so on. Many senior high schools allow students to choose their track or to choose the level they wish to take in each subject, but in plans of this kind counselors tend to steer students into the level of classes to which they would have been assigned if the school were not allowing students a choice (Rosenbaum, 1978).

Another form of ability grouping common in secondary schools involves assigning students to ability-grouped classes for all academic subjects, but allowing for the possibility that students will be placed in a high-ranking group for one subject and a low-ranking group for another. In practice, scheduling constraints often make this type of grouping similar to plans in which all courses are taken within the same track. In some cases schools ability group for some subjects and not for others; for example, students may be in ability-grouped math and English classes but in heterogeneous social studies and science classes. Ability grouping usually involves higher and lower sections of the same course, but sometimes consists of assignment to completely different courses, as when ninth graders are assigned either to Algebra I or to general math. When high achievers are assigned to markedly different courses usually offered to older students (as when seventh graders take algebra), this is called acceleration.

More commonly, high achievers may be assigned to "honors" or "advanced placement" sections of a given course, and low achievers may be assigned to special "remedial" sections.

Although between-class ability grouping is by far the most common type of ability grouping in secondary schools, forms of within-class grouping are also occasionally seen. These are plans in which students are assigned to homogeneous instructional groups within their classes. Within-class ability grouping, such as use of reading or math groups, is the most common form of grouping at the elementary level (McPartland et al., 1987). Complex plans, such as those that involve grouping across grade lines, flexible grouping for particular topics, and part-time grouping, are also occasionally seen in secondary schools. In general, a wider range of grouping plans are used in middle/junior high schools than in senior high schools.

Arguments for and against ability grouping have been essentially similar for 70 years. For example, Turney (1931), summarizing writings of the 1920s, listed advantages and disadvantages of ability grouping. The advantages were as follows:

1. It permits pupils to make progress commensurate with their abilities.
2. It makes possible an adaption of the technique of instruction to the needs of the group.
3. It reduces failures.
4. It helps to maintain interest and incentive, because bright students are not bored by the participation of the dull.
5. Slower pupils participate more when not eclipsed by those much brighter.
6. It makes teaching easier.
7. It makes possible individual instruction to small slow groups.

The following were the disadvantages:

1. Slow pupils need the presence of the able students to stimulate them and encourage them.
2. A stigma is attached to low sections, operating to discourage the pupils in these sections.
3. Teachers are unable, or do not have time, to differentiate the work for different levels of ability.
4. Teachers object to the slower groups.

A research symposium, school board meeting, or PTA meeting on the topic of ability grouping in 1990 is likely to bring up much the same arguments on both sides, with two important additions: the argument that ability grouping discriminates against minority and lower-class students (e.g., Braddock, 1990; Rosenbaum, 1976), and the argument that students in the low tracks receive a lower pace and lower quality of instruction than do students in the higher tracks (e.g., Gamoran, 1989; Oakes, 1985).

In essence, the argument in favor of ability grouping is that it will allow teachers to adapt instruction to the needs of a diverse student body and give them an opportunity to provide more difficult material to high achievers and more support to low achievers. The challenge and stimulation of other high achievers are believed to be beneficial to high achievers (see Feldhusen, 1989). Arguments opposed to ability grouping focus primarily on the perceived damage to low achievers, who receive a slower pace and lower quality of instruction, have teachers who are less experienced or able and who do not want to teach low-track classes, face low expectations for performance, and have few positive behavioral models (e.g., Gamoran, 1989; Oakes, 1985; Persell, 1977; Rosenbaum, 1980). Because of the demoralization, low expectations, and poor

behavioral models, students in the low tracks are believed to be more prone to delinquency, absenteeism, dropout, and other social problems (Crespo & Michelna, 1981;
Wiatrowski, Hansell, Massey, & Wilson, 1982). With few college-bound peers, students
in low tracks have been found to be less likely to attend college than other students
(Gamoran, 1987). Ability grouping is perceived to perpetuate social class and racial
inequities because lower class and minority students are disproportionally represented
in the lower tracks. Ability grouping is often considered to be a major factor in the
development of elite and under-class groups in society (Persell, 1977; Rosenbaum,
1980). Perhaps most important, tracking is believed to work against egalitarian, democratic ideals by sorting students into categories from which escape is difficult or impossible.

There are important differences between the pro-grouping and anti-grouping positions that go beyond the arguments themselves. Arguments in favor of ability grouping focus on *effectiveness*, saying in effect that as distasteful as grouping may be, it so
enhances the learning of students (particularly but not only high achievers) that its
use is necessary. In contrast, arguments opposed to grouping focus at least as much
on *equity* as on effectiveness and on democratic values as much as on outcomes. In
one sense, then, the burden of proof is on those who favor grouping, for if grouping
is not found to be clearly more effective than heterogeneous placement, none of the
pro-grouping arguments apply. The same is not true of anti-grouping arguments, which
provide a rationale for abolishing grouping that would be plausible even if grouping
were found to have no adverse effect on achievement.

Research on the achievement effects of ability grouping has taken two broad forms.
One type of research compares the achievement gains of students who are in one or
another form of grouping to those of students in ungrouped, heterogeneous placements. Another type of research compares the achievement gains made by students
in high-ability groups to those made by students in the low groups.

Reviews of the grouping versus nongrouping literature have consistently shown
that grouping has little or no impact on overall student achievement in elementary
and secondary schools (e.g., Borg, 1965; Esposito, 1973; Findley & Bryan, 1971; Good
& Marshall, 1984; Heathers, 1969; Kulik & Kulik, 1982). Primarily on the basis of his
own empirical research, Borg (1965) claimed that ability grouping had a slight positive effect on the achievement of high achievers and a slight negative effect on low
achievers, but Kulik and Kulik (1987) found no such trend.

In contrast, researchers who have compared gains made by students in different
tracks have generally concluded that controlling for ability level, socioeconomic status, and other control variables, being in the top track accelerates achievement and
being in the low track significantly reduces achievement (Alexander, Cook, & McDill,
1978; Dar & Resh, 1986; Gamoran & Berends, 1987; Gamoran & Mare, 1989; Oakes,
1982; Persell, 1977; Sorensen & Hallinan, 1986). In fact, many researchers and theorists in the sociological tradition maintain that tracking is a principal source of social
inequality in society and that it causes or greatly magnifies differences along lines of
class and ethnicity (e.g., Braddock, 1990, Jones, Erickson & Crowell, 1972; Schafer &
Olexa, 1971; Vanfossen, Jones, & Spade, 1987).

One area of research has investigated the quality of instruction offered to students
in high- and low-ability groups, usually concluding that low-ability group classes
receive instruction that is significantly lower in quality than that received by students
in high-track classes (e.g., Evertson, 1982; Gamoran, 1989; Oakes, 1985; Trimble &
Sinclair, 1987). However, it is difficult to compare "quality of instruction" in high-
and low-track classes. For example, teachers typically cover less material in low-track
classes (e.g., Oakes, 1985). Is this an indication of poor quality of instruction or an
appropriate pace of instruction? Students in low-track classes are more off-task than

those in high-track classes (e.g., Evertson, 1982). Is this due to the poor behavioral models and low expectations in the low-track classes, or would low achievers be more off-task than high achievers in any grouping arrangement? Evidence that low-track classes are often taught by less experienced or less qualified teachers or that they manifest other objective indicators of lower-quality instruction could justify the conclusion that regardless of measurable effects on learning, students in the lower tracks do not receive equal treatment, but such evidence is rare.

In addition to synthesizing research on overall effects of ability grouping on the achievement of high-average- and low-achieving secondary students, this review will attempt to reconcile research comparing achievement gains in different tracks with research comparing grouped and ungrouped settings.

Review Methods

This review uses a procedure called "best-evidence synthesis" (Slavin, 1986), which incorporates the best features of meta-analytic and traditional reviews. Best-evidence syntheses specify clear, well-justified methological and substantive criteria for inclusion of studies in the main review and describe individual studies and critical research issues in the depth typical of good-quality narrative reviews. However, whenever possible, effect sizes are used to characterize study outcomes, as in meta-analyses (Glass, McGaw, & Smith, 1981). Systematic literature search procedures, also characteristic of meta-analysis, are similarly applied in best-evidence syntheses.

Criteria for Study Inclusion

The studies on which this review is based had to meet a set of a priori criteria with respect to relevance to the topic and methodological adequacy. First, all studies had to involve comprehensive ability grouping plans that incorporated most or all students in the school. This excludes studies of special programs for the gifted or other high achievers as well as studies of special education, remedial programs, or other special programs for low achievers. Studies of within-class ability grouping are included, but studies of such grouping-related programs as individualized instruction, mastery learning, cooperative learning, and continuous-progress groupings are excluded.

Studies had to be available in English, but otherwise no restrictions were placed on study location or year of publication. Every attempt was made to locate dissertations and other unpublished documents in addition to the published literature.

Methodological requirements for inclusion. Criteria for inclusion of studies in the main review were essentially identical to those used in an earlier review of elementary ability grouping (Slavin, 1987). These were as follows:

1. Ability-grouped classes were compared to heterogeneously grouped classes. This requirement excluded a few studies that correlated "degree of heterogeneity" with achievement gain (e.g., Millman & Johnson, 1964; Wilcox, 1963). Studies that compared achievement gains for students in different tracks but not to heterogeneous classes (e.g., Alexander et al., 1978) were excluded from the main review but are discussed in a separate section.

2. Achievement data from standardized or teacher-made tests were presented. This excluded many anecdotal reports and studies that used grades as the dependent measure. Teacher-made tests, used in a very small number of studies, were accepted only if there was evidence that they were designed to assess objectives taught in all classes.

3. Initial comparability of samples was established by use of random assignment

or matching of students or classes. When individual students in intact schools or classes were matched, evidence had to be presented that the intact groups were comparable.

4. Ability grouping had to be in place for at least a semester.

5. At least three ability-grouped and three control classes were involved.

The criteria outlined above excluded very few studies comparing comprehensive ability grouping plans to heterogeneous placements. Every study located that satisfied criteria 1, 2, and 3 also satisfied criteria 4 and 5. Excluding studies of special programs for high achievers (e.g., Atkinson & O'Connor, 1963), all but two of the studies included in meta-analyses by Kulik and Kulik (1982, 1987) were also included in the present review. The exceptions were a study by Adamson (1971) that had substantial IQ differences favoring the ability-grouped school and one by Wilcox (1963) that compared more and less heterogeneous tracked classes.

One major category of studies included in the present review but excluded by the Kuliks includes studies that did not present data from which effect scores could be computed (e.g., Borg, 1965; Ferri, 1971; Lovell, 1960; Postlethwaite & Denton, 1978). These studies are discussed in terms of the direction and statistical significance of their findings.

Literature Search Procedures

The studies included here were located in an extensive search. Principal sources included the Education Resources Information Center (ERIC), *Dissertation Abstracts International*, and citations made in other reviews, meta-analyses, and primary sources. Every attempt was made to obtain a complete set of published and unpublished studies that met the criteria outlined above.

Computation of Effect Sizes

Effect sizes were generally computed as the difference between the experimental and control means divided by the control group's standard deviation (Glass et al., 1981). In the ability grouping literature, the heterogeneous group is almost always considered the control group, and this convention is followed in the present article; positive effect sizes are ones that favored ability grouping, whereas negative effect sizes indicated higher means in the heterogeneous groups. The standard deviation of the heterogeneous group is also preferred as the denominator because of the possibility that ability grouping may alter the distribution of scores. However, when means or standard deviations were omitted in studies that otherwise met the inclusion criteria, effect sizes were estimated when possible from ts, Fs, exact p values, sums of squares in factorial designs, or other information, following procedures described by Glass et al. (1981).

Several of the studies included in this review presented data comparing gain scores without reporting actual pre- or posttest means. Standard deviations of gain scores are typically lower than those of raw scores (to the degree that pre-post correlations exceed +0.5), so effect sizes computed on gain scores are often inflated. If pre-post correlations are known, effect sizes from all scores can be transformed to the scale of posttest values. However, because none of the studies using gain scores also provided pre-post correlations, a pre-post correlation of +0.8 was assumed (following Slavin, 1987). Using a formula from Glass et al. (1981), this correlation produces a multiplier of 0.632, which was used to deflate effect size estimates from gain score data. The purpose of this procedure and others was to attempt to put all effect size estimates in the same metric, the unadjusted standard deviation of the heterogeneous classes. However, because this multiplier is only a rough approximation, effect sizes from studies using gain scores should be interpreted with even more caution than that which is warranted for effect sizes in general.

Another deviation from usual meta-analytic procedure used in the present view involved adjustments of posttest scores for any pretest differences. This was done either by subtracting pretest means from posttests (if the same tests were used), by converting pre- and posttest means to z scores and then subtracting (if different tests were used), or by using covariance-adjusted scores. However, even when such adjustments were made, affecting the numerator of the effect size formula, the denominator remained the unadjusted posttest standard deviation.

One effect size is reported for each study (see Bangert-Drowns, 1986). When multiple subsamples, subjects, or tests were used, medians were computed across the data points. For example, if four measures were used with three subgroups (e.g., high, middle, and low achievers), the effect size for the study as a whole would be the median of the 12 (4 x 3) resulting effect sizes. Whenever possible, findings were also broken down by achievement level (high, average, low), and separate effect sizes were computed for each major subject.

In pooling findings across studies, medians rather than means were used, principally to avoid giving too much weight to outliers. However, any measure of central tendency in a meta-analysis or best-evidence synthesis should be interpreted in light of the quality and consistency of the studies from which it was derived, not as a finding in its own right.

Research on Ability Grouping in Secondary Schools

A total of 29 studies of tracking or streaming in secondary schools met the inclusion criteria listed earlier. The studies, their major characteristics, and their findings are listed in Table 1.

The studies listed in Table 1 are organized in three categories according to their research designs. Six studies used random assignment of students to ability-grouped or heterogeneous classes. Nine studies took groups of students; matched them individually on IQ, composite achievement, and other measures; and then assigned one of each matched pair of students to an ability-grouped class and one to a heterogeneous class. The quality of these randomized or matched experimental designs is very high, and the findings of the 15 studies using such designs must be given special weight. The remaining 14 studies investigated existing schools or classrooms that used or did not use ability grouping, and then either selected matched groups of students from within each type of school or used analyses of covariance or other statistical procedures to equate the groups. The difficulty inherent in such designs is that any differences between schools that are systematically related to ability grouping would be confounded with the practice of ability grouping per se. For example, a secondary school that used heterogeneous grouping might have a staff, principal, or community more concerned about equity, affective development, or other goals than would a "matched" school that used ability grouping. However, several of the correlation studies used very large samples and longitudinal designs, and these provide important additional information not obtainable from the typically smaller and shorter experimental studies.

Within each category, studies are listed in descending order of sample size. All other things being equal, therefore, studies near the top of Table 1 should be considered as better evidence of the effects of ability grouping that studies near the end of the table. However, the nature and quality of the studies are discussed in more detail in the following sections.

Table 1 Studies of secondary tracking

Article	Grades	Location	Sample size	Duration	Design	Effect Sizes By achievement	By subject	Total
			Randomized experimental studies					
Marascuilo & McSweeney, 1972	8-9	Berkeley, CA	603 students	2 yrs.	Students randomly assigned to 3-group AG or hetero social studies classes. Compared students on teacher-made and standardized tests.	HI +.14 AV –.37 LO –.43	Social studies –.22	–.22
Drews, 1963	9	Lansing, MI	4 schools, 432 students	1 yr.	Students randomly assigned to 3-group AG or hetero English classes compared on standardized tests.	HI –.16 AV +.01 LO –.01	Reading –.11 Language .00	–.05
Fick, 1963	7	Olathe, KS	1 school, 168 students	1 yr.	Students randomly assigned to 3-group AG or hetero "core" classes. Both classes taught by same teacher. Iowa Tests of Basic Skills used as posttests.	HI +.01 AV .00 LO –.04	Reading –.01 Language .00	–.01
Peterson, 1966	7-8	Chisholm, MN	1 school, 152 students	1 yr.	Students randomly assigned to AG or hetero classes. AG based on composite achievement, grades. Compared on standardized tests.	HI +.05 AV –.44 LO –.06	Reading +.02 Language –.01 Math –.25 Social studies .07	–.04
Ford, 1974	9	New York, NY (Low SES minority)	80 students	1 sem.	Students randomly assigned to 2-group AG or hetero math classes. Same teachers taught both types of classes. Students compared on Metropolitan Achievement Test.		Math (0)	(0)

Study	Grade	Location	Sample	Duration	Description			
Bicak, 1962	8	Minneapolis, MN (lab school)	1 school, 75 students	1 sem.	Students randomly assigned to 2-group AG or hetero science classes at university lab school	HI −.39 LO −.10	Science −.25	−.25
					Matched experimental studies			
Lovell, 1960	10	Panama City, FL	500 students	1 yr.	Matched students assigned to 5-group AG or hetero English, biology, and algebra classes	HI (+) AV (0) LO (0)	Language (+) Math (0) Biology (0)	(0)
Billett, 1928	9	Painesville, OH	408 students	1 yr.	In 3 successive years, matched students assigned to 3-group AG or hetero English classes. Compared gains on standardized tests.	HI −.11 AV +.03 LO +.18	English +.04	+.04
Platz, 1965	9	—[a]	298 students	1 sem.	Matched students assigned to 3-group AG or hetero science classes. Students compared on standardized science test.	HI +.24 AV −.10 LO +.22	Science +.22	+.22
Bailey, 1968	9	St.Louis, MO	255 students	1 yr.	Matched students assigned to 2-group AG or hetero algebra classes. Same teachers taught both types of classes. Students compared on gains on standardized algebra measure.	HI +.18 LO −.24	Math −.03	−.03
Thompson, 1974	11	Suburban VA	240 students	1 yr.	Compared students in 2 schools, one AG in social studies, one hetero. Students matched. Compared gain scores on teacher-made tests.	HI −.50 HI AV −.47 LO AV −.54	Social studies −.48	−.48
Barton, 1964	9	Rural UT	204 students	1 yr.	Matched students assigned to 4-group AG or hetero English classes, compared on California Achievement Test gains.	HI +.22 HI AV −.03 LO AV −.13 LO −.20	Reading +.06 Language −.13	−.04

Table 1 (continued)

Article	Grades	Location	Sample size	Duration	Design	By achievement	Effect Sizes By subject	Total
Willcutt, 1969	7	Bloomington, IN (lab school)	156 students	1 yr.	Matched students assigned to 4-group flexible AG in math or to hetero. Grouping changed 8 times in the year.		Math −.15	−.15
Holy & Sutton, 1930	9	Marion, OH	148 students	1 sem.	Matched students assigned to AG, hetero algebra classes. Same teacher taught all classes.		Math +.28	+.28
Martin, 1927	7	New Haven, CT	83 students	1 yr.	Matched students assigned to 3-group AG or hetero.	HI +.12 AV −.06 LO +.23	Reading +.17 Math +.13 Language +.03 Social studies .00 Science −.04	+.10
Correlational studies								
Kerckhoff, 1986	5-10	Britain	8,500 students	5 yrs.	Longitudinal study of students throughout Britain who attended streamed or unstreamed secondary schools.		Reading +.02 Math +.03	+.03
Fogelman, Essen, & Tibbenham, 1978	6-10	Britain	5,923 students	4 yrs.	Retrospective study compared students who had been in streamed, partially streamed, or heterogeneous schools throughout secondary school, controlling for Grade 5 general ability.		Reading +.02 Math +.03	+.03

Study	Location	Grades	Sample	Duration	Description	Ability level effects	Subject effects	Overall
Borg, 1965	Utah	6-9 7-10 8-11 9-12	2,934 students	4 yrs.	Longitudinal study of students in districts using AG compared to students in neighboring district using heterogeneous grouping, controlling for pretests.	HI (0) AV (0) LO (0)	Math (0) Science (0)	(0)
Ferri, 1971	Britain	5-6	28 schools, 1,716 students	2 yrs.	Streamed and nonstreamed schools matched on 7+ (Grade 2) reading, followed 4 years in junior school, 2 years in secondary.	HI (0) AV (0) LO (0)	Math (0) English (0)	(0)
Breidenstine, 1936	Soudersburg, PA	7-9	11 schools, 860 students	1 yr.	Compared students in 4 AG, 7 hetero schools matched on IQ.		Composite achievement	−.19
Purdom, 1929	—[a]	9	700 students	1 sem.	Matched students in AG, hetero English and algebra classes compared in achievement.	HI −.02 AV −.08 LO +.07	English −.02 Algebra .00	+.01
Postlethwaite & Denton, 1978; Newbold, 1977	Britain	5-7	1 school, 450 students	2 yrs.	Students within one secondary school assigned to streamed or unstreamed halls. Achievement assessed on national examinations.	HI (0) AV (0) LO (0)	Math (0) English (0) Social studies (0) French (0)	(0)
Bachman, 1968	Portland, OR	7	15 schools, 23 teachers, 404 students	1 yr.	Math classes in schools using AG compared to hetero classes, controlling for IQ.	Math (0)	Math (0)	(0)
Kline, 1964	St. Louis, MO	9-12	4 schools	4 yrs.	Retrospective study of successive cohorts of students, one in 3- or 4-group AG, one hetero, in 4 schools. Compared on standardized tests after 4 years of AG or hetero placement.	V. HI −.02 HI +.08 AV .00 LO −.02	Reading −.05 Language +.07 Math +.01	+.01

Table 1 (continued)

Article	Grades	Location	Sample size	Duration	Design	Effect Sizes By achievement	By subject	Total
Stoakes, 1964	7	Cedar Rapids, IA	3 schools	1 yr.	Matched mentally advanced and slow-learning students compared in schools using AG or hetero assignment. Compared on standardized tests.	HI (0) LO (0)	Reading (0) English (0) Math (0)	(0)
Martin, 1959	6-8	Nashville, TN	3 schools	2 yrs.	Retrospectivestudy compared gains on Stanford Achievement Tests for 2 AG and 1 hetero school from Grades 6-8.	HI (0) AV (0) LO (0)	Reading (0) Language (0) Math (0)	(0)
Chiotti, 1961	9	Issaquah, WA	3 schools	1 yr.	Matched students in 3-group AG and hetero schools compared in math achievement.	HI +.14 AV +.06 LO +.35	Math +.18	+.18
Fowlkes, 1931	7	Glendale, CA	2 schools	1 sem.	Students in school using 3-group AG based on IQ matched with students in hetero school. Compared gains on Stanford Achievement Tests.	HI −.45 AV −.18 LO −.05	Reading −.04 Language −.17 Math −.17 Social studies −.21	−.20
Cochran, 1961	8	Kalamazoo, MI	1 school	1 yr.	Compared studentsgrouped separately for English, math, to previous year (hetero) students matched in IQ, age, sex, achievement.	HI (0) AV (0) LO (0)	Math(0) English (0)	(0)

Note: AG = ability grouping; hetero = heterogeneous assignment; HI = high achieving students; AV = average achieving students; LO = low achieving students.
[a] Not available.

Overall Findings

Across the 29 studies listed in Table 1, the effects of ability grouping on student achievement are essentially zero. The median effect size (ES) for the 20 studies from which effect sizes could be estimated was –.02, and none of the 9 additional studies found statistically significant effects. Counting the studies with nonsignificant differences as though they had effect sizes of .00, the median effect size for all 29 studies would be .00. Results from the 15 randomized and matched experimental studies were not much different; the median effect size was –.06 for the 13 studies from which effect sizes could be estimated. In 9 of these 13 studies (including all 5 of the randomized studies) results favored the heterogeneous groups, but these effects are mostly very small.

There are few consistent patterns in the study findings. Most of the studies involved Grades 7-9, with ninth graders sometimes in junior high schools and sometimes in senior high schools. No apparent trend is discernible within this range. Above the ninth grade the evidence is too sparse for firm conclusions. Lovell (1960) found that high achieving tenth graders performed significantly better in ability-grouped English classes, but there were no effects in biology or algebra and no effects for average or low achievers. In a 4-year study of students in Grades 912, Borg (1965) found significant positive effects of ability grouping for average and low achievers in math but no differences in science or for high achievers. Cohorts followed from Grades 7-10 and 8-11 showed no significant differences on any measure for any ability level. In contrast, Thompson (1974), in a study of 11th grade social studies, found the largest effects favoring heterogeneous grouping (ES = –.48), whereas Kline (1964), in another 4-year study of students in Grades 9-12, found no differences.

Twelve of the 29 studies tracked students for all subjects according to one composite ability or achievement measure. The remaining 17 studies grouped students on the basis of performance in one or more specific subjects. However, there were no differences in the outcomes of these different forms of ability grouping. In addition, there were no consistent patterns in terms of the number of ability groups to which students were assigned (the great majority of studies used 3). Study duration had no apparent impact on outcome. Studies that used adjusted gain scores produced the same effects as other studies, and the use of the adjustment of gain scores described above made no difference in outcomes.

There was no discernible pattern of findings with respect to different subjects, with one possible exception. Studies by Marascuilo and McSweeney (1972), Thompson (1974), and Fowlkes (1931) found relatively strong effects favoring heterogeneous grouping in social studies, and three additional studies by Peterson (1966), Martin (1927), and Postlethwaite and Denton (1978) found no differences or slight effects in the same direction. This is not enough evidence to conclusively point to a positive effect of heterogeneous grouping in social studies, but it is important to note that all three of the randomized or matched experimental studies found differences in this direction.

There were no consistent effects according to study location. All four of the British studies found no differences between streamed and unstreamed classes; a large, longitudinal Swedish study by Svensson (1962), not shown in Table 1 because it lacked adequate evidence of initial equality, also found no differences between streamed and unstreamed classes. Urban, suburban, and rural schools had similar outcomes. The one study that involved large numbers of minority students, a randomized experiment in a New York City high school by Ford (1974), found no differences between ability-grouped and heterogeneous math classes.

Studies conducted before 1950 were no more likely than more recent studies to find achievement differences. On this topic, it is interesting to note that experimental-control studies of ability grouping have not been done in recent years. The only study of the 1980s, by Kerckhoff (1986), was done by a sociologist who focused his attention on differences between students in different streams. This study is described in more detail below. Otherwise, the most recent experimental-control comparisons were done in the early 1970s.

Differential Effects According to Achievement Levels

One of the most important questions about ability grouping in secondary schools concerns the degree to which it differentially affects students at different achievement levels. As noted earlier, many researchers and reviewers, particularly those working the sociological tradition, have emphasized the *relative* impact of grouping for different groups of students far more than the average effect for all students.

Twenty-one of the 29 studies presented in Table 1 presented data on the effects of ability grouping on students of different ability levels. Most studies divided their samples into three categories (high, average, and low achievers), but some used two or four categories.

Across the 15 studies from which effect sizes could be computed, the median effect size was +.01 for high achievers, −.08 for average achievers, and −.02 for low achievers. Effects of this size are indistinguishable from zero, and if all the nonsignificant differences found in studies from which effect sizes could not be computed are counted as effect sizes of .00, the median effect size for each level of student becomes .00. In addition, only one of seven studies from which effect sizes could not be computed (Lovell, 1960) found significantly positive effects of ability grouping for high achievers, and none of these studies found significant effects in either direction for average and low achievers. The randomized and matched experimental studies provided slightly more support for the idea that ability grouping has a differential effect; the median effect sizes for high, average, and low achievers were +.05, -.10, and -.06, respectively. It is interesting to note that the study by Borg (1965), which is often cited to support the differential effect of ability grouping on students of different ability levels, in fact provides very weak support for this phenomenon. Across two measures given to members of four 4-year cohorts that principally included secondary years, significant effects favoring ability grouping were found for high achievers in one of eight comparisons, for average achievers in three of eight, and for low achievers in one of eight. Only in a cohort that included Grades 4 to 7 were there significant effects favoring heterogeneous grouping for low achievers.

It might be expected that differential effects of track placement would build over time and that longitudinal studies would show more of a differential impact than l-year studies. The one multiyear randomized study, by Marascuilo and McSweeney (1972), did find that over a 2-year period, students in the top social studies classes gained slightly more than similar students in heterogeneous classes (ES = +.14), whereas middle (ES = −.37), and low (ES = −.43) groups gained significantly less than their ungrouped counterparts. However, across seven multiyear correlational studies of up to 5 years' duration, not one found a clear pattern of differential effects.

A few studies provided additional information on differential effects of ability grouping by investigating effects of grouping on high or low achievers only. For example, Torgelson (1963) randomly assigned low achieving students in Grades 7-9 to homogeneous or heterogeneous classes. Across several performance measures, the median effect size was +.13 (nonsignificantly favoring ability grouping). Similarly, Borg and Prpich (1966) randomly assigned low achieving 10th graders to ability-grouped or heterogeneous English classes and found that there were no differences in one cohort. In a second cohort, differences favoring ability grouping on a writing measure were found, but there were no differences on eight other measures.

Studies of ability grouping of high achievers are difficult to distinguish from studies of special programs for the gifted. Well-designed studies of programs for the gifted generally find few effects of separate programs for high achievers unless the programs include acceleration (exposure to material usually taught at a higher grade level) (Fox, 1979; Kulik & Kulik, 1984). That is, grouping per se has little effect on the achievement of high achievers. An outstanding illustration of this is a dissertation by Mikkelson (1962), who randomly assigned high achieving seventh and eighth graders to ability-grouped or heterogeneous math classes. The seventh grade homogeneous classes were given enrichment, but the eighth graders were accelerated, skipping to ninth grade algebra. No effects were found for the seventh graders. The accelerated eighth graders, or course, did substantially better than similar students who were not accelerated on an algebra test, and they did no worse on a test of eighth grade math. Taken together, research comparing ability-grouped to heterogeneous placements provides little support for the proposition that high achievers gain from grouping whereas low achievers lose. However, there is an important limitation to this conclusion. In most of the studies that compared tracked to untracked grouping plans (including all of the randomized and matched experimental studies), tracked students took different levels of the same courses (e.g., high, average, or low sections of Algebra I). Yet much of the practical impact of tracking, particularly at the senior high school level, is on determining the nature and number of courses taken in a given area. The experimental studies do not compare students in Algebra I to those in Math 9, or students who take 4 years of math to those who take 2. The conclusions drawn in this section are limited, therefore, to the effects of between-class grouping *within the same courses*, and should not be read as indicating a lack of differential effects of tracking as it affects course selection and course requirements.

Other Forms of Ability Grouping

The studies discussed above and summarized in Table 1 evaluated the most common forms of ability grouping in secondary schools – full-time, between-class ability grouping for one or more subjects. However, a few studies have evaluated other grouping plans.

The most widely used form of grouping in elementary schools, within-class ability grouping, has also been evaluated in a few studies involving middle and junior high schools. Campbell (1965) compared the use of three math groups within the class to heterogeneous assignment in two Kansas City junior high schools. There were no differences between the two programs in achievement. Harrah (1956) compared five types of within-class grouping in Grades 7-9 in West Virginia and found ability grouping to be no more successful than other grouping methods. Note that these findings conflict with those of studies of within-class ability grouping in mathematics in the upper elementary grades, which tended to support the use of math groups (Slavin, 1987).

Vakos (1969) evaluated the use of a combination of heterogeneous and homogeneous instruction in 11th grade social studies classes in Minneapolis. Students were grouped by ability 2 days each week but heterogeneously grouped the other 3 days. No achievement differences were found. Zweibelson, Bahnmuller, and Lyman (1965) evaluated a similar mixed approach to teaching ninth grade social studies in New Rochelle, New York, and also found no achievement differences. Chiotti (1961) compared a flexible plan for grouping junior high school students across grade lines for mathematics to both ability-grouped and heterogeneous grouping plans, and again found no differences in achievement. A cross-grade grouping arrangement similar to the Joplin Plan (Slavin, 1987) was compared to within-class grouping in reading by Chismar (1971) in Grades 4-8. Significantly positive effects of this program were found in Grades 4 and 7 but not 5, 6, and 8.

Reconciling Track/No-Track and High-Track/Low-Track Studies

As noted earlier, two very different traditions of research have dominated research on ability grouping. One involves comparisons of ability-grouped to heterogeneous placements. The other involves comparisons of the progress made by students in different ability groups or tracks. Whereas there has been little experimental research comparing ability-grouped to heterogeneous placements since the early 1970s, research comparing the achievement of students in different tracks largely began in the 1970s and continues to the present.

The findings of high-track/low-track studies of ability grouping conflict with those emphasized in this review in that they generally find that even after controlling for IQ, socioeconomic status, pretests, and other measures, students in high tracks gain significantly more in achievement than do students in low tracks, especiallym in mathematics (see Gamoran & Berends, 1987, for a review). How can these findings be reconciled with those of the experimental studies?

One important difference between experimental and correlational studies of ability grouping is that, as mentioned earlier, correlational studies (especially at the senior high school level) often include not only the effects of being in a high, average, or low class, but also the effects of differential course taking. Students in academic tracks may score better than those in general or vocational classes because they take more courses or more advanced courses. The experimental studies comparing grouped and ungrouped classes are all studies of grouping per se, holding course taking and other factors constant. The correlational studies examine tracking as it is in practice, where track placement implies differences in course requirements, course taking patterns, and so on. Also, experimental track versus no-track studies are rare beyond the ninth grade, whereas most correlational studies comparing students in high versus low tracks involve senior high schools. The lack of track versus no-track studies at the senior high school level is hardly surprising given the nearly universal use of some form of tracking at that level. However, tracking usually has a different meaning in senior than in junior high school. Whereas junior high school tracking mostly involves different levels of courses (e.g., high English vs. low English), senior high tracking is more likely to involve completely different patterns of coursework (e.g., metal shop vs. French III). Also, the problem of dropouts becomes serious in senior high school; a study of 12th graders unavoidably excludes the students who may have suffered most from being in the low track and left school (see Gamoran, 1987). This could reduce observed differences between high- and low-track students.

There is limited evidence, however, that differences in course taking or grade level account for the different conclusions of the track/no-track and high-track/low-track studies. Four-year longitudinal studies in U.S. senior high schools by Kline (1964) and Borg (1965) found no differential effects of track placement for high, average, and low achievers (as compared to similar students in untracked placements). Presumably, course-taking patterns in these senior high school studies varied by track. A correlational study by Alexander and Cook (1982) found that although taking more courses in senior high school did increase achievement (controlling for background factors), different course-taking patterns in different tracks did not account for track differences in achievement. Gamoran (1987) found that track effects on math and science achievement were explained in part by the fact that students in the academic tracks take more math and science courses and in particular, more *advanced* courses in these areas. However, no such patterns were seen on reading, vocabulary, writing, or civics achievement measures. Gamoran noted the difficulty of disentangling track and course taking, which are highly correlated in math and science (and, of course, both track and course taking are strongly correlated with ability, socioeconomic status, and other factors). It is certainly logical to expect correlational studies of senior high school tracking to find different effects of different track placements because of different course-taking patterns, but because of confounding of tracking, course-taking, and student background factors, that is difficult to determine conclusively.

Another likely explanation for different findings of track/no-track and high-track/low-track studies involves the difficulty of statistically controlling for large differences. Students in higher tracks tend to achieve at much higher levels than those in lower tracks (both before and after taking secondary courses), and statistically controlling for these differences is probably not sufficient to completely remove the influence of ability or prior performance on later achievement. Further, studies in higher tracks are also likely to be higher in such attributes as motivation, internal locus of control, academic self-esteem, and effort, factors that are not likely to be controlled in correlational studies.

To understand the difficulty of controlling for large initial differences between students, imagine an experiment in which a new instructional method was to be evaluated. The experimenter selects a group of students who have high test scores and high IQ scores and are nominated by their teachers as being hard working, motivated, and college material. This group becomes the experimental group, and the remaining students serve as the control group. To control for the differences between the groups, prior composite achievement and socioeconomic status are used as covariates or control variables.

In such an experiment, no one would doubt that regardless of the true effectiveness of the innovative treatment, the experimental group would score far better than the control group, even controlling for prior achievement and socioeconomic status. No journal or dissertation committee would accept such a study. Yet this "experiment" is essentially what is being done when researchers compare students in different tracks. When there are significant pretest differences, use of statistical controls through analysis of covariance or regression are considered inadequate to equate the groups. Most often, the statistical controls will undercontrol for true differences (Lord, 1960; Reichardt, 1979). Yet high- and low-track students usually differ in pretests or IQ by one to two standard deviations, an enormous systematic difference for which no statistical procedure can adequately control.

The only study that compared both tracked to untracked schools and high-track to low-track students was a 5-year longitudinal study by Kerckhoff (1986) in Britain. This study illustrates the problem of controlling for large differences. For example, in mathematics, boys in the high track of three-group ability grouping programs gained about 11 z score points from a test given at age 11 to one given at age 16, whereas students in a remedial track gained 18 z score points. Yet the regression coefficient comparing the high-track to ungrouped students was +2.34, indicating performance about 42% of a standard deviation above "predicted" performance. In contrast, the remedial-track boys had a regression coefficient (in comparison to ungrouped students) of −.72, indicating performance about 13% of a standard deviation below "predicted" performance, despite the fact that the remedial students actually gained more than the top-track students. The reason for this is that the remedial students started out (at age 11) scoring 1.64 standard deviations below the ungrouped students, whereas top-track students started out 1.02 standard deviations above the ungrouped students, a total difference between top-track and remedial students of 2.66. No regression or analysis of covariance can adequately control for such large pretest differences. Because of unreliability in the measures and less-than-perfect within-group correlations of pre- and posttests, "predicted" scores based on pretests and other covariates will (other things being equal) be too low for high achievers and too high for low achievers.

Another factor that can contribute to overestimates of the effects of curriculum track on achievement in studies lacking heterogeneous comparison groups is fan spread. Put simply, high achievers usually gain more per year than do low achievers, so over time the gap between high and low achievers grows. This increasing gap cannot be unambiguously ascribed to ability grouping or other school practices, as it occurs under virtually all circumstances. A student who is performing at the 16th percentile in the 6th grade and is still at the 16th percentile in 12th grade will be further "behind" the 12th grade mean in grade equivalents, for example (Coleman & Karweit, 1972). An additional factor that can contribute to spurious findings indicating a benefit of being in the high track is that factors other than test scores factor into placement decisions. For example, a study by Balow (1964) found that on math tests not used for group placement, there was enormous overlap between students in supposedly homogeneous seventh-grade math classes. More than 72% of the students scored between the lowest score in the top group and the highest score in the bottom group. Among these students in the "area of overlap," students who were in the top group gained the most in math achievement over the course of the year, whereas those in the low group gained the least.

On its surface this study provides support to the "self-fulfilling prophecy" argument. Yet consider what is going on. Imaging two students with identical scores, one assigned to the high group and one to the low group. Why were they so assigned? Random error is a possibility, but all the systematic possibilities weigh in the direction of higher performance for the student assigned to the high group. Because teacher judgement was involved, teachers may have accurate knowledge of student motivation, self-esteem, behavior, or other factors to enable them to predict who will do well and who will not. The actual assignments were done on different tests than those used in the Balow study; it is likely that students who scored low on Balow's pretests but were put in the high groups scored high on the test used for placement, and then regressed to a higher mean on Balow's posttest.

What this discussion is meant to convey is not that different tracks do or do not have a differential impact on student achievement, but that comparisons of students in existing tracks cannot tell us one way or another. To learn about the differential impacts of track placement, there are two types of research that might be done. One would be to randomly assign students at the margin to different tracks something that has never been done. The other is to compare similar students randomly assigned to ability-grouped or ungrouped systems. This has been done several times, and, as noted earlier in this review, there is no clear trend indicating that students in high-track classes learn any more than high achieving students in heterogeneous classes, or that students in low-track classes learn any less than low achieving students in heterogeneous classes.

Why Is Ability Grouping Ineffective?

The evidence summarized in Table 1 and discussed in this review is generally consistent with the conclusions of earlier reviews comparing homogeneous and heterogeneous grouping (e.g., Kulik & Kulik, 1982, 1987; Noland, 1985), but runs counter to two quite different kinds of "common sense." On one hand, it is surprising to find that assignment to the low-ability group is not detrimental to student learning. A substantial literature has indicated the low quality of instruction in low groups (e.g., Evertson, 1982; Gamoran, 1989; Oakes, 1985), and a related body of research has documented the negative impact of ability grouping on the motivations and self-esteem of students assigned to low groups (e.g., Cottle, 1974; Schafer & Olexa, 1971; Trimble & Sinclair, 1987). How can the effect of ability grouping on low achieving students be zero, as this review concludes?

On the other hand, another kind of "common sense" would argue that, at least in certain subjects, ability grouping is imperative in secondary schools. How can an 8th grade math teacher teach a class composed of students who are fully ready for algebra and students who are still not firm in subtraction and multiplication? How does an English teacher teach literature and writing to a class in which reading levels range from 3rd to 12th grade? Yet study after study, including randomized experiments of a quality rarely seen in educational research, finds no positive effect of ability grouping in any subject or at any grade level, even for the high achievers most widely assumed to benefit from grouping.

The present review cannot provide definitive answers to these questions. However, it is worthwhile to speculate on them.

One possibility is that the standardized tests used in virtually all of the studies discussed in this review are too insensitive to pick up effects of grouping. This seems particularly plausible in looking at tests of reading, because reading has not generally been taught as such in secondary schools. However, standardized tests of mathematics do have a great deal of face validity and curricular relevance, and these show no more consistent a pattern of outcomes. Marascuilo and McSweeney (1972) used both teacher-made and standardized measures of social studies achievement and found similar results with each.

Another possibility is that it simply does not matter whom students sit next to in a secondary class. Secondary teachers use a very narrow range of teaching methods, overwhelmingly using some form of lecture or discussion (Goodlad, 1983). In this setting, the direct impact of students on one another may be minimal. If this is so, then any impacts of ability grouping on students would have to be mediated by teacher characteristics or behaviors or by student perceptions and motivations.

Studies contrasting teaching behaviors in high- and low-track classes usually find that the low tracks have a slower pace of instruction and lower time on-task (e.g., Evertson, 1982; Oakes, 1982). Yet, as noted earlier, the meaning and impact of these differences are not self-evident. It may be that a slower pace of instruction is appropriate with lower-achieving students, or that pace is relatively unimportant because a higher pace with lower mastery is essentially equivalent to a lower pace with higher mastery. Higher time on-task should certainly be related to higher achievement (Brophy & Good, 1986), but the comparisons of time on task between high and low tracks are misleading. What would be important to compare is time on task *for low achievers* in homogeneous and heterogeneous classes, because low achievers may simply be off-task more than high achievers regardless of their class placement. In this regard, it is important to note that Evertson, Sanford, and Emmer (1981) found time on-task to be lower in extremely heterogeneous junior high school classes than in less heterogeneous ones because teachers had difficulty managing the more heterogeneous classes.

The lesson to be drawn from research on ability grouping may be that unless teaching methods are systematically changed, school organization has little impact on student achievement. This conclusion would be consistent with the equally puzzling finding that substantial reductions in class size have little impact on achievement (Slavin, 1989); if teachers continue to use some form of lecture/discussion/seatwork/quiz, then it may matter very little in the aggregate which or how many students the teachers are facing. In contrast, forms of ability grouping that were found to make a difference in the upper elementary grades – the Joplin Plan (cross-grade grouping in reading to allow for whole-class instruction) and within-class grouping in mathematics (Slavin, 1987) – both significantly change time allocations and instructional activities within the classroom.

Alternatives to Ability Grouping

If the effects of ability grouping on student achievement are zero, then there is little reason to maintain the practice. As noted earlier in this article, arguments in favor of ability grouping depend on assumptions about the effectiveness of grouping, at least for high achievers. In the absence of any evidence of effectiveness, these arguments cannot be sustained.

Yet there is also no evidence that simply moving away from traditional ability grouping practices will in itself enhance student achievement, and there are legitimate concerns expressed by teachers and others about the practical difficulties of teaching extremely heterogeneous classes as the secondary level. How can schools moving away from traditional ability grouping use this opportunity to contribute to student achievement?

One alternative to ability grouping often proposed (e.g., Oakes, 1985) is the use of cooperative learning methods, which involve students working in small, heterogeneous learning groups. Research on cooperative learning consistently finds positive effects of these methods if they incorporate two major elements: group goals and individual accountability (Slavin, 1990). That is, the cooperating groups must be rewarded or recognized on the basis of the sum or average of individual learning performances. Cooperative learning methods of this kind have been used successfully at all grade levels, but there is less research on them in Grades 10-12 than in Grades 2-9 (see Newmann & Thompson, 1987). Cooperative learning methods have also had consistently positive impacts on such outcomes as self-esteem, race relations, acceptance of mainstreamed academically handicapped students, and ability to work cooperatively (Slavin, 1990).

One category of cooperative learning methods may be particularly useful in middle schools moving toward heterogeneous class assignments. These methods are Cooperative Integrated Reading and Composition (Stevens, Madden, Slavin, & Farnish, 1987) and Team Assisted Individualization—Mathematics (Slavin & Karweit, 1985; Slavin, Madden, & Leavey, 1984). Both of these methods are designed to accommodate a wide range of student performance levels in one classroom, using both homogeneous and heterogeneous within-class groupings. These programs have been successfully researched in Grades 3-6 but are often used up to the eighth grade level.

Other alternatives to between-class ability grouping have also been found to be successful in the upper elementary grades (see Slavin, 1987) and could probably be effective in middle schools as well. These include within-class ability grouping in mathematics (e.g., teaching two or three math groups within a heterogeneous class) and the Joplin Plan in reading. The Joplin Plan involves regrouping students for reading across grade levels but according to reading level, so that no within-class reading groups are necessary. However, although these alternatives to between-class grouping are promising because of their success in the upper elementary grades, the few studies of within-class ability grouping at the junior high school level have not found this practice to be effective (Campbell, 1965; Harrah, 1956), and the one middle school study of the Joplin Plan found only inconsistent positive effects (Chismar, 1971). (For descriptions of secondary schools implementing alternatives to traditional ability grouping, see Slavin, Braddock, Hall, & Petza, 1989.)

Limitations of This Review

It is important to note several limitations of the present review. Perhaps the most important is that in none of the studies reviewed here were there systematic observations made of teaching and learning. Observational studies and outcome studies have proceeded on parallel tracks; it would be important to be able to relate evidence of outcomes to changes in teacher behaviors or classroom characteristics. In particular, it would be important to know the degree to which teachers in ability-grouped schools actually differentiate instruction. For example, are teachers of high-track classes more likely to provide enrichment (e.g., greater depth on the same objectives) or acceleration (e.g., coverage of more material usually taught at a later grade level)? How do teachers of low-track classes adapt instruction to the needs of their students? How do teachers of untracked, heterogeneous classes accommodate the wide range of performance levels in their classes? What level and pace of instruction is provided in untracked, heterogeneous classes? Most important, how do variations from teacher to teacher in instructional behaviors in high, low, and heterogeneous classes relate to the outcomes of ability grouping for students of different ability levels?

Another limitation, mentioned earlier, is that almost all studies reviewed here used standardized tests of unknown relationship to what was actually taught. It may be, for example, that positive effects of ability grouping for high achievers could be missed by standardized tests because what these students are getting is enrichment or higher-order skills not assessed on the standardized measures, or that negative effects for low achievers are missed because teachers of low-track classes are hammering away at the minimum skills that are assessed on the standardized tests but ignoring other content. Future research on ability grouping needs to closely examine possible outcomes of grouping on more broadly based and sensitive measures.

A third limitation is the age of most of the studies reviewed. It is possible that schools, students, or ability grouping have changed enough since the 1960s and 1970s to make conclusions from these and older studies tenuous.

As noted earlier, the results reported in this review mainly concern the effects of grouping per se, with little regard for the effects of tracking on such factors as course taking. Effects of tracking on differential course taking are most important in senior high schools. There is a need for additional research comparing tracked to untracked situations at the senior high school level, particularly research designed to disentangle the effects of tracking from those of differential course taking.

In addition, it would add greatly to the understanding of ability grouping in secondary schools to have evaluations or even descriptions of a wider range of alternatives to traditional ability grouping. The few studies of within-class grouping, cross-grade groupings, and flexible grouping plans are not nearly adequate to explore alternatives. Cooperative learning, often proposed as an alternative to ability grouping, has frequently been found to increase student achievement in ability-grouped as well as ungrouped secondary classes (Newmann & Thompson, 1987; Slavin, 1990), yet no study has compared cooperative learning in heterogeneous classes to traditional instruction in homogeneous ones. Descriptions of creative alternatives to ability grouping currently exist only at the anecdotal level (Slavin et al., 1989).

Conclusions

Although there are limitations to the scope of this review and to the studies on which it is based, there are several conclusions that can be advanced with some confidence. These are as follows:

1. Comprehensive between-class ability grouping plans have little or no effect on the achievement of secondary students, at least as measured by standardized tests. This conclusion is most strongly supported in Grades 7-9, but the more limited evidence that does exist from studies in Grades 10-12 also fails to support any effect of ability grouping.

2. Different forms of ability grouping are equally ineffective.

3. Ability grouping is equally ineffective in all subjects, except that there may be a negative effect of ability grouping in social studies.

4. Assigning students to different levels of the same course has no consistent positive or negative effects on students of high, average, or low ability.

For the narrow but extremely important purpose of determining the impact of ability grouping on standardized achievement measures, the studies reviewed here are exemplary. Six randomly assigned individual students to ability-grouped or heterogeneous classes, and nine more individually matched students and then assigned them to one or the other grouping plan. Many of the studies followed students for 2 or more years. If there had been any true effect of ability grouping on student achievement, this set of studies would surely have detected it.

For practitioners, the findings summarized above mean that decisions about whether or not to ability group must be made on bases other than likely effects on achievement. Given the antidemocratic, antiegalitarian nature of ability grouping, the burden of proof should be on those who would group rather than those who favor heterogeneous grouping, and in the absence of evidence that grouping is beneficial, it is hard to justify continuation of the practice. The possibility that students in the low groups are at risk for delinquency, dropout, and other social problems (e.g., Rosenbaum, 1980) should also weigh against the use of ability grouping. Yet schools and districts moving toward heterogeneous grouping have little basis for expecting that abolishing ability grouping will in itself significantly accelerate student achievement unless they also undertake changes in curriculum or instruction likely to improve actual teaching.

There is much research still to be done to understand the effects of ability grouping in secondary schools on students achievement. Studies using more sensitive achievement measures, studies of grouping at Grades 10-12, studies of a broader range of alternatives to grouping, and studies relating observations to outcomes of grouping are areas of particular need. Enough research has been done comparing the effects of ability grouping on standardized achievement tests for students assigned to high, middle, and low tracks, at least up through the ninth grade. It is time to move beyond these simple comparisons to consider more fully how secondary schools can adapt instruction to the needs of a heterogeneous student body.

CHAPTER 9

Achievement Effects of the Nongraded Elementary School

Background and Commentary

The review reprinted in this chapter, from the *Review of Educational Research*, was written in collaboration with Roberto Gutiérrez, then a graduate student in sociology at Johns Hopkins. In the course of assembling research for my reviews on ability grouping I came across many studies of nongraded elementary school plans of the 1960's and early 1970's. I excluded these studies from my 1987 review, as the nongraded elementary school is too complex and too important a topic to be submerged in a review of ability grouping. However, I always intended to return to this topic. When Roberto Gutiérrez arrived at Johns Hopkins (from his native Bogotá, Colombia) with an interest in the intersection of education and sociology, I suggested he might want to pick up where I left off on a review of the achievement outcomes of the nongraded elementary school. This he did, with enormous skill and energy.

At the time I was reviewing research on elementary ability grouping, the nongraded elementary school was only of historical interest. By the early 1970's, the nongraded elementary movement had largely been incorporated in the open classroom movement, and then both utterly disappeared in the back-to-the-basics movement of the mid-1970's. Amazingly, however, the nongraded elementary school made a reappearance in the early 1990's, probably in part as a reaction to high rates of retention in the primary grades in the 1980's, and in part as a movement consistent with current conceptions of "developmentally appropriate practice" in early childhood education.

We hoped to provide a base of research evidence to inform the rapidly evolving reintroduction of nongraded organization. In our review we found that it mattered a great deal what form of nongraded organization was being used. Those that most resembled the open classroom tended to have few if any benefits for student achievement, while those that used nongraded plans as a flexible means of grouping students for instruction appropriate to their needs tended to be much more effective.

My perception is that the nongraded programs of the 1990's have resembled the less effective nongraded plans of the early 1970's rather than the more limited but more effective plans of the 1960's. At present it would be difficult to separate the effects of nongraded organization from those of use of whole language in reading and other "developmentally appropriate" practices that emphasize discovery, experimentation, and relatively unstructured forms of cooperative learning. In any case, I am unaware of any studies comparing modern nongraded plans to traditional control groups.

One interesting question, never satisfactorily answered in the first cycle of enthusiasm for nongraded primary plans, is the degree to which schools actually use nongrading as a means of either accelerating student progress through the primary grades or of allowing struggling or immature students additional time in the primary grades. The possibility of allowing time in the primary grades to vary was part of the rationale for the early nongraded designs expressed by Goodlad and Anderson (1963), among others, but limited evidence from that time tended to show that schools rarely altered students' progression through the early grades. Today, with most elementary schools operating under considerable accountability pressure, it seems more likely that schools might hold low-achieving students in the primary block for an additional year, hoping to increase test scores by testing students who are a year older (see Slavin & Madden, 1991). I am unaware of any evidence about this from today's nongraded schools.

A nongraded elementary program is one in which children are flexibly grouped according to performance level, not age, and proceed through the elementary school at their own rates. Popular in the 1950s, 1960s, and early 1970s, the nongraded plan is returning today. This article reviews research on the achievement effects of nongraded organization. Results indicated consistent positive achievement effects of simple forms of nongrading generally developed early: cross-grade grouping for one subject (median ES = +.46) and cross-grade grouping for many subjects (median ES = +.34). Forms of nongrading making extensive use of individualization were less consistently successful (median ES = +.02). Studies of Individually Guided Education (IGE), which used nongrading and individualization, also produced inconsistent effects (median ES = +.11). The article concludes that nongraded organization can have a positive impact on student achievement if cross-age grouping is used to allow teachers to provide more direct instruction to students but not if it is used as a framework for individualized instruction.

Greek mythology tells us of the cruel robber, Procrustes (the stretcher). When travelers sought his house for shelter, they were tied to an iron bedstead. If the traveler was shorter than the bed, Procrustes stretched him until he was the same length as the bed. If he was longer, his limbs were chopped off to make him fit. Procrustes shaped both short and tall until they were equally long and equally dead.

[Graded systems of school organization] trap school-age travelers in much the same fashion as Procrustes' bed trapped the unwary. (Goodlad & Anderson, 1959, p. 1)

So begins the book that launched one of the most interesting innovations in the history of education: *The Nongraded Elementary School,* by John Goodlad and Robert Anderson. The nongraded elementary school movement was an important force in North American education in the 1960s and early 1970s, even if its major elements were only implemented in a small proportion of schools. The challenge to the traditional age-graded classroom posed by the nongraded concept is one that still has relevance today. More importantly, the nongraded elementary school itself is reappearing in U.S. schools. Recently, the states of Kentucky and Oregon have promoted a shift to nongraded programs, and many districts and schools elsewhere are moving in this direction (Willis, 1991).

A great deal of research has been done to evaluate various forms of the nongraded elementary school, but there are few comprehensive reviews on this topic. McLoughlin (1967), reviewing studies done up to 1966, concluded that most found no differences between graded and nongraded programs in reading, arithmetic, and language arts performance. In contrast, Pavan (in press), who limited her review of achievement to studies reported between 1968 and 1990, concluded that most comparisons favored the nongraded plan. However, both of these reviews were quite limited. Both simply counted statistically significant findings favoring graded or nongraded programs, paying little attention to the particular forms of nongrading used, the methodological quality of the studies, or the size of the effects.

The purpose of this article is to describe the nongraded elementary school in its earlier incarnations, to systematically review research on the academic achievement effects of nongraded schooling, and to draw inferences from this research for applications of the nongraded ideal in today's schools.

What Is a Nongraded Elementary School?

The term *nongraded* (or ungraded) elementary school covers a wide range of school and classroom organizational arrangements. Central to the concept is the elimination of traditional grade level designations. Students are grouped according to their level of academic performance, not their ages. Sometimes this grouping is done for just one subject, sometimes it is done for many subjects, and sometimes students are placed in self-contained multiage classrooms according to their reading performance or general ability. For example, a nongraded reading class might contain six-, seven-, and eight-year-old students, all reading at what would ordinarily be considered the second-grade level. Students are allowed to proceed through the grades at their own rates. Some may take longer than usual to complete the elementary grades, while others may complete elementary school in less time than usual. Because a school has classes at many levels, a child who spurts ahead or falls behind can easily be moved to another class appropriate to his or her level. As a result, no child is ever retained or skipped a whole grade at once.

Frequently, the nongraded program applies only to the primary grades (1-3 or K-3) and is called a *nongraded primary school*. This is the main form that is returning today (Willis, 1991). The idea is that all students will have a certain level of academic performance on entering fourth grade but may have taken more or less time to reach this level.

In their original conception, nongraded elementary schools usually incorporated a curriculum structure called *continuous progress,* in which the skills to be learned in such subjects as reading and mathematics are organized into a hierarchical series of levels covering all the grades involved in the plan (usually 1-3 or 1-6). For example, the reading curriculum ordinarily taught in grades 1-3 might be organized in four levels per grade, for a total of 12 levels for the entire nongraded period leading to grade four. In a continuous-progress model, students pick up each year where they left off the previous year. For example, a low achiever who has only completed Level 5 at the end of his or her second year would start with Level 6 at the beginning of his third year (rather than being retained in second grade, as might occur in a graded school). A high achiever who mastered Level 12 at the end of his or her second year might simply go to fourth grade a year early. Goodlad and Anderson (1963) recommended having such a hierarchical structure only for hierarchical subjects (such as reading and math) and using interest groupings, age groupings, or other criteria for subjects (such as social studies and science) that depend to a lesser degree on prerequisite skills.

Beyond the use of flexible, multiage groupings, actual operationalizations of the nongraded elementary school model have varied enormously. At one end of a continuum of complexity, nongraded organization is essentially equivalent to the Joplin

Plan (Floyd, 1954; Slavin, 1987). This is an arrangement in which students are grouped across grade lines for just one subject, almost always reading. For example, at a common reading period all students might move to a class composed of students at the same performance level in reading drawn from different classes and grade levels; a second grade, first-semester reading class might have first, second, and third graders in it. Students move through a continuous-progress sequence of reading levels that cover the material students are expected to learn in all grades involved in the plan. They move as rapidly as they are able to go, taking as much time as they need to master the material. Groupings are reassessed frequently and changed if student performance warrants it.

The main effect of the use of the Joplin Plan is to reduce the number of reading groups taught by each teacher, often to one (i.e., whole-class instruction), thereby reducing the difficulties inherent in managing multiple groups and reducing the need for students to do follow-up activities independently of the teacher.

The Joplin Plan can be described as a nongraded reading program that still maintains an age-graded organization for other subjects. Studies of the Joplin Plan, which was popular in the 1950s and 1960s, do not make it clear what happened when students reached the end of the elementary grades and were reading at a level quite different from their grade level.

In the 1960s, nongraded programs began to resemble more closely the model described by Goodlad and Anderson (1963), which suggested flexible multiage grouping for most or all academic subjects, with continuous-progress curricula for such subjects as reading and mathematics.

When it was first described and implemented in the 1950s and early 1960s, nongraded organization primarily involved changes in grouping patterns, not instructional methods. Teachers in the earlier implementations still overwhelmingly taught students in groups using traditional methods and curricula. Starting in the late 1960s, however, the nongraded plan often absorbed another innovation becoming popular at that time, individualized instruction. Increasingly, descriptions of nongraded schools began to include the extensive use of learning stations, learning activity packets, and other individualized, student-directed activities. In many cases, these individual activities were also combined with tasks students completed in small groups which primarily worked independently of the teacher. Another typical attribute of these forms of nongrading was team teaching. For example, two to six teachers might occupy a section of the school and take joint responsibility for a large group of students, flexibly grouping and regrouping them throughout the day. As time went on, programs of this kind were increasingly implemented in schools without classroom walls and tended to be called open schools rather than nongraded elementary schools (see Giaconia & Hedges, 1982), and, in an introduction to the 1987 reprinting of their 1963 book, Goodlad and Anderson acknowledge the essential commonality between the two approaches.

A good summary of the goals and elements of a fully realized operationalization of the nongraded ideal is adapted by Goodlad and Anderson (1963/1987, pp. *xv-xviii*) from the dissertation of Barbara Pavan (1972), who was the principal of one of the earliest model nongraded elementary schools and continues to be an important advocate of this approach. It is presented below.

I. Goals of Schooling

1. The ultimate school goal is to develop self-directing autonomous individuals.
2. The school should help develop individual potentialities to the maximum.

3. Each individual is unique and is accorded dignity and respect. Differences in people are valued. Therefore the school should strive to increase the variability of individual differences rather than stress conformity.
4. Development of the child must be considered in all areas: aesthetic, physical, emotional, and social, as well as intellectual.
5. Those involved in the school enterprise are colearners, especially teachers and students.
6. The school atmosphere should allow children to enjoy learning, to experience work as pleasurable and rewarding, and to be content with themselves.

II. Administrative-Organizational Framework

A. Vertical Grouping
7. Each individual works in varied situations where he or she will have opportunities for maximum progress. There are no procedures for retention or promotion, nor any grade levels.
8. A child's placement may be changed at any time if it is felt to be in the best interests of the child's development considering all five phases of development: aesthetic, physical, intellectual, emotional, and social.

B. Horizontal Grouping
9. Grouping and subgrouping patterns are extremely flexible. Learners are grouped and regrouped on the basis of one specific task and are disbanded when that objective is reached.
10. Each child should have opportunities to work with groups of many sizes, including one-person groups, formed for different purposes.
11. The specific task, materials required, and student needs determine the number of students that may be profitably engaged in any given educational experience.
12. Children should have frequent contact with children and adults of varying personalities, backgrounds, abilities, interests, and ages.

III. Operational Elements

A. Teaching Materials – Instructional
13. A wide variety of textbooks, trade books, supplemental materials, workbooks, and teaching aids must be available and readily accessible in sufficient quantities.
14. Varied materials must be available to cover a wide range of reading abilities.
15. Alternate methods and materials must be available at any time so that the child may use the learning style and materials most suitable to his or her present needs and the task at hand (including skill building, self-teaching, self-testing, and sequenced materials).
16. A child is not really free to learn something he or she has not been exposed to. The teacher is responsible for providing a broad range of experiences and materials that will stimulate many interests in the educational environment.

B. Curriculum (Knowledge)
17. The unique needs, interests, abilities, and learning rates, styles, and patterns of each child will determine his or her individual curriculum. Conformity and rigidity are not demanded.
18. The curriculum should be organized to develop the understanding of concepts and methods of inquiry more than specific content learning.

19. Process goals will be stressed: the development of the skills of inquiry, evaluation, interpretation, application – the skills of learning to learn.
20. Sequence of learning must be determined by each individual student and his or her teacher, because:
    (a) no logical or inherent sequence is in the various curriculum areas,
    (b) no predetermined sequence is appropriate to all learners,
    (c) individual differences in level of competence and in interest are constantly in flux.
21. Each child will formulate his or her own learning goals with guidance from his or her teachers.

C. Teaching Methods
22. Different people learn in different ways.
23. Learning is the result of the student's interaction with the world he or she inhabits. Individuals learn by direct experience and manipulation of their environment: Therefore, the child must be allowed to explore, to experiment, to mess around, to play, and to have the freedom to err.
24. The process is more important than the product. How the child learns is emphasized.
25. All phases of human growth – aesthetic, physical, intellectual, emotional, and social – are considered when planning learning experiences for a child.
26. The teacher is a facilitator of learning. He or she aids in the child's development by helping each one to formulate goals, diagnose problem areas, suggest alternative plans of action; by providing resource materials; and by giving encouragement, support, or prodding as needed.
27. Children should work on the level appropriate to present attainment and should move as quickly as their abilities and desires allow them to.
28. Successful completion of challenging experiences promotes greater confidence and motivation to learn than fear of failure.
29. Learning experiences based on the child's expressed interests will motivate the child to continue and complete a task successfully much more frequently than teacher-contrived techniques.

D. Evaluation and Reporting
30. Children are evaluated in terms of their past achievements and their own potential, not by comparison to group norms. Expectations differ for different children.
31. Evaluation by teacher and/or child is done for diagnostic purposes and results in the formulation of new education objectives.
32. Evaluation must be continuous and comprehensive to fulfill its diagnostic purpose.
33. A child strives mainly to improve his or her performance and develop potential rather than to compete with others.
34. Teachers accept and respond to the fact that growth patterns will be irregular and will occur in different areas at different times.
35. Individual pupil progress forms are used to record the learning tasks completed, deficiencies that need new assignments to permit mastery, and all other data that will show the child's progress in relation to past achievements and potential or that will help the teacher in suggesting possible future learning experiences for the individual.
36. Evaluating and reporting will consider all five areas of the child's development: aesthetic, physical, intellectual, emotional, and social.

Of course, few nongraded programs have incorporated all of the features identified by Pavan (1972). Those that are most central to the nongrouping concept (and most likely to be implemented in practice under this name) are the ones relating to vertical and horizontal grouping, in particular the abolition of grade levels and of promotion and retention.

Rationales for Nongrading

The major rationale for a nongraded approach is to provide an alternative to both retention and social promotion (i.e., promoting students regardless of performance). In the view of Goodlad and Anderson (1963) and many who followed them (e.g., Shepard & Smith, 1989), retention is harmful to students, is applied inconsistently, and fails to take into account developmental inconsistencies (e.g., late bloomers), especially among young children. A retained child repeats a whole year of content he or she failed to learn the first time. Spending a year failing to learn a body of curriculum and then spending a second year going over the same curriculum seems to be a poor practice for low achievers. Advocates of nongrading would argue that it is far better to allow such students to move more slowly through material with a high success rate and never have to repeat unlearned content. As noted earlier, nongraded elementary programs use a continuous progress plan, in which a hierarchical curriculum (such as reading or mathematics) is divided into some number of units across the grades, and then students can take as much time as they need to complete the units. A low achiever moving slowly through a continuous progress curriculum may take as many years to reach the fourth grade as a similar low achiever in a graded structure who is retained at some point in grades 1-3, but advocates of nongrading would argue that the continuous progress plan is less stigmatizing, less psychologically damaging, and more instructionally sensible than retention.

Nongraded organization also offers an alternative to traditional forms of ability grouping. Goodlad and Anderson (1963) point out how nongrading can be an improvement on both between-class ability grouping (e.g., high, middle, and low self-contained second grades) and within-class ability grouping (e.g., reading groups). The problem with between-class ability grouping, they argue, is that grouping on any one criterion (such as reading performance or general ability) cannot group students well for any particular skill. For example, a class grouped according to reading skill will have a very broad range of math levels and will even be quite diverse in performance levels on any particular reading task. As a result, the costs of ability grouping in terms of stigmatizing low achievers are not compensated for by any practically meaningful reduction in heterogeneity. Formation of reading groups within heterogeneous classes is similarly flawed in their view. In order to create homogeneous groups, teachers must have many reading groups, and therefore much class time must be spent on follow-up activities of little instructional value.

In the nongraded plan, students are flexibly grouped for major subjects (especially reading and math) across class and age lines, so that the resulting groups are truly homogeneous on the skills being taught. Further, by creating multiage groups from among all students in contiguous grade levels, it is possible for teachers to create entire reading or math classes at one or, at most, two levels, so that they need not devote much class time to follow-up.

Finally, the nongraded plan is proposed as a solution to the problem of split grades. In many schools with, for example, a class size of 25 and 38 students in each of grades two and three, principals would create one second-grade class, one third-grade class, and one second- and third-grade combination class. In a graded structure, teaching the second- and third-grade class is difficult, as the two portions of the class may be taught completely separately. A nongraded organization, by eliminating the designation of students as second or third graders, solves this problem.

The rationale for the reemergence of the nongraded plan today is similar to that of the 1950s. In the 1980s, retention rates increased dramatically in elementary schools, especially those in large cities (Levine & Eubanks, 1986-1987). This was partly a result of accountability pressures, which focus on the performance of students according to grade level, not age, thereby rewarding districts for such policies as imposing grade-to-grade promotion standards and holding back low achievers (see Allington & McGill-Franzen, 1992; Slavin & Madden, 1991). However, in more recent years, a reaction against high retention rates has taken place, influenced in particular by the work of Shepard and Smith (1989) which documents the negative long-term effects of retention in the elementary grades. Unwilling to return to social promotion (and still under accountability pressures which discourage it), many school districts are currently experimenting with a variety of means of holding standards constant while allowing time spent in the early grades to vary. Among these is the growing use of means of adding a year between kindergarten and second grade for at-risk children – such as, developmental kindergarten, junior kindergarten, transitional first grade, or prefirst programs. However, research on the long-term impacts of these approaches has questioned their value (see Karweit & Wasik, in press). The nongraded primary has been rediscovered as a means of avoiding both retention and social promotion, just as it was in the 1950s.

Another rationale for the nongraded primary school still important today is a reaction against traditional ability grouping. Between-class ability grouping (e.g., high, middle, and low second grades) has been used by a minority of elementary schools, but use of reading groups has been almost universal until very recently (McPartland, Coldiron, & Braddock, 1987). At present, many schools are seeking alternatives to the use of set reading groups (see Barr, 1990), and the nongraded program appears to be a means of doing away with reading groups while still allowing teachers to accommodate instruction to individual needs.

An important factor today in the move toward the nongraded primary that was not a rationale in the 1950s is the trend toward "developmentally appropriate" practices in the early grades. Developmentally appropriate practices are instructional approaches that allow young children to develop skills at their own pace. For example, the National Association for the Education of Young Children (1989) published a position statement, entitled *Appropriate Education in the Primary Grades*, that described developmentally appropriate education for children ages 5-8. Among the prescriptions were the following:

> Each child is viewed as a unique person with an individual pattern and timing of growth.... Children are allowed to move at their own pace in acquiring important skills.... For example, it is accepted that not every child will learn how to read at age 6, most will learn by 7, and some will need intensive exposure to appropriate literacy experiences to learn to read by 8 or 9. (p. 4)

The NAEYC position paper also supported integrated curriculum and instruction, extensive use of projects and learning stations, cooperative learning, and other strategies quite consistent with the nongraded primary plans of the late 1960s and early 1970s (and with the open classroom of the same period). A book by Katz, Evangelou, and Hartman (1991), published by NAEYC, makes a case for mixed age grouping that emphasizes developmentally appropriate activities and downplays grouping by ability or performance level.

Individually Guided Education

One important outgrowth of the nongraded concept was Individually Guided Education (IGE), developed and researched by Klausmeier and his colleagues at the University of Wisconsin (Klausmeier, Rossmiller, & Saily, 1977). IGE, in its Wisconsin version or in the one developed by the Kettering Foundation (through I/D/E/A), was a very ambitious, comprehensive restructuring of elementary education. It used a nongraded grouping strategy, in which students were flexibly grouped according to instructional needs rather than age. As in any nongraded elementary school, students could take as much time as they needed to complete the objectives prescribed for each subject. However, IGE affected all aspects of school organization and instruction, not only grouping. Individual plans were prepared for each student, and students were constantly assessed to determine their continuing placements. Instruction could be delivered one-on-one by teachers or peers, to small groups, or (rarely) to large groups. Extensive use was made of learning stations at which students could perform experiments, work on individualized units, or do other individual or small-group activities independently of the teacher. Comprehensive instructional models were developed and implemented in reading, mathematics, social studies, and science. Students were organized into multiage Instruction and Research (I & R) units of 100 to 150 students with (ideally) a unit leader, three to five staff teachers, an aide, and a teacher intern. This team planned and carried out the instruction students received in all subjects. Often, individual teachers would become experts in a given subject and take responsibility for that subject with the entire unit. A building-level Instructional Improvement Committee worked to establish objectives and policies for the school as a whole.

Review Methods

This review synthesizes the findings of research comparing the achievement effects of nongraded and traditional organizations in the elementary grades (K-6). The review method used is best evidence synthesis (Slavin, 1986), which combines elements of meta-analysis (Glass, McGaw, & Smith, 1981) with those of narrative reviews. Briefly, a best evidence synthesis requires locating all research on a given topic, establishing well-specified criteria of methodological adequacy and germaneness to the topic, and then reviewing this "best evidence" with attention to the substantive and methodological contributions of each study. Whenever possible, study outcomes are characterized in terms of effect sizes, the difference between the experimental and control means divided by the control group's standard deviation. Details of the review procedures are described in the following sections.

Literature Search Procedures

Every effort was made to obtain every study ever reported that met the broad substantive inclusion criteria described below. Principal sources included the Education Resources Information Center (ERIC), *Dissertation Abstracts International,* and the reference lists of earlier reviews and of the primary studies themselves. Most of the studies located were doctoral dissertations. These were obtained from University Microfilms International in Michigan or from the Library of Congress in Washington, DC, which maintains microfilm copies of all U.S. dissertations. In a few cases where unpublished documents could not be found or where clarifications were needed about important studies, authors were contacted directly.

Substantive Inclusion Criteria

Studies were included in an initial search if they could be identified as evaluating nongraded, ungraded, multiage, or Individually Guided Education programs in grades K-6. Studies spanning elementary and middle grades were included, but only data up to grade six were considered.

Methodological Inclusion Criteria

Studies were included if they met the following methodological criteria, which are identical to those applied in earlier reviews of ability grouping by Slavin (1987, 1990):

1. Some objective measure of achievement was used. Because of their subjective nature, grades were not included as achievement variables. In practice, all achievement outcomes were assessed using standardized measures.
2. Initial comparability of the nongraded and graded samples was established by means of random assignment of students, matching of schools or classes, or matching of individual students within classes or schools. In studies using matching, evidence had to be presented to indicate that either the groups were initially equivalent (within 20% of a standard deviation) or that they were not equivalent in which case pretest data had to be presented to allow for adjustment of posttest scores for pretest differences. Studies that used gain scores or analyses of covariance to control for initial differences between nongraded and graded programs are listed in separate portions of each table, as statistical adjustment for pretest differences cannot be assumed to completely control for their influence on posttests (see Reichart, 1979). Results of these studies should be interpreted cautiously.
3. The nongraded program was in place for at least a semester. All studies located met this standard; in fact, only two studies were less than a year in duration.

Very few studies which used any achievement measure to compare nongraded and graded programs were excluded on the basis of these inclusion criteria. Examples of studies excluded are ones which involved nongraded secondary schools (e.g., Chalfant, 1972); studies without any evidence that nongraded and control groups were initially equivalent and without adjustments for pretests (e.g., Ingram, 1960); and studies of school organization plans related to, but not the same as, the nongraded program (e.g., Heathers, 1967; Maresh, 1971).

Studies were not excluded if they met the above criteria but failed to present data that would allow for computation of effect sizes. Instead, such studies were discussed in the review and were included in all tables with an indication of the direction and statistical significance of any differences (see below).

Computation of Effect Sizes

Whenever possible, effect sizes were computed in a manner similar to earlier reviews of ability grouping by Slavin (1987, 1990). In general, effect sizes were computed as the difference between the nongraded and graded programs' means divided by the graded program's standard deviation. When reports omitted means or standard deviations, effect sizes were estimated from ts, Fs, ps, or other statistics, using procedures described by Glass, McGaw, and Smith (1981). However, one important departure from the Glass et al. procedures was used when appropriate. If pretest scores were available, posttests were adjusted for them using ANCOVA or raw gain scores. However, denominators in the effect-size computations were always unadjusted individual-level posttest standard deviations. The purpose of these procedures was to avoid situations in which one treatment exceeded another at pretest and posttest to the same degree yet the posttest difference was coded as meaningful. See Slavin (1987) for more on this adjustment procedure and other details of effect size computation.

For some purposes, effect sizes were pooled across studies. Whenever this was done, medians (not means) were computed on all studies from which effect-size estimates could be derived. Pooling effect sizes within well-defined categories of studies can provide a useful summary of the size and direction of effects, but the pooled estimate should always be evaluated in light of the methodological quality and the consistency of results of the individual studies narratively described in the text.

Categories of Nongraded Programs

As noted earlier, nongraded elementary schools have varied widely in their particulars. This variation is not surprising, given that the original conception of the nongraded idea did not pretend to touch on all aspects of school organization and instruction:

> Nongrading is a scheme for organizing schools vertically. It does not account for the many problems of organizing schools horizontally. (Goodlad & Anderson, 1963, p. 210)

In looking at studies of nongraded elementary schools over time, an interesting pattern emerges. The earlier studies tended to apply nongrading to only one subject, usually reading. As time went on, studies began to include more than one subject but still to maintain traditional curriculum and instruction; later still, nongraded programs began to incorporate much more radical changes in curriculum and instruction, along with increased use of team teaching, individualized instruction, learning stations, peer tutoring, cooperative learning, and so on. Individually Guided Education (IGE) represented a full flowering of this form of nongrading.

It is possible to distinguish four distinct categories of nongraded programs, and this review considers each type separately. These are as follows:

1. *Nongraded Programs Involving Only One Subject (Joplin-Like Programs)*
   Nine studies, all reported in the 1950s or 1960s, evaluated nongraded plans that only involved one subject. The subject was reading in eight studies, math in one.
2. *Nongraded Programs Involving Multiple Subjects (Comprehensive Programs)*
   Fourteen studies, reported from the late 1950s or 1960s to the early 1980s, evaluated nongraded plans incorporating two or more subjects (and often including all academic subjects). This category adheres most closely to the original conception put forward by Goodlad and Anderson (1963), in that the nongraded programs emphasize continuous progress and flexible, multiage grouping but do not emphasize individualized instruction.
3. *Nongraded Programs Incorporating Individualized Instruction*
   Eleven studies, all but one reported in the brief period from 1969 to 1973, evaluated nongraded programs that emphasized individualized instruction, learning stations, learning activity packages, programmed instruction, and/or tutoring.
4. *Individually Guided Education (IGE)*
   Ten studies evaluated implementations of Individually Guided Education (IGE), described earlier. This was the latest group of studies, with reports appearing over the period from 1972 to 1985.

5. *Studies Lacking an Explicit Description of the Nongraded Program*
In addition to the four categories discussed above, 12 studies failed to state what was actually implemented in the nongraded programs they evaluated. These were generally ex post facto studies, often with large samples, in which the researchers simply accepted principals' words that their schools were nongraded. Given the considerable diversity among implementations that were described, it would be foolish to assume anything about what the independent variable in these studies really was. However, this category is included for the sake of completeness.

Research on Nongraded Programs

The following sections discuss the research on each of the categories of nongraded programs described above. The five sections contain tables summarizing the major characteristics and findings, first of randomized studies, then matched equivalent studies, and finally matched studies lacking evidence of initial equality. Within these categories, the larger studies are listed first. The text usually follows the same order. In general, then, studies listed and discussed earlier in each section can be considered higher in methodological quality than those that come later.

In each table, effect sizes are presented for each measure or subgroup used, and then an overall effect size is presented. Asterisks by effect sizes indicate that the differences were statistically significant, according to the authors. When effect sizes could not be computed, outcomes are characterized as favoring nongraded (+), no difference (0), or favoring graded (–), with asterisks if the differences were significant. A key to all symbols and abbreviations used in all tables appears in an appendix.

Joplin-Like Nongraded Programs

Table 1 summarizes the research on nongraded programs that have as a distinctive feature the homogeneous grouping of students according to performance level in only one subject. These plans can be labeled Joplin-like programs because they share with the Joplin Plan the idea of regrouping students for just one subject (usually reading), ignoring grade levels or ages. Nine studies, all done during the 1960s, are included in this category. Most of them were described under the Joplin Plan arrangement in an earlier synthesis of ability grouping and student achievement in elementary schools (Slavin, 1987). These studies appeared early in the nongraded movement, suggesting that the earlier implementations were more conservative (affecting only one subject) than those which appeared later.

Results from five of the nine studies found strong positive effects for the nongraded plans, three studies reported no differences between them and graded plans, and one significantly favored the graded program.

Jones, Moore, and Van Devender (1967) randomly assigned students and teachers to nongraded or traditional classes. Students in the nongraded classes were assigned to heterogeneous classes but regrouped across grade levels for reading. They proceeded through nine reading levels and were continually regrouped on the basis of their reading performance. Within each reading class, teachers had multiple reading groups and used traditional basal readers and instructional methods. The results of this study supported the efficacy of the nongraded program. After 3 semesters, reading scores for experimental students on three standardized scales were considerably higher than for control students (ES = +.72, or about +.41 grade equivalents). After 3 years in the program, experimental-control differences had diminished but were still moderately positive (ES = +.33).

Three studies compared Joplin-like nongraded classes to matched control classes and presented evidence of initial comparability. In the largest of these, Halliwell (1963) evaluated a nongraded primary that was virtually identical to the Joplin Plan. Students in first through third grades were regrouped for reading only and remained in heterogeneous classes the rest of the day. Spelling was also included in the regrouped classes for second and third graders. The article was unclear as to whether within-class grouping was used in regrouped reading classes, but there was some indication that reading groups were not used. Results indicated considerably higher reading achievement in nongraded classes than in the same school the year before nongrading was introduced (ES = +.53). Scores were higher for nongraded students at every grade level, but by far the largest differences were for first graders, who exceeded earlier first grade classes by +.94 grade equivalents (ES = +1.22). It is important to note that mathematics achievement, measured at the second- and third-grade levels, also increased significantly more in the nongraded classes than in previous years (ES = +.51). Because mathematics was not part of the nongraded program, this finding suggests the possibility that factors other than the nongraded plan might account for the increases in student academic achievement. However, the author notes that teachers claimed to have been able to devote more time to mathematics because the nongraded program required less time for reading, spelling, and language instruction than they had spent on these subjects in previous years.

A study by Skapski (1960) also evaluated the use of a Joplin-like nongraded organization for reading only. The details of the nongraded program were not clearly described, but it appears that reading groups were not used in regrouped classes and that curricula and teaching methods were traditional. Two comparisons were made. First, the reading scores of students in the nongraded program were compared to the same students' arithmetic scores, on the assumption that because arithmetic was not involved in the nongraded plan any differences would reflect an effect of nongrading. Results of this comparison indicated that second- and third-grade-aged students achieved an average of 1.1 grade equivalents higher in reading than in arithmetic. Further, scores of third graders who had spent 3 years in the nongraded program were compared to students in two control schools matched on IQ. Results indicated that the nongraded students achieved at a much higher level in reading than did control students (ES = +.57) but that there were no differences in arithmetic (which was not involved in the nongraded program). Differences were particularly large for students with IQs of 125 or higher (ES = +.91) but were still quite substantial for students with IQs in the range 88-112 (ES = +.52).

Only one study evaluated the use of a Joplin-like program in mathematics. This was a study by Hart (1962). Experimental students were regrouped for arithmetic instruction across grade lines and were taught as a whole class. Students were frequently assessed on arithmetic skills and reassigned to different classes if their performance indicated that a different level of instruction was needed. Experimental students who had spent 3 years in the nongraded arithmetic program were matched on IQ, age, and SES with students in similar schools using traditional methods. It was not stated whether control classes used within-class ability grouping for arithmetic instruction. Results indicated an advantage of about one half of a grade equivalent for the experimental group (ES = +.46).

Five studies matched Joplin-like nongraded classes with graded ones and dealt statistically with initial differences among students. In a study in Catholic schools in the Archdiocese of St. Louis, Bockrath (1958) analyzed the largest sample. She conducted three studies in one: first, a comparison of the fourth-grade reading test scores of 1953 and 1956 in the 366 schools of the archdiocese (12,450 students); second, the same comparison for a stratified sample of 50 of these schools (3,596 students); and third,

Table 1 Nongraded programs involving only one subject (Joplin-like programs)

Article	Grades	Location	Sample size	Duration of program	Design	Test	Effect sizes By achievement	By subject	Total
Randomized studies									
Jones, Moore, & Van Devender (1967)	2-3	Shamokin, Pennsylvania	52 (26 NG, 26 G) (1 school)	1.5 yrs. 3 yrs. (follow-up)	Students and teachers randomly assigned to NG/Joplin or heterogeneous graded classes for reading only.	SAT, LCRT		Rdg. (1.5 yrs.) +.72* Rdg. (3 yrs.) +.33	+ .33 .33
Matched studies with evidence of initial equality									
Hlliwell (1963)	1-3	New Jersey	295 (146 NG, 149 G) (1 school)	1 yr.	Compared NG/Joplin in reading and spelling to previous yr. heterogeneous grouping in the graded program. Students had comparable IQ at the beginning of the study.	CAT (gr. 1), MAT (gr. 2-3)		Rdg. +.53*	+.53*
Skapski (1960)	3	Burlington, Vermont	110 (3 schools)	3 yrs.	Students matched on IQ. Compared NG/Joplin in reading only to heterogeneous grouping in a graded program	SAT	Su. +.91 Hi. +.48 Av. +.52	Rdg. +.57**	+.57**
Hart (1962)	4	Hillsboro, Oregon	100 (50 NG, 50 G) (1 school)	3 yrs.	Students matched on sex, mental maturity, age, and SES. Compared NG/Joplin in arithmetic only to heterogeneous grouping in a graded program.	CAT		Math +.46-	+.46
Matched studies lacking evidence of initial equality									
Bockrath (1958)	4	Archdiocese of St. Louis	(1974 NG, 1622 G) (50 schools)	3 yrs.	Comparison between 1956 students' reading achievement with 1953 students' scores. IQ used to adjust score medians. Stratified sample by size and locatbn of schools.	CAT	Hi. + Lo. +	Rdg. +.51**	+.51**

Study	Grades	N	Duration	Description	Test	Results	Reading	Effect
Jacquette (1959)	1-6	3517 (1554 NG, 1963 G) (4 schools)	5 yrs.	Schools matched on rdg. achievement and IQ. Pretest used to compute gain scores.	CAT, GPRT		Rdg. +.03	+.03
Moore (1963)	1-2	621 (292 NG, 329 G) (4 schools)	1 yr.	Schools matched on SES and curriculum. Change scores used to control pretest achievement significant differences. Compared NG/Joplin in reading only to conventional graded plans.	MAT	Hi. -.22 Av. -.43*** Lo. -.29	Rdg. -.41**	-.41**
Enevoldsen (1961)	1-3	420 (210 NG, 210 G) (7 schools)	2 yrs. (2 sch.) 3 yrs. (1 sch.)	Schools matched on SES. IQ used as a covariate. Compared NG/Joplin in reading only to graded programs.	CAT		Rdg. 0	0
Kierstead (1963)	3-8	277 (111 NG, 166 G) (2 schools)	1 yr.	Students equated and classified by IQ and pretest. Pretest used to compute gain scores. Compared NG/Joplin in reading only to ability grouping in a graded plan.	ITBS	Hi. -.01 GE Av. +.08 GE Lo. -.14 GE	Rdg. -.02 GE	-.02 GE

Note. A key to the abbreviations used is provided as an appendix to this article.
* p < .05
** p < .01

a 3-year study of one of the archdiocesan schools to examine how the nongraded primary functioned (106 students). In 1956, students had been in the nongraded program for 3 years, while the 1953 pupils had received graded instruction. Besides a 1
point difference in mean IQ, the students differed in entrance age. An effort was made
by the author to adjust the fourth-grade reading score medians in relation to IQ, but
only a narrative description took into account the new entrance age adopted for the
second group of students (a mean increase of 2 months for first-year primary entrants
in 1953). The results clearly favored the nongraded plan (ES = +.51), which was characterized by the creation of flexible skill-level groupings.

The only study of a Joplin-like program that found a clear advantage for a graded
plan was presented by Moore (1963). First- and second-grade students' reading and
arithmetic achievement scores were compared for nongraded and conventionally
graded schools. Following comparable instructional practices, second-grade students'
reading scores in the graded and nongraded schools did not differ significantly (ES =
−.12). However, a substantial negative difference (i.e, favoring the graded program)
was found among first graders' reading scores (ES = −.70). As the author surveyed teachers with respect to which reading material their pupils were using, he found that
first-grade control students were reading approximately one basal reading text higher
than the experimental group. Arithmetic scores did not differ for first-and second-grade
students, but no description is given of the instruction in this subject.

Enevoldsen (1961) did another study that found schools which differed in label
rather than in organizational structure. The graded and nongraded schools chosen
from the same public system had similar nongraded reading programs. Consequently,
no significant difference was found in students' reading achievement. Two additional
studies of relatively low quality, by Jacquette (1959) and Kierstead (1963), reported
effect sizes that were close to zero. Both studies stated that learning levels established
in the nongraded programs followed very closely the sequential pattern of graded reading skills, and as a result few differences in outcomes were found.

Summary and discussion: Joplin-like nongraded programs. Overall, the findings of
methodologically adequate studies of this type of nongraded program were consistent. All studies exhibiting good methodological quality (randomized and matched
studies with evidence of initial equality) found substantial positive results in favor of
the nongraded program. The median effect size for the four best quality studies was
+.50; for all seven studies from which effect sizes could be estimated, it was +.46. The
matched studies lacking evidence of initial equality that do not report positive results
were characterized by similar reading programs; the biggest difference between them
appeared to be their label. Two features were important in almost all of the successful nongraded programs evaluated: flexibility in pupil grouping, with frequent assessment of mastery at each level; and increased amounts of teaching time for the homogeneous instructional groups. Because each teacher had to manage fewer groups, there
was less need for independent follow-up activities, such as worksheets in reading.
Perhaps this last characteristic is one of the most important elements that favors students in a nongraded program: More homogeneous groups allow teachers to define
more specific objectives for instruction, and children receive a greater amount of direct
teaching.

Comprehensive Nongraded Programs

Some studies described plans in which more than one subject was nongraded. These
14 studies, summarized in Table 2, were conducted from the late 1960s to the beginning of the 1980s. Only three evaluations presented small (and nonsignificant) neg-

ative total effect sizes, while 8 of the 10 that presented results favoring the nongraded plans reported statistically significant differences.

Among the eight studies with evidence of initial equality, Brody's (1970) had the largest sample size. It evaluated a nongraded program in which first and second graders had to pass a series of sequential steps in several subjects at 90% mastery and were placed in groups according to their mastery of specific skills (regardless of grade level). Vertical advancement of students was strongly emphasized. At the time of assessment, first graders had been in this program 1 year, and second graders 2 years. Both groups of nongraded students gained significantly more than did students in graded classes (ES = +.20). Effects were particularly large in mathematics (ES = +.73). This study was somewhat flawed by the fact that, before matching, the nongraded students were 5.4 points higher in IQ than their graded counterparts.

The only matched equivalent study to find no differences in achievement between nongraded and graded programs was done by Otto (1969), in a laboratory school at the University of Texas. Unlike many other studies of nongraded plans, this study fully described the nongraded intervention, designed to be a full-scale implementation of the Goodlad and Anderson (1963) nongraded plan. Unfortunately, experimental and control groups did not differ on many elements held to be essential to the nongraded program. Teachers of the nongraded classes did assign students to instructional groups across grade lines, had students use materials suited to different levels, and provided less whole-class instruction than did teachers in the graded program. However, the nongraded classes did not use more subgroups than graded classes and did not reduce the heterogeneity of subgroups. Because the experiment took place in a laboratory school, it may be that control classes were of high quality and control teachers may have used many aspects of nongrading in their classes.

It is interesting to note that two other studies conducted in university laboratory schools by Muck (1966) and Ross (1967) also failed to find differences between nongraded and graded classes. The principal differences between the graded and nongraded programs in the Muck study concern the sequence in which the curriculum was taught and the policy that children in the nongraded program remain with the same teacher for 3 years while three different teachers faced students in the graded plan. These are not key issues in the usual characterization of the nongraded approach.

Perrin (1969) also found slight differences in favor of the nongraded plans (ES = +.11). As he analyzed the data at the end of each year in a 3-year study, it became clear that, as time passed, the results started to differ significantly. Perrin evaluated a nongraded program in which a minimum skills chart was used to trace the progress of each child. These basic skills in reading, language, and arithmetic were divided into levels, and children moved through them at their own paces.

Buffie (1962) compared the scores children obtained during their last year in a graded or nongraded primary program. In the graded plan, pupils worked on the same program at the same time in all subjects except reading. In the nongraded plan, grouping within as well as between classrooms was done in reading, arithmetic, and spelling. Team teaching practiced in the nongraded schools also differentiated the plans. The results favor the nongraded plan (total ES = +.34), especially on the language subtest (ES = +.67).

Another study that pointed out sharp differences in the instructional practices of the two groups compared was done by Guarino (1982). Using an index for nongradedness (Pavan, 1972), he tested the congruence between labels and structures. The main distinction was that grouping in the nongraded program was intended to provide an appropriate level of instruction for all students and was guided by frequently administered diagnostic tests to discover deficiencies in skill areas. High and low achievers in the graded program had problems in receiving instruction appropriate to their

Table 2 Nongraded programs involving multiple subjects (comprehensive programs)

Matched studies with evidence of initial equality

Article	Grades	Location	Sample size	Duration of program	Design	Test	Effect sizes		Total
							By achievement	*By subject*	
Brody (1970)	1-2	Pennsylvania	603 (362 NG, 241 G) (3 schools)	2 yrs.	Students matched on IQ.	SAT	Hi. + Lo. +	Rdg. +.20 Math +.73**	+.46**
Otto (1969)	3-5	Austin, Texas	450 (2 upper middle-class lab schools)	2 yrs.	Students matched on pretest achievement.	MAT, ITBS		Rdg. 0 Math 0	0
Perrin (1969)	1-3	Little Rock, Arkansas	288 (144 NG, 144 G) (13 schools)	3 yrs.	Schools matched on SES. Students matched on IQ, age, sex, and race.	MAT (1-2), ITBS (3)		Rdg. +.08 Math +.14	+.11
Buffie (1962)	3	Cedar Falls, Iowa	234 (117 NG, 117 G) (8 schools)	3 yrs.	Schools matched on SES, enrollment, class size, and teachers' experience. Students matched on sex, age, and intelligence.	ITBS	Hi.+.39 Lo. +.19	Rdg. +.19 Math +.17 Lang. +.67**	+.34**
Guarino (1982)	2-5	New Jersey	162 (81 NG, 81 G) (2 schools)	5 yrs.	Schools matched on SES amd ethnic mix. Students matched on age, sex, and IQ.	CAT		Rdg. +.49** Math +.19	+.34*
Ramayya (1972)	6	Darmouth, Nova Scotia	160 (80 NG, 80 G)	6 yrs.	Students matched on sex, IQ, and SES.	TBS		Rdg. +.41* Math +.25 Lang. +.59*	+.42
Muck (1966)	1-3	Buffalo, New York	148 (1 lab school)	3 yrs.	Students matched on mental maturity.	MAT, ITBS		Rdg. +.04 Math -.36 Lang. +.15	-.06
Machiele (1965)	1	Urbana, Illinois	100 (50 NG, 50 G) (1 school)	1 yr.	Students matched on IQ, mental age, and chronological age. Compared students in NG program to students in previous yr.	CAT		Rdg. +.61** Math +.38	+.49*

Matched studies lacking evidence of initial equality

Study	Grades	Location	N	Duration	Test	Description	Results	Overall
Zerby (1960)	3	Morristown, Pennsylvania	394 (187 NG, 207 G) (2 schools)	3 yrs.	CAT	Schools matched on SES. IQ score used to compute achievement beyond anticipated achievement level.	Rdg. +.10 Math +.57**	+.34*
Chastain (1961)	4-6	Rangey, Colorado	360 (120 NG, 120 G) (1 school)	1 yr.	MAT	Students matched on sex and IQ. Pretest used as a covariate.	Rdg. +.01 Math −.09	−.04
Lawson (1973)	1, 3, 5	Kokomo, Indiana	338 (6 schools)	1, 3, & 5 yrs.	CAT	IQ used as a covariate.	Rdg. +**	+**
Ross (1967)	1-3	Bloomington, Indiana	314 (128 NG, 186 G) (1 lab school)	6 months	MAT	Pretest and IQ used as covariates. Students nonrandomly assigned to NG and G programs in the school.	Rdg. +.06 GE Math + .06 GE	+.06 GE
Morris (1968)	1-3, 5	Montgomery County, Pennsylvania	117 (57 NG, 60 G) (1 school)	3 yrs.	ITBS, SAT	IQ used as a covariate. Compared students in NG program to students in previous year. Intervention stopped after 3 years.	After 3 yrs. +* After 5 yrs. +**	+**
Gumpper, Meyer & Kaufman (1971)	1-4	Pennsylvania	(2 schools)	1 yr.	DLRT, SAT	Schools matched on SES, enrollment, teachers' characteristics, SAT and students' previous academic achievement. Pretest used to compute gain scores.	Gr. 1 Rdg. 0 Math (+) Gr.2-4 Rdg. (−) Math −** Lang. −**	(−)

Note. A key to the abbreviations used is provided as an appendix to this article.
* p < .05
** p < .01

special needs. The standardized scores differed significantly in favor of the nongraded program (total ES = +.34), especially in the reading subtests (ES = +.49) and for the older students.

Ramayya (1972) reported positive results for the sixth-year students in a nongraded school (total ES = + .42). For 6 years, these students attended reading and arithmetic classes that were reorganized into several levels. This study confirms the findings reported by Perrin (1969), Brody (1970), Buffie (1962), and Guarino (1982): The longer the duration of the nongraded program, the greater its favorable impact on student academic achievement.

Among the matched studies lacking evidence of initial equality, the largest was a study by Zerby (1960). Instructional practices were similar in the two programs he compared. Reading and arithmetic texts differed between programs, and the nongraded program provided the students with the same teacher for all 3 years, while different teachers every year instructed children in the graded plan. Despite the resemblance in instructional practices, the results significantly favored the nongraded program (total ES = +.34), especially in arithmetic (ES = +.57).

Lawson (1973) and Morris (1968) conducted studies that had several similarities. Reading and mathematics programs were organized by levels, and regrouping allowed teachers to teach classes of students all at one level in each subject. In the nongraded plans studied, team teaching was described by Lawson, while Morris emphasized the fact that teachers did not face more than two different ability groups per class. Both studies found positive results that increased with time. After 3 and 5 years, significant differences favored the nongraded programs.

After only 1 year of intervention, Chastain (1961) evaluated the academic achievement of students in an intermediate school that shifted from a graded structure to achievement-level grouping in reading classes and finally to a nongraded plan. No differences were found in the reading achievement of students belonging to either plan; a negative difference in arithmetic achievement became smaller in the second year of homogeneous grouping (first year of the nongraded plan).

Another study that evaluated what could be considered a comprehensive nongraded program was conducted by Gumpper, Meyer, and Kaufman (1971). Test scores from children attending the first 4 years of school in a continuous-progress program and in a modified self-contained graded program were compared. Some ability grouping was used for mathematics and reading classes at the same grade level in the control school. Students changed classes several times during each day, breaking some of the atmosphere of the self-contained classroom situation. In the experimental school, children were grouped homogeneously according to achievement for language arts and mathematics. When the nongraded program was introduced, problems with ability grouping occurred, and teachers had to deal with as many as three different levels of children at the same time. The fact that the posttest was administered at the end of the first year of the nongraded program was Gumpper's main explanation for the positive results for first graders and the negative results for second-, third-, and fourth-grade students. Rather than a true difference between graded and nongraded plans, Gumpper believes that the lower achievement gains were more a function of problems of reorientation for older students in the continuous-progress school.

Summary and discussion: Comprehensive nongraded programs. Findings from this group of studies consistently favored the nongraded program. Almost all of its positive results were significant; not one study found significant differences in favor of the graded plan. The median effect size for the matched equivalent studies was +.34, and it was the same for all nine studies from which effect sizes could be estimated. Among those studies that did not report any significant difference, three were conducted in university laboratory schools, and another three found equivalence in the first year of the program but started to see favorable changes in subsequent years. In the case of laboratory schools, control classes were similar to experimental ones, or they appeared to be very high quality classes. Perhaps for these reasons, significant differences did not appear in those circumstances. Across many studies, greater duration of the program was associated with higher positive differences. Other common characteristics of academically successful nongraded plans were subjects organized by levels, use of texts written in accordance with those levels, and regrouping of students in multiage environments that allowed teachers to reduce the heterogeneity of their instructional groups.

Nongraded Programs Incorporating Individualized Instruction

Many studies of nongraded programs included indications that individualized instruction was an important part of the nongraded plan. These individualized approaches included one-to-one tutoring, programmed instruction, and learning activity packages. Two examples of the types of individualization adopted are as follows:

> Most students would be on contracts of work . . . [that] might last from one to five days with the student coming to the teachers only in particular moments of difficulty. (Bowman, 1971, p. 46)

> The Individually Prescribed Instruction mathematics program . . . was used in the model school. This individualized system of instruction provided each student with the opportunity to work on undeveloped skills, to obtain a diagnosis of new learnings, and to receive a prescription for the next sequence of material to be mastered. Math specialists, instructional aides, and volunteer aides were available to pupils on a one-to-one basis. (Jeffreys, 1970, p. 30)

All but one of the 11 studies of this type were published in a brief period from 1969 to 1973, with a median of 1971. This is considerably later than the time frame in which the studies summarized in Table 1 appeared. The median publication date for the Joplin-like programs is 1962, and, for the comprehensive (multiple subjects) programs, it is 1969. What this progression suggests is that individualized instruction increasingly became part of the nongraded elementary school in the late 1960s and early 1970s at a time when individualization was gaining popularity in North American schools in general.

Table 3 summarizes the characteristics and findings of the 11 studies of nongraded programs including individualized instruction. Only one of these, conducted by Higgins (1980), randomly assigned students to nongraded or graded classes. In that study, reading was the only subject of interest. The physical arrangements and instructional practices in both settings were essentially different. For the most part, instruction for children in the nongraded classrooms was individualized. The graded classes were grouped by ability, and discussions took place in each group to check comprehension. The scores obtained by each group of students on the Metropolitan Reading Test did not show any significant difference (ES = +.02).

Sie (1969) studied pupils from second, third, and fourth grades who were in two schools, one with a traditional graded plan and the other with a nongraded plan. The students were matched according to their SAT scores. Both schools shared similar group instruction in the areas of work study skills, social studies, and science. The principal difference between them was that the nongraded school emphasized individualized instruction in reading and arithmetic, while the graded one used some form of

Table 3 Nongraded programs incorporating individualized instruction

Article	Grades	Location	Sample size	Duration of program	Design	Test	Effect sizes By achievement	Effect sizes By subject	Total
Randomized studies									
Higgins (1980)	3-5	Baton Rouge, Louisiana	246 (75 NG, 171 G) (3 schools)	1 yr.	Students randomly assigned to nongraded/combination or traditional reading classes. Pretest used to compute gain scores.	MRT	Hi. (+) Lo. (+)	Rdg. +.02	+.02
Matched studies with evidence of initial equality									
Sie (1969)	2-4	Ames, Iowa	124 (67 NG, 67 G) (2 schools)	1 yr.	Schools matched on SES. Students matched on SAT scores. Pretest used to compute gain scores.	SAT		Rdg. +.03 Math +.14 Lang. –.11	+.02
Jeffreys (1970)	3, 5	Howard County, Maryland	88 (44 NG, 44 G) (2 schools)	1 yr.	Schools matched on SES. Students matched on pretest achievement measure. Pretest scores and parent occup. status used as covariates.	ITBS		Rdg. +.08 Math –.13	–.03
Wilt (1470)	4	Chicago suburb, Illinois	84 (32 NG, 52 G) (2 schools)	4 yrs.	Students matched on IQ and age.	ITBS		Rdg +.49 GE* Math +.10 GE Lang. –.27 GE	+.11 GE
Matched studies lacking evidence of initial equality									
Ward (1969)	1-2	Fort Worth, Texas	797 (376 NG, 421 G) (4 schools)	2 yrs.	Schools matched on SES, race, and available resources. IQ, age, and readiness scores used as covariates.	MAT		Rdg. + Math (+)	+
Burchyett (1972)	3-5	Grand Blanc, Michigan	535 (332 NG, 203 G) (2 schools)	2 yrs.	Schools matched on SES. Pretest used as a covariate.	STEP		Rdg –.06 Mat –.10	–.08
Bowman (1971)	1-6	Burlington, North Carolina	457 (313 NG, 144 G) (2 schools)	1 yr. (2 schools)	IQ used as a covariate. Pretest used to compute gain scores.	MAT		Rdg. +.27* Math +.28*	+.28*

Study	Grades	Location	N	Duration	Description	Test	Results by group	Results by subject	Effect size
Case (1970)	5	Montgomery County, Maryland	269 (131 NG, 138 G) (4 schools)	1 yr.	Schools matched on SES. Students matched on age, sex, race, and SES (higher IQ scores for control group). Pretest used to compute gain scores.	SAT	Hi. +.18 Av. +.14 Lo. – .01	Rdg. +.01 Math +.16	+.09
Killough (1971)	1-8	Houston, Texas	267 (132 NG, 135 G) (4 schools)	3 yrs.	Schools matched on SES and ethnic distribution. IQ used as a covanate. Pretest used to compute gain scores.	SRAAS		Rdg. +* Math +*	+*
Givens (1972)	5	Saint Louis, Missouri	100 (50 NG, 50 G) (1 lab, 1 control school)	1 yr.	Students randomly selected from two populations of students that received either individualized or traditional instruction. Ex post facto experimental design. Pretest used to compute gain scores.	ITBS	Hi. (+) Av. 0 Lo. (+)	Rdg. –.11 Math +.10	.00
Walker (1973)	1-12	Kentucky	96 (32 NG, 64 G)	12 yrs.	Schools rated on an eight dimension scale, the Nongradedness Assessment Scale. Longitudinal study to determine the long term effects of NC and G primary school years (1-3). Rate of progress used as a covariate.	CAT		Rdg. +.24 Math +.14 Lang. +.17	+.18
Snake River School District (1972)	1-3	Blackfoot, Idaho	78 (39 NG, 39 G) (2 schools)	1 yr.	Students matched on SES. Pretest used to compute gain scores.	SRAAS		Rdg. .00 Math .00	.00

Note. A key to the abbreviations used is provided as an appendix to this article.
* p < .05
** p < .01

within-class grouping for reading. Of 24 gain scores computed for the SAT subtests, the nongraded students scored significantly better in one, arithmetic computation, while the graded students performed significantly better on paragraph meaning and language subtests. The overall effect size was near zero.

Jeffreys (1970) evaluated the academic achievement of children in a nongraded program characterized by an open-space building design, team teaching, and individualized instruction with that of children in a traditional, graded plan. In the nongraded school, reading ability levels were used to group pupils for language arts, and skill levels were used to group them homogeneously in science. Students we grouped heterogeneously for social studies and health. In addition to spending more time in individualized and small group settings (math and spelling instruction followed an individualized system), nongraded students were found to initiate verbal interaction with teachers and to be involved in after-school activities a significantly greater number of times. However, no significant differences were found on the Iowa Test of Basic Skills.

Another evaluation of a nongraded program operating in an open area and using team teaching was done by Wilt (1970). The author administered a teacher questionnaire to identify differences between both schools studied with respect to their internal structure, operating procedures, and teacher and student flexibility. Teachers in both schools supported the basic concept of individualized instruction, and it appears that those in the nongraded program used it somewhat more. There is no mention of its use in any specific subject; apparently, it was used whenever the need arose. Consequently, criteria for grouping in the nongraded plan were more diverse (interest, academic achievement, student-teacher relationship) than in the graded program, where homogeneous grouping according to performance level prevailed. Despite the differences in vertical and horizontal organization, students' scores on the Iowa Test of Basic Skills did not differ significantly.

Among the matched studies lacking evidence of initial equality, Ward (1969) investigated the largest sample. He compared the academic achievement of children in four different schools, two of them implementing nongraded programs and two following graded plans. Results favored the experimental group in each of the 72 comparisons (although only 16 of these were statistically significant differences). Ward notes that the nongraded schools differed mainly in the larger amount of time used by their teachers and pupils in reading and arithmetic. The nongraded approach exhibited a more flexible use of time and provided "the kind of 'atmosphere' which is conducive to the individualization of instruction" (Ward, 1969, p. 168).

Burchyett (1972) found the largest differences in favor of the conventional school among the studies reporting use of individualized instruction (total ES = −.08). None of these, however, were statistically significant for all of the three grades studied. He compared children attending a nongraded, multiaged, team-taught school with children attending self-contained classrooms in a graded organization. Unfortunately, the author did not specify which areas of instruction were approached on an individual basis, which were characterized by multiage grouping, and how organizational patterns differed in the schools under study.

One of the two studies in Table 3 reporting significant differences in favor of the nongraded program was a study by Bowman (1971). He compared pupils from first to sixth grade in a conventional graded school with students in a nongraded program from another school. The latter used individualized instruction, team teaching, flexible grouping, and learning centers. Individual work was emphasized in reading and mathematics; contracts for work on individualized units were agreed on by teachers and pupils. Grouping across class and grade lines was the organizational arrangement for all other subjects, although curriculum changes were also undertaken for social studies, music, and art (science was an exception due to the lack of time to plan new units adaptable to a multiage situation). Strong positive effects on the academic achievement of intermediate students were found (ES = +.52), but the nongraded plan did not have similar effects on the academic achievement of primary students (ES = +.06).

Killough (1971) reported another study with significant positive effects of a nongraded plan implemented in an open-space school. In this study, children benefited from being in a nongraded program from first grade through the junior high school years. After 3 years in the program, pupils had significantly higher cognitive achievement gains than control students. Details of the intervention were not described.

Walker (1973) also studied the long-term effects of graded and nongraded primary programs. After rating the degree of nongradedness of each of the six programs he studied, his conclusions after comparing those that truly followed the nongraded principles with all the other plans resembled those of Bowman (1971). Scores began to differ significantly when children remained in a nongraded school after the primary unit. Walker found the greatest gap in academic achievement favoring these children at the fifth-grade level. From his descriptions of the six programs studied, they differed mainly in the graded materials and terminology used in four schools and absent in two. Individualized instruction and grouping across grades were present in most schools, although with varying emphasis in reading and mathematics.

A sharper contrast between self-contained classes and nongradedness with individualized instruction was evident in a study by Case (1970). It compared three conventional elementary schools to one nongraded middle school. The elementary schools used ability grouping and concentrated primarily on the development of reading and mathematics skills (instruction was given to smaller homogeneous groups), while pupils in the nongraded program were encouraged to do independent study in these subjects. The study reported no significant differences between the gain scores of each group of students (total ES = +.09).

Givens (1972) evaluated fifth graders in two different schools. The demonstration school featured team teaching, multiage grouping of pupils, an open space concept, and individualization of instruction. Local universities contributed interns and student teachers to the schools' staff. No further description of the differences in instructional practices was provided. The standardized test scores did not favor any of the schools (total ES = .00). A zero effect size was also obtained in a small study conducted in the Snake River (Idaho) School District (1972).

Summary and discussion: Nongraded programs incorporating individualized instruction. Considered together, the results of research on these nongraded programs were remarkably consistent. No significant differences appeared in most studies. A median effect size of essentially zero (ES = +.02) was found across the nine studies from which effect sizes could be computed. These findings suggest that nongraded programs using individualized instruction were equivalent to graded plans in terms of academic achievement. As the nongraded plans became more complicated in their grouping arrangements, they apparently lost the comparative advantage of Joplin-like or comprehensive nongraded programs.

There is one interesting trend in the data on nongraded programs using individualized instruction: More positive effects were obtained with older than with younger children. It may be that students need a certain level of maturity or self-organizational skills to profit from a continuous-progress program that includes a good deal of independent work. Another indication of this is the observation that the longer the duration of the program, the better the results.

Individually Guided Education (IGE) Programs

Ten studies, most of which were done in the 1970s, met the inclusion criteria for this review. All of these evaluated the University of Wisconsin IGE program, not the Kettering model (Klausmeier, Quilling, Sorenson, Way, & Glasrud, 1971). Since nongradedness is a characteristic of IGE schools, comparisons between them and non-IGE schools cast light on the effects of programs with a stress on individualized instruction. Although the degree of implementation of ME processes varied from one research setting to another, IGE concepts, components, and practices were clearly established by its developers at the University of Wisconsin. As an ideal nongraded plan, the IGE program takes into account individual differences and uses specialized curriculum materials in reading, mathematics, and other subjects. But the IGE program is far more complicated than a usual nongraded program. Most reports do not provide any description of the type of intervention actually experienced by the experimental schools; it is implied that their organization follows the structure set in the implementation guidelines of 1971.

Schneiderhan (1973) did the only randomized study of IGE programs. She compared two experimental programs, Individually Guided Instruction (IGI) and Individually Guided Education (IGE), with a traditionally graded program. Fourth-through sixth-grade students from one school were randomly selected to follow an IGI or a traditional program. Their ITBS scores and the scores from a third group of students in a second school with an IGE program were used in the comparison. Both IGE and IGI programs were characterized by nongradedness, team teaching, multiage grouping, differentiated staffing, and open-space environments. It can be inferred from the short descriptions provided that the only difference between them was the additional components an IGE program has: an Instructional Improvement Committee and a Systemwide Program Committee. There were no consistent significant differences between the academic achievement of the three groups. It is possible that only 1 year in the IGI program after several years (at least 3) in the conventional graded plan could not dramatically affect children's performance.

Price (1977), in the largest matched study lacking evidence of initial comparability, started by measuring the level of implementation of the 35 processes emphasized in IGE schools. Then he compared students' academic achievement in schools that were high implementers of the IGE philosophy with students' achievement in schools that were low implementers. Schools identified as high implementers, compared to low implementers, were associated with either higher or comparable pupil outcomes. Significant differences were found for reading achievement, but these cannot be associated with specific classroom practices and organization because no description of them was provided. The findings of this study are biased by the fact that high implementers are probably better schools than low implementers, and this superiority is likely to be related to student achievement differences.

Five other studies showed no consistent differences among the schools compared. Except for Kuhlman's (1985), none of them carefully described the organizational arrangements adopted by each program. Biernacki (1976) evaluated an IGE multi-unit program implemented by an inner city elementary school. The academic achievement of selected students attending grades three through six from the multi-unit school and from a self-contained graded school was compared, and no significant differences were found (total ES = +.17). Klaus (1981) tried not only to compare achievement of students in IGE and non-IGE schools but also to determine the effect of the number of years spent by a student in each program on his or her performance. His sample consisted of students who attended the same elementary schools and remained in the system through 11th grade. The IGE program ended at the sixth-grade level. Overall student change scores from fourth to sixth grade showed no significant effects of the program (total ES = +.05), and no significant differences were found in the achievement of 11th-grade students. This study reported larger differences favoring the scores of fourth graders in ME schools. These differences became smaller in subsequent grades. In the only study that favors non-IGE schools, Flowers (1977) studied students' academic achievement in open-space IGE schools and in traditional schools. He found no significant differences between their standardized test scores (total ES = −.25). Also, Henn's (1974) comparison of IGE and non-IGE programs did not yield significant differences in the academic achievement of students in the two programs (total ES = +.03). Kuhlman (1985) compared four types of school organization: conventional graded schools, traditional alternative schools (also graded), open alternative schools, and IGE schools. From the author's descriptions, it is clear that the last two types of schools emphasized individualized instruction and used a nongraded approach. In regression analyses, the variable for school organization did not yield a significant coeffficient. Three more studies, besides the one conducted by Price (1977), found significant differences in favor of IGE schools. Bradford (1972) studied an IGE school characterized by multilevel programs for reading, mathematics, and spelling by an effective use of a multimedia approach and community resources for classroom enrichment and by experiences with dramatization as a classroom technique. Preassessment tools were used to assemble students in small groups, in an independent study program, or in a one-to-one relationship. Students' gains were close to statistical significance in reading and significantly greater in mathematics in the IGE school than in self-contained classrooms.

Burtley's (1974) comparison of IGE and traditional programs favored the former with respect to academic achievement (total ES = +.48). The author reported substantial differences in teachers' behavior and in instructional programs between the two plans compared. Teachers in the IGE school assumed more professional responsibilities and engaged in team teaching. Children in this school were exposed to several instructional modes: large or small multiage groups, one-to-one tutoring, independent study, and study in dyads. Another important difference was that an effort was made to avoid teaching subjects in isolation from each other; consequently, skills were frequently reinforced in the IGE program.

Finally, a small study done by Soumokil (1977) reported the largest total effect size (+.80). As in the Price (1977) study, the research started by assessing the differences between an IGE elementary program in one school with a standard program in another. After confirming the character of each school's label, the author proceeded to compare students' standardized test scores. IGE students scored significantly higher.

Summary and discussion: IGE programs. Overall, research findings on IGE schools resemble results obtained by other studies on nongraded programs incorporating individualized instruction (Table 3). The median effect size across six studies from which effect sizes could be computed was near zero (ES = +.11). Nevertheless, four studies reported significant differences in favor of IGE schools, and all of these were evaluations of schools that clearly differ from one another. Schools closer to a full implementation of IGE concepts seemed to supply students with a wider range of instructional possibilities for their specific needs: small groups, one-to-one tutoring, or independent work. This finding supports the argument that selective use of individualized instruction can yield positive results for children's academic performance.

Studies Lacking an Explicit Description of the Nongraded Program

The last group of studies, summarized in Table 5, included six studies dated in the 1960s and six in the 1970s, all of which lacked an explicit description of the type of

Table 4 Individually Guided Education (IGE)

Article	Grades	Location	Sample size	Duration of program	Design	Test	Effect sizes By achievement	Effect sizes By subject	Total
Randomized studies									
Schneiderhan (1973)	4 6	Roseville, Minnesota	484 (206 IGE, 88 IGI, 190 G) (2 schools)	1 yr.	Students randomly selected to individually guided instruction or traditional programs in the same school. IQ amd pretest scores used as covariates.	ITBS	(IGI & control) (IGE & IGI & control)	Rdg. 0, Math 0, Lang. 0 Rdg. (−), Math 0 Lang. (+)	0
Matched studies lacking evidence of initial equality									
Price (1977)	4, 6	Iowa	1081 (637 Hi., 444 Lo.) (14 schools)	3 yrs.	Comparison of high and low implementers of 35 processes employed in IGE. Schools matched on size, SES, and location. IQ used as a covariate.	ITBS		Rdg. +*, Math (+)	+*
Biernacki (1976)	3-6	Toledo, Ohio	479 (174 NG, 305 G) (2 schools)	6 months	Schools matched on SES, race, and similar achievement in grade equivalents in reading and math for students in Grade 6. Students randomly selected from chosen schools. Pretest used to compute gain scores.	MAT		Rdg. +.13, Math +.20	+.17
Klaus (1981)	4-6	LaCrosse, Wisconsin	433 (219 NG, 214 G)	3 yrs.	Pretest used to compute gain scores. IQ used as a covariate.	ITBS, SRAAS		Rdg. +.07, Math +.12, Lang. −.05	+.05
Bradford (1972)	1-3	Detroit, Michigan	394 (299 NG, 93 G) (2 schools)	1 yr.	Students matched on sex, SES, and reading and math achievement. Pretest used to compute gain scores. IQ used as a covariate.	MAT		Rdg. +, Math +*	+*

Burtley (1974)	2-3	Woodberry, Illinois	302 (167 NG, 135 G)	2 yrs.	Schools matched on SES, ethnicity, size, and enrollment. Pretest used to compute gain scores.	MAT	Rdg. +.40* Math +.55*	+.48*
Flowers (1977)	3	Westminster, Colorado	221 (99 NG, 122 G)	3 yrs.	Test school matches available among the remaining schools within the district based on SES. Students classified by SES.	SAT	Rdg. −.25 Lang. −.25	−.25
Kuhlman (1985)	2, 4, 6	Kansas	200 (50 OS, 50 IGE, 100 G)	2 yrs.	Students randomly selected from chosen schools. SES and number of parents used as covariates.	KCT	Rdg. 0 Math 0	0
Soumokil (1977)	3, 5	Columbia, Missouri	102 (2 schools)	2 yrs. (Gr. 3) 3 yrs. (Gr. 5)	Pretest and IQ used as covariates.	ITBS	Rdg. +.79** Math +.80-**	+.80**
Henn (1974)	4	Ohio	(24 schools) (Enrollment: 7,072 NG, 6,958 G)	2 yrs.	Schools matched on SES, available resources, and teachers' qualifications.	MAT, OST	Rdg. +.05 Math +.01	+ 03

Note. A key to the abbreviations used is provided as an appendix to this article.
* p < .05
** p < .01

intervention applied to experimental schools and the characteristics of control schools. Two doubts confront any reader of these reports: To what extent was the nongraded label a good description of practices in the experimental situations, and what characteristics did control schools have that really made them fit a conventional description. The value of these studies was perhaps in putting to rest the idea that simply giving a school an innovative label, in this case *nongraded,* would have had some effect on student learning. These studies were included for the sake of completeness, but little can be learned from them.

In the largest matched equivalent study, Hickey (1962) found that students in nongraded primaries in seven Catholic schools learned significantly more after 3 years than did students in similar graded schools (total ES = +.46).

Among all other studies, only the smallest (Engel & Cooper, 1971) reported positive significant differences (total ES = +1.10). This study also differed from all others in its assurance that the schools compared belonged to extremes in an index of nongradedness.

Two medium-sized matched studies that lacked evidence of initial equality reported significant differences in favor of graded schools. Vogel and Bowers (1972) conducted a 1-year study of students in kindergarten through sixth grade. Pupils in graded schools scored higher on the standarized test (SAT) but significantly lower on measures of conceptual maturity (according to the Harris Draw-A-Person Test).

The second study that found significantly higher achievement in graded schools was also one of the lowest in methodological quality. Carbone (1961) compared the achievement of students in traditional graded schools to that in schools mentioned by Goodlad and Anderson (1959) as nongraded, controlling for IQ scores. The students involved were in grades four, five, and six – which is to say, 1, 2, or 3 years (respectively) after their experience in the nongraded primary. Further, there were substantial IQ differences between the two sets of students, and teacher questionnaires indicated very few differences between the two sets of teachers in reported classroom practices.

Summary and discussion: Programs not adequately described. This was the only group of research studies in which a trend was not evident. The median effect size was near zero (ES = +.02), but two studies found significantly positive effects, and two found significantly negative ones. The most serious limitation of these studies was the lack of descriptions that could have helped to interpret the findings. A closer look at the four studies that presented significant differences made the argument in favor of the nongraded programs more convincing. Both positive studies had greater methodological quality: Both had evidence of initial equality (students were matched on IQ), and Engel and Cooper (1971) even tested the validity of schools' labels. In contrast, each of the negative studies had some serious problems: inconsistency of findings or flaws in the experimental setting.

Interactions With Study Features

In addition to differentiating results according to categories of programs, assessments were made of the interaction of nongraded versus graded organization with other features of the studies. One finding, which buttresses the main conclusions of this review, was that program effects for the Joplin-like and comprehensive models were particularly strong and consistent in the higher quality studies – that is, the randomized and matched equivalent experiments. However, there were no consistent patterns with respect to effects at different grade levels (1-3 vs. 4-6) or subjects (reading, math, or language arts). Longer implementations (more than 1 year) were only inconsistently associated with larger effects.

Program effects declined according to the year of publication of the studies, but this is of course confounded with program types; the Joplin-like and comprehensive models without individualization were mostly studies reported in the 1960s, while most studies of nongraded programs emphasizing individualization, including studies of IGE, were reported in the 1970s.

Does Nongrading Accelerate or Decelerate Student Progress?

One of the principal rationales for nongrading is that it allows students to spend more time in the grades involved, if necessary, until they can reach a high level of performance or to spend less time if they are able to go more quickly than other students. Surprisingly, only one study actually assessed the degree to which nongraded students took nonnormative amounts of time to complete the primary or elementary grades. This was a study by McLoughlin (1970), which compared students in graded and nongraded primary programs in eight New York State school districts. The nongraded programs used flexible cross-class and cross-age grouping, teaching to homogeneous groups, and continuous-progress curricula; they would therefore probably fall into the comprehensive category defined earlier. The comparisons were made in 1964-1965 and again in 1965-1966. In 1964-1965, an average of 4.4% of students took an extra year to complete the primary grades in the nongraded schools; 4.6% took an extra year in the graded ones. In 1965-1966, 2.9% of nongraded students took an extra year, while 7.3% of graded students were retained. No students were accelerated in either type of program in 1964-1965, and 1/10 of one percent were accelerated in the nongraded schools in 1965-1966.

Put another way, 95.6% of the nongraded students made normative progress through the primary grades in 1964-1965, and 97.0% in 1965-1966. What this means is that, at least in the time and places studied by McLoughlin (1970), nongrading was not being used as a means of altering the amount of time students spent in the primary grades. On the contrary, in 1965-1966 slightly more students were decelerated by retention in the graded schools than the number that took the opportunity to spend more time in the primary grades offered by the nongraded structure.

McLoughlin (1970) also checked to see whether schools that had been implementing the nongraded plan over a longer period had more students who made nonnormative progress than newer nongraded programs. There was a slight (nonsignificant) trend, but it was in the opposite direction. First-year nongraded programs had somewhat more students making nonnormative progress than did schools that had implemented nongrading for 2-7 years.

If McLoughlin's findings (1970) apply to other implementations of the nongraded concept, this result has important methodological and substantive consequences. Methodologically, there is a concern in studies of nongrading that, if nongraded students take more time to complete the primary grades, their test scores will be artificially increased. That is, if third-year students in a nongraded school were older on average than third graders in a graded school, this could explain any test-score advantage of the nongraded programs. It would be important to know this is not the case. Substantively, McLoughlin's findings (1970) may be seen as questioning one assumption of many advocates of nongrading, who often paint a picture of the low-achieving child proceeding happily and successfully through the grades, never particularly aware that he or she is taking 4 years to accomplish what his or her classmates are completing in 3 years. Yet students (and, more particularly, their parents) can count, and they know who their classmates were when they entered school. The pressures to have students make normative progress may be as strong in nongraded programs as in graded programs.

Table 5 *Studies lacking an explicit description of the nongraded program*

Article	Grades	Location	Sample size	Duration of program	Design	Test	Effect sizes By achievement	By subject	Total
Matched studies lacking evidence of initial equality								*By subject*	
Hickey (1962)	3	Diocese of Pittsburgh	1348 (745 NG, 603 G) (14 schools)	3 yrs.	Schools matched on SES. Students matched on IQ.	MAT	Hi. +.31* Av. +.18	Rdg. +.24* Math +.68** Lo. -.01	+.46**
Lair (1975)	3	Richardson, Texas	463 (183 NG, 280 G) (12 schools)	3 yrs.	Random selection of 6 G and 2 NG schools. Students matched on readiness for learning scores.	CLTBS	Su. -.12 Hi. -.21 Av. -.36	Lang. -.09	-.09
Aigner (1961)	4	Bellevue, Washington	428 (214 NG, 214 G)	3 yrs.	Groups equated with the School and College Abilities Total test.	STEP	Hi. 0 Av. 0 Lo. 0	Rdg. (-) Math (-)	(-)
Mycock (1966)	1	Manchester, England	108 (4 schools)	1 yr.	Schools matched on size, resources, and staff ratio and quality. Students matched on age, sex, and intelligence.	NARA, PT		Rdg. 0 Math 0	0
Reid (1973)	4	Alabama	100 (50 NG, 50 G)	3 yrs.	Students matched on age, sex, and mental ability.	SAT	Hi. -.11 Av. -.01 Lo. -.05	Rdg. +.01 Math -.05 Lang. +.01	-.01
Williams (1966)	3	Hammond, Indiana	76 (38 NG, 38 G) IQ.	3 yrs.	Students matched on age, sex, and	SAT	Hi. +1.29* Lo. -1.30*	Rdg. -.46- Math -.23	-.34
Engel & Cooper (1971)	6	Darmouth, Nova Scotia	40 (20 NG, 20 G) (2 schools)	6 yrs.	Schools selected according to an index for nongradedness. Students matched on IQ.	CAT		Rdg. +1.20** Lang. +1.02**	+1.10**
Herrington (1973)	6	Dade County, Florida	951 (16 schools)	1 yr.	Schools randomly selected from SES-ranked lists. Classes randomly selected. Pretest used as covariate.	SAT		Rdg. (+) Math +	(+)
Vogel & Bowers (1972)	K-6	Evanston, Illinois	473 (224 NG, 249 G)	1 yr.	Teachers matched on sex, training, experience, and age level taught. Pretest used as covariate.	SAT		Composite -**	-**
Hopkins, Oldridge, & Williamson (1965)	3-4	Los Angeles County, California	330 (139 NG, 191 G) (4 schools)	3 yrs.	IQ used as covariate.	CRT		Rdg. +.02	+.02
Carbone (1961)	4-6		244 (122 NG, 122 G) (6 schools)	At least 3 yrs.	Schools matched on SES. Casses randomly selected. Students matched on sex and age. IQ used as covariate.	ITBS		Rdg. -** Math -** Lang. -**	-**
Remacle (1970)	5-6	Brookings, South Dakota	128 (64 NG, 64 G) (1 school)	2 yrs. (Gr. 5) 1 yr. (Gr. 6)	Random selection of students in control group. IQ used as covariate.	ITBS		Rdg. +.24 GE Math +.37 GE Lang. +.33 GE	+.31 GE

Note. A key to the abbreviations used is provided as an appendix to this article.
* p < .05
** p < .01

Yet the main thrust of Goodlad and Anderson's rationale (1959) is not affected by McLoughlin's findings (1970). It is still plausible that deceleration in a continuous-progress curriculum is preferable to retention. Further, in a flexible nongraded program it may be that students who would otherwise fall behind can be identified and given extra assistance so that they may catch up with their peers. The nongraded plan might be seen not as a way to give low achievers more time but rather as a way to use time and other resources more flexibly. A student who is not reading at the end of first grade might well be reading at the end of second grade if he or she receives extra help (and does not suffer the humiliation of repeating first grade).

Discussion

As the nongraded elementary plan reappears in schools of the 1990s, it is important to learn about the history of this movement of 30 years ago. Most important, one needs to understand the achievement effects of nongraded organization and to understand the conditions under which achievement was or was not enhanced by this innovation.

A review of research on the nongraded elementary school is particularly needed today because there was little consensus on its effects in its own time. Only two reviewers examined portions of the literature, and they came to opposite conclusions. Pavan (1973, 1977, in press) concluded that the evidence favored the nongraded primary, while McLoughlin (1967) stated that most research showed no differences between graded and nongraded plans. Slavin and I conclude that, when their review methods are applied to a much larger set of studies, the evidence could be interpreted as confirming both Pavan's and McLoughlin's conclusions, contradictary though they are.

Table 6 summarizes the outcomes of the 57 studies that met the inclusion standards. Looking only at the box score of significant and nonsignificant positive and negative findings, one can read the results as supporting either McLoughlin's (1967) negative conclusion or Pavan's (1973, 1977, in press) positive one. McLoughlin argued that, because nonsignificant findings outnumbered significant positive ones, the effects of the nongraded primary were equivocal. Twenty-five years later, the proportions of significantly positive findings are like those he reported; only 20 of the 57 studies were significantly positive. Pavan came to the opposite conclusion in her reviews, noting that significant positive findings far outnumbered significant negative ones. This is also true in the present review; only 3 studies significantly favored graded programs, while 20 favored nongraded ones.

However, the conclusions of the present review, which uses a best evidence synthesis, conform to neither McLoughlin's nor Pavan's conclusions. Instead, the evidence presented here supports a conclusion that the effects of nongraded programs depend on the type of program being implemented. Using median effect sizes rather than box scores, one sees that the positive effects of nongraded organization are most consistent and strongest when the program focuses on the vertical organization of the school and when nongrading is used as a grouping strategy but not as a framework for individualized instruction.

Four categories of nongraded programs were examined, in addition to one group of studies in which the nature of the nongraded program could not be determined. Studies in two of these categories clearly supported the nongraded plans. These are the Joplin-like programs, in which students are grouped across age lines in just one subject (usually reading), and the comprehensive programs, which involve cross-age grouping in many subjects but still rely on teacher-directed instruction. The median effect sizes for studies in these categories were clearly positive (+.46 for Joplin-like pro-

Table 6 Summary of effects by type of nongraded plan

Type of program	Total studies	Significant positive	Nonsignificant positive	No difference	Nonsignificant negative	Significant negative	Median effect size
Joplin-Like	9	4	2	1	1	1	+.46 (7)
Comprehensive	14	8	2	1	3	0	+.34 (9)
Individualized	12	2	6	2	2	0	+.02 (9)
IGE	10	4	3	2	1	0	+.11 (6)
Unspecified	12	2	3	1	4	2	+.01(6)

Note. Number of studies in which an effect size could be computed is presented in parentheses.

grams, +.34 for comprehensive), and the best designed evaluations were the ones most likely to show the positive effects.

In contrast, nongraded programs that incorporated a great deal of individualized instruction (and correspondingly less teacher-directed instruction), including Individually Guided Education (IGE), were less consistently associated with achievement gain. This is not to say that these approaches reduce student achievement; rather, their effects are very inconsistent, generally neither helping nor hurting student achievement, with more studies finding positive than negative effects (especially in the case of IGE). Poorly described nongraded programs also had median effects very near zero, perhaps because experimental and control groups may not have differed in anything essential except label.

What accounts for the relatively consistent positive effects of the Joplin-like and comprehensive nongraded plans and the less consistent effects of programs incorporating individualization? At this remove of time from the flowering of the nongraded ideal, one can only speculate, but there are many more recent developments in educational research that suggest some possibilities.

The most obvious reason that incorporating a great deal of individualization might have reduced the effectiveness of the nongraded elementary school is suggested by research on individualized instruction itself, which has generally failed to support this innovation (e.g., see Bangert, Kulik, & Kulik, 1983; Horak, 1981; Miller, 1976; Rothrock, 1982). Correlational evidence from process-product studies of more and less effective teachers has consistently found that student learning is enhanced by direct instruction from teachers, as contrasted with extensive reliance on individualization, seatwork, and written materials (see Brophy & Good, 1986). Further, to the degree that the nongraded elementary school came to resemble the open school, the research finding few achievement benefits to this approach (e.g., Giaconia & Hedges, 1982) takes on increased relevance.

In its simplest forms, the nongraded elementary school has many likely benefits. By grouping students across age lines, it may allow teachers to reduce the number of within-class reading and math groups they teach at any given time, thereby reducing the need for independent seatwork and follow-up. In fact, in several of the evaluations of Joplin-like programs, it was noted that cross-age groupings made within-class groupings (i.e., reading groups) unnecessary, so teachers could spend the class period teaching the entire class, with no need for seatwork unless they saw a specific need for it.

Another factor in the success of simple nongraded plans is the likelihood that flexible cross-age grouping allows teachers to fully accommodate instruction to the needs of each child in a particular subject while still delivering instruction to groups. Goodlad and Anderson's (1959, 1963) criticism of traditional ability grouping is that it does not truly reduce heterogeneity in the specific skill being taught. Grouping students within classes or within grades (in all but the largest elementary schools) does not provide enough opportunity to have group instruction closely tailored to student needs. Flexible cross-age grouping does provide such an opportunity, so the instructional costs of grouping (in terms of disruption, movement, and stigma for children in low groups) can perhaps be outweighed by the greater opportunity to adapt instruction to the precise needs of students and to continue to adapt to students' needs by examining and changing groupings at frequent intervals (see Slavin, 1987).

If the effectiveness of nongraded organization is due to increased direct instruction delivered at students' precise instructional level, then it is easy to see how a move to greater individualization would undermine these effects. Individualized instruction, learning stations, learning activity packets, and other individualized or small group activities reduce direct instruction time with little corresponding increase in appropriateness of instruction to individual needs (in comparison to the simpler cross-age grouping plans).

It is difficult to assess the impact of one of the key rationales given for the nongraded plan throughout its history, the opportunity to allow at-risk students to take as much time as they need to complete the primary or elementary grades without the use of retention. An early study by McLoughlin (1970) found that self-described nongraded programs did not generally take advantage of the opportunity to let low achievers take more time, but one does not know if McLoughlin's findings would apply to most nongraded programs implemented now or in the past. Clearly, however, the effectiveness of the simpler nongraded programs does not depend on the opportunity to accelerate or decelerate student progress, since most studies found positive effects in the first year of implementation, before any acceleration or deceleration could take place.

This discussion is, as noted earlier, completely speculative. There is much more we would have liked to know about how nongraded programs were actually implemented in the 1950s, 1960s, and 1970s. The return of the nongraded idea in the 1990s may, however, answer many questions. But assessments of current forms of nongrading as well as component analyses are necessary to understand which elements of nongrading account for the program's effects, and studies combining qualitative and quantitative methods are necessary to understand both what really changes in nongraded schools and what differences these changes make in student achievement.

Is Earlier Research on the Nongraded Elementary School Relevant Today?

How relevant is research on the nongraded elementary school to education today? Many of the problems that the nongraded elementary school was designed to solve still exist, and the reemergence of nongraded programs appears to be due in large part to concern about these problems, especially the tension between retention and social promotion and rejection of traditional forms of ability grouping. Yet there are also many differences between education today and that of 30 years ago. The general perception that both individualized instruction (e.g., Bangert et al., 1983; Horak, 1981) and the open classroom (e.g., Giaconia & Hedges, 1982) failed in their attempt to increase student achievement means that it is unlikely that the nongraded elementary schools of the 1990s will, like those of the early 1970s, embrace these methods. As a result, it is more likely that the nongraded programs of the 1990s will resemble the simpler forms found in this review to be instructionally effective. Yet there are other developments in North American education today that will certainly influence the forms taken by the nongraded programs, their effects on achievement, and their ultimate impact on educational practice. The movement toward developmentally appropriate early childhood education and its association with nongrading means that the nongraded primary school of the 1990s will often incorporate 4- and 5-year-olds (earlier forms rarely did so) and that instruction in nongraded primary programs will probably be more integrated and thematic, and less academically structured or hierarchical, than other schools. A proposal for nongraded primary programs of this type was recently made by Katz et al. (1991). In other words, like in the early 1970s, the effectiveness of the nongraded school organization plan may become confounded with innovative instructional methods. Whether these instructional methods will have positive or negative effects on ultimate achievement is currently unknown.

The ultimate impact of the nongraded ideal will also have much to do with rapidly unfolding changes in assessment and accountability. One reason for the increase in retention, prefirst, and other extra-year programs in the late 1970s and early 1980s was greatly increased accountability pressures in U.S. schools. Retaining more students has a strong (though short-lived), positive impact on achievement test scores reported by grade (not age), because the children taking the tests are older (see Allington & McGill-Franzen, 1992; Slavin & Madden, 1991). There is currently widespread concern about high retention rates (Shepard & Smith, 1989), yet returning to social promotion would greatly reduce test scores in districts currently retaining many students. If the nongraded elementary school emerges as a means of giving low achievers more time in the elementary grades, it may be favored by the current policies of reporting test scores by grade (for the same reasons that they favor retention). On the other hand, if high-stakes accountability systems begin to report achievement by age (e.g., as does the National Assessment of Educational Progress), this advantage may not become a factor.

Clearly, there is a need for much more research on the nongraded elementary school as it is being implemented today. Because of scientific conventions of the time, most of the earlier research reviewed here was strong in experimental design (most studies used random assignment or careful matching of experimental and control groups) but weak in description of the independent variable – that is, the characteristics of the nongraded and graded schools. Research done today must be strong in both dimensions. Goodlad and Anderson (1963) emphasized that the nongraded plan addresses only vertical organization, not instruction. Yet, as this review has shown, differences in instructional methods between nongraded and graded schools may account for differences (or nondifferences) in outcomes.

Research on the nongraded elementary school offers a fascinating glimpse into the history and ultimate fate of a compelling innovation. The return of this idea after nearly 20 years of dormancy is fascinating as well. This review concludes that the evidence from the first cycle of research on the nongraded elementary school supports use of simpler forms of the model and certainly supports the need and potential fruitfulness of further experimentation. Yet there is a cautionary note in this review as well. Good ideas can be undermined by complexification over time. A constant cycle of experimentation, research, evaluation, revision, and continued experimentation is necessary to build compelling ideas into comprehensive, effective plans for school organization and instruction.

APPENDIX *Abbreviations used in the tables*

+	Results clearly favor nongraded programs
(+)	Results generally favor nongraded programs
0	No trend in results
(–)	Results generally favor graded programs
–	Results clearly favor graded programs
AG	Ability Grouping
CAT	California Achievement Test
CRT	California Reading Test
CTBS	Canadian Test of Basic Skills
CLTBS	California Test of Basic Skills
DLRT	Durrell Listening-Reading Test
GPRT	Gates Primary Reading Test
IGE	Individually Guided Education
IGI	Individually Guided Instruction
ITBS	Iowa Test of Basic Skills
G	Graded Program
GE	Grade Equivalent
KCT	Kansas Competency Test
LCRT	Lee Clark Reading Test
MAT	Metropolitan Achievement Test
MRT	Metropolitan Reading Test
NARA	Neale Analysis for Reading Ability
NG	Nongraded Program
OS	Open-Area School
OST	Ohio Survey Test
PT	Piaget-Type Test
Rdg.	Reading
SAT	Stanford Achievement Test
SRAAS	Science Research Associates Achievement Series
STEP	Sequential Test of Educational Programs

CHAPTER 10

Mastery Learning Reconsidered

Background and Commentary

Mastery learning was probably the dominant instructional innovation of the late 1970's and early- to mid-1980's. The ideas behind mastery learning are very consistent with the talent development philosophy, described in Chapter 1, that guides our work at the Center for Research on the Education of Students Placed at Risk (see, for example, Bloom, 1981). Mastery learning begins with the assertion that all children can learn to high levels, but that some will simply need more time or additional assistance to reach these high levels. Success for All, described in Chapter 4, used procedures that would be familiar to any proponent of mastery learning; first graders who are struggling in reading are given one-to-one tutoring closely tied to the curriculum of their regular reading class, and are frequently assessed to determine their progress toward well-specified standards of excellence. In fact, the philosophical congruence between our work and that of mastery learning proponents led me to follow research on mastery learning very closely, and even to create and evaluate a model combining cooperative learning and mastery learning (see Slavin & Karweit, 1984).

However, in the mid-1980's I began to be concerned that reviews of research on mastery learning were seriously misrepresenting the evidence. I was particularly concerned by article published by Bloom (1984) implying up to "two sigma" effects, which would mean that mastery learning students score two standard deviations higher than controls, and an article by Walberg (1984) claiming average effects of almost one sigma (one standard deviation). These claims were far beyond the effects I was seeing in the many classroom experiments I had been accumulating for some time.

In looking into the mastery learning literature I soon discovered what was going on. There were many studies, primarily done by Bloom's students, that were really intended not to evaluate an instructional method but to test the limits of educability of students at different levels of performance, to validate the idea that with enough time and assistance, all children could learn a given objective. These were essentially laboratory studies of artificial, specifically selected content (typically new material with few if any prerequisite skills or knowledge required), in which enormous amounts of corrective instruction might be given to students who did not meet criterion on a formative test. This corrective instruction was often in the form of one-to-one tutoring by University of Chicago graduate students. The entire study might occupy one to two weeks, or less. The procedures used in these studies could never have been applied in real classrooms over extended time periods; even if limitless funds were available for corrective tutoring, some students would have to spend so much time on each objective that their other subjects would suffer. Further, the control groups in these studies did not receive the additional instructional time or resources received by the mastery learning groups, so total instructional time provided was sometimes two or three times more in experimental groups than control groups.

Clearly, making inferences from these brief laboratory studies to the forms of mastery learning actually being disseminated to educators across the world was completely inappropriate. In my review I simply required a four-week duration for a study to be included. This limitation was fairly arbitrary, but it did provide an easily understood (and replicated) means of separating the laboratory studies, which were uniformly brief, from studies of methods that could in principle have been used over an extended time period.

In writing my review my hope was to counter the excessive claims and encourage a more realistic focus of research on conditions under which mastery learning or related methods might be effective. However, I did not intend to discourage continued research or application of well-developed mastery learning strategies. Whatever my differences with mastery learning as a set of techniques, I have sympathy for its objectives and philosophy and respect for the work done by many mastery learning researchers.

Neither mastery learning nor any other method is a panacea for educational practice. The details and contexts matter. I hope the ideas behind mastery learning will continue to lead to programs capable to accelerating the achievement of all students, even as I stand by my conclusion that the forms of group-based mastery learning evaluated in the 1970's and 1980's did not have the extraordinary effects claimed by their enthusiasts.

Several recent reviews and meta-analyses have claimed extraordinarily positive effects of mastery learning on student achievement, and Bloom (1984a, 1984b) has hypothesized that mastery-based treatments will soon be able to produce "2-sigma" (i.e., 2 standard deviation) increases in achievement. This article examines the literature on achievement effects of practical applications of group-based mastery learning in elementary and secondary schools over periods of at least 4 weeks, using a review technique, "best-evidence synthesis," which combines features of meta-analytic and traditional narrative reviews. The review found essentially no evidence to support the effectiveness of group-based mastery learning on standardized achievement measures. On experimenter-made measures, effects were generally positive but moderate in magnitude, with little evidence that effects maintained over time. These results are discussed in light of the coverage versus mastery dilemma posed by group-based mastery learning.

The term "mastery learning" refers to a large and diverse category of instructional methods. The principal defining characteristic of mastery learning methods is the establishment of a criterion level of performance held to represent "mastery" of a given skill or concept, frequent assessment of student progress toward the mastery criterion, and provision of corrective instruction to enable students who do not initially meet the mastery criterion to do so on later parallel assessments (see Block & Anderson, 1975; Bloom, 1976). Bloom (1976) also includes an emphasis on appropriate use of such instructional variables as cues, participation, feedback, and reinforcement as elements of mastery learning, but these are not uniquely defining characteristics; rather, what defines mastery learning approaches is the organization of time and resources to ensure that most students are able to master instructional objectives.

There are three primary forms of mastery learning. One, called the Personalized System of Instruction (PSI) or the Keller Plan (Keller, 1968), is used primarily at the postsecondary level. In this form of mastery learning, unit objectives are established for a course of study and tests are developed for each. Students may take the test (or parallel forms of it) as many times as they wish until they achieve a passing score. To do this, students typically work on self-instructional materials and/or work with peers to learn the course content, and teachers may give lectures more to supplement than to guide the learning process (see Kulik, Kulik & Cohen, 1979). A related form of mastery learning is *continuous progress* (e.g., Cohen, 1977), where students work on individualized units entirely at their own rate. Continuous progress mastery learning programs differ from other individualized models only in that they establish mastery criteria for unit tests and provide corrective activities for students who do not meet these criteria the first time.

The third form of mastery learning is called *group-based mastery learning,* or Learning for Mastery (LFM) (Block & Anderson, 1975). This is by far the most commonly used form of mastery learning in elementary and secondary schools. In group-based mastery learning the teacher instructs the entire class at one pace. At the end of each unit of instruction a "formative test" is given, covering the unit's content. A mastery criterion, usually in the range of 80-90% correct, is established for this test. Any students who do not achieve the mastery criterion on the formative test receive corrective instruction, which may take the form of tutoring by the teacher or by students who did achieve at the criterion level, small group sessions in which teachers go over skills or concepts students missed, alternative activities or materials for students to complete independently, and so on. In describing this form of mastery learning, Block and Anderson recommend that corrective activities be different from the kinds of activities used in initial instruction. Following the corrective instruction, students take a parallel formative or "summative" test. In some cases only one cycle of formative test-corrective instruction-parallel test is used, and the class moves on even if several students still have not achieved the mastery criterion; in others, the cycle may be repeated two or more times until virtually all students have gotten a passing score. All students who achieve the mastery criterion at any point are generally given an "A" on the unit, regardless of how many tries it took for them to reach the criterion score.

The most recent full-scale review of research on mastery learning was published more than a decade ago by Block and Burns (1976). However, in recent years two meta-analyses of research in this area have appeared, one by Kulik, Kulik, and Bangert-Drowns (1986) and one by Guskey and Gates (1985, 1986). Meta-analyses characterize the impact of a treatment on a set of related outcomes using a common metric called "effect size," the posttest score for the experimental group minus that for the control group divided by the control group's standard deviation (see Glass, McGaw, & Smith, 1981). For example, an effect size of 1.0 would indicate that, on the average, an experimental group exceeded a control group by one standard deviation; the average member of the experimental group would score at the level of a student in the 84th percentile of the control group's distribution.

Both of the recent meta-analyses of research on mastery learning report extraordinary positive effects of this method on student achievement. Kulik et al. (1986) find mean effect sizes of 0.52 for pre-college studies and 0.54 for college studies. Guskey and Gates (1985) claim effect sizes of 0.94 at the elementary level (grades 1-8), 0.72 at the high school level, and 0.65 at the college level. Further, Walberg (1984) reports a mean effect size of 0.81 for "science mastery learning" and Lysakowski and Walberg (1982) estimate an effect size for "cues, participation, and corrective feedback," principal components of mastery learning, at 0.97. Bloom (1984b, p. 7) claims an effect size of 1.00 "when mastery learning procedures are done systematically and well" and has predicted that forms of mastery learning will be able to consistently produce achievement effects of "2 sigma" (i.e., effect sizes of 2.00). To put these effect sizes in perspective, consider that the mean effect size for randomized studies of one-to-one adult tutoring reported by Glass, Cohen, Smith, and Filby (1982) was 0.62 (see Slavin, 1984a). If the effects of mastery learning instruction approach or exceed those for one-to-one tutoring, then mastery learning is indeed a highly effective instructional method.

The purpose of the present article is to review the research on the effects of group-based mastery learning on the achievement of elementary and secondary students in an attempt to understand the validity and the practical implications of these findings. The review uses a method for synthesizing large literatures called "best-evidence synthesis" (Slavin, 1986a), which combines the use of effect size as a common metric of treatment effect with narrative review procedures. Before synthesizing the "best evidence" on practical applications of mastery learning, the following sections discuss the theory on which group-based mastery learning is based, how that theory is interpreted in practice, and problems inherent in research on the achievement effects of mastery learning.

Mastery Learning in Theory and Practice

The theory on which mastery learning is based is quite compelling. Particularly in such hierarchically organized subjects as mathematics, reading, and foreign languages, failure to learn prerequisite skills is likely to interfere with students' learning of later skills. For example, if a student fails to learn to subtract, he or she is sure to fail in learning long division. If instruction is directed toward ensuring that nearly all students learn each skill in a hierarchical sequence, then students will have the prerequisite skills necessary to enable them to learn the later skills. Rather than accepting the idea that differences in student aptitudes will lead to corresponding differences in student achievement, mastery learning theory holds that instructional time and resources should be used to bring all students up to an acceptable level of achievement. To put it another way, mastery learning theorists suggest that rather than holding instructional time constant and allowing achievement to vary (as in traditional instruction), *achievement level* should be held constant and *time* allowed to vary (see Bloom, 1968; Carroll, 1963).

In an extreme form, the central contentions of mastery learning theory are almost tautologically true. If we establish a reasonable set of learning objectives and demand that every student achieve them at a high level *regardless of how long that takes,* then it is virtually certain that all students will ultimately achieve that criterion. For example, imagine that students are learning to subtract two-digit numbers with renaming. A teacher might set a mastery criterion of 80% on a test of two-digit subtraction. After some period of instruction, the class is given a formative test, and let's say half of the class achieves at the 80% level. The teacher might then work with the "nonmasters" group for one or more periods, and then give a parallel test. Say that half of the remaining students (25% of the class) pass this time. If the teacher continues this cycle indefinitely, then all or almost all students will ultimately learn the skill, although it may take a long time for this to occur. Such a procedure would also accomplish two central goals of mastery learning, particularly as explicated by Bloom (1976): to reduce the variation in student achievement and to reduce or eliminate any correlation between aptitude and achievement. Since all students must achieve at a high level on the subtraction objective but students who achieve the criterion early cannot go on to new material, there is a ceiling effect built into the procedure that will inherently cause variation among students to be small and will correspondingly reduce the correlation between mathematics aptitude and subtraction performance. In fact, if we were to set the mastery criterion at 100% and repeat the formative test-corrective instruction cycle until all students achieved this criterion, then the variance on the subtraction test would be zero, as would the correlation between aptitude and achievement.

However, this begs several critical questions. If some students take much longer than others to learn a particular objective, then one of two things must happen. Either corrective instruction must be given outside of regular class time, or students who achieve mastery early on will have to spend considerable amounts of time waiting for their classmates to catch up. The first option, extra time, is expensive and difficult to arrange, as it requires that teachers be available outside of class time to work with the nonmasters and that some students spend a great deal more time on any particular subject than they do ordinarily. The other option, giving enrichment or lateral extension activities to early masters while corrective instruction is given, may or may not be beneficial for these students. For all students mastery learning poses a dilemma, a choice between content coverage and content mastery (see Arlin, 1984a; Mueller, 1976; Resnick, 1977). It may often be the case that even for low achievers, spending the time to master each objective may be less productive than covering more objectives (see, for example, Cooley & Leinhardt, 1980).

Problems Inherent in Mastery Learning Research

The nature of mastery learning theory and practice creates thorny problems for research on the achievement effects of mastery learning strategies. These problems fall into two principal categories: unequal time and unequal objectives.

Unequal time. One of the fundamental propositions of mastery learning theory is that learning should be held constant and time should be allowed to vary, rather than the opposite situation held to exist in traditional instruction. However, if the total instructional time allocated to a particular subject is fixed, then a common level of learning for all students is likely to require taking time away from high achievers to increase it for low achievers, a leveling process that would in its extreme form be repugnant to most educators (see Arlin, 1982, 1984b; Arlin & Westbury, 1976; Fitzpatrick, 1985; Smith, 1981).

To avoid what Arlin (1984b) calls a "Robin Hood" approach to time allocation in mastery learning, many applications of mastery learning provide corrective instruction during times other than regular class time, such as during lunch, recess, or after school (see Arlin, 1982). In short-term laboratory studies, the extra time given to students who need corrective instruction is often substantial. For example, Arlin and Webster (1983) conducted an experiment in which students studied a unit on sailing under mastery or nonmastery conditions for 4 days. After taking formative tests, mastery learning students who did not achieve a score of 80% received individual tutoring during times other than regular class time. Nonmastery students took the formative tests as final quizzes and did not receive tutoring.

The mastery learning students achieved at twice the level of nonmastery students in terms of percent correct on daily chapter tests, an effect size (ES) of more than 3.0. However, mastery learning students spent more than twice as much time learning the same material. On a retention test taken 4 days after the last lesson, mastery students retained more than nonmastery students (ES = .70). However, nonmastery students retained far more *per hour of instruction* than did mastery learning students (ES = −1.17). Similarly, Gettinger (1985) found that students who were given enough time to achieve a 100% criterion on a set of reading tasks achieved only 15.5% more than did students who were allowed an average of half the time allocated to the 100% mastery group.

In recent articles published in *Educational Leadership* and the *Educational Researcher,* Benjamin Bloom (1984a, 1984b) noted that several dissertations done by his graduate students at the University of Chicago found effect sizes for mastery learning of 1 sigma or more (i.e., one standard deviation or more above the control group's mean). In all of these, corrective instruction was given outside of regular class time, increasing total instructional time beyond that allocated to the control groups. The additional time averaged 20-33% of the initial classroom instruction, or about 1 day per week. For example, in a 2-week study in Malaysia by Nordin (1979), an extra period for corrective instruction was provided to the mastery learning classes, while control classes did other school work unrelated to the units involved in the study. A 3-week study by Anania (1981) set aside one period each week for corrective instruction. In a study by Leyton (1983), students received 23 periods of corrective instruction for every 2-3 weeks of initial instruction.

In discussing the practicality of mastery learning, Bloom (1984a, p. 9) states that "the time or other costs of the mastery learning procedures have usually been very small." It may be true that school districts could in theory provide tutors to administer corrective instruction outside of regular class time; the costs of doing so would hardly be "very small," but cost or cost-effectiveness is not at issue here. But as a question of experimental design, the extra time often given to mastery learning classes is a serious problem. It is virtually unheard-of in educational research outside of the mastery learning tradition to systematically allocate an experimental group more instructional time than a control group, except in studies of the effects of time itself. Presumably, any sensible instructional program would produce significantly greater achievement than a control method that allocated 20-33% less instructional time. Studies that fail to hold time constant across treatments essentially confound treatment effects with effects of additional time.

It might be argued that mastery learning programs that provide corrective instruction outside of regular class time produce effects that are substantially greater *per unit time* than those associated with traditional instruction. However, computing "learning per unit time" is not a straightforward process. In the Arlin and Webster (1983) experiment discussed earlier, mastery learning students passed about twice as many items on immediate chapter tests as did control students, and the time allocated to the mastery learning students was twice that allocated to control. Thus, the "learning per unit time" was about equal in both groups. Yet on a *retention* test only 4 days later, the items passed per unit time were considerably higher for the control group. Which is the correct measure of learning per unit time, that associated with the chapter tests or that associated with the retention test?

Many mastery learning theorists (e.g., Block, 1972; Bloom, 1976; Guskey & Gates, 1985) have argued that the "extra time" issue is not as problematic as it seems, because the time needed for corrective instruction should diminish over time. The theory behind this is that by ensuring that all students have mastered the prerequisite skills for each new unit, the need for corrective instruction on each successive unit should be reduced. A few very brief experiments using specially constructed, hierarchically organized curriculum materials have demonstrated that over as many as three successive 1-hour units, time needed for corrective instruction does in fact diminish (Anderson, 1976; Arlin, 1973; Block, 1972). However, Arlin (1984a) examined time-to-mastery records for students involved in a mastery learning program over a 4-year period. In the first grade, the ratio of average time to mastery for the slowest 25% of students to that for the fastest 25% was 2.5 to 1. Rather than decreasing, as would have been predicted by mastery learning theorists, this ratio increased over the

4-year period. By the fourth grade, the ratio was 4.2 to 1. Thus, while it is theoretically possible that mastery learning procedures may ultimately reduce the need for corrective instruction, no evidence from long-term practical applications of mastery learning supports this possibility at present.

It should be noted that many studies of mastery learning do hold total instruction time more or less constant across experimental and control conditions. In discussing the "best evidence" on practical applications of mastery learning, issues of time for corrective instruction will be explored further.

Unequal objectives. An even thornier problem posed by research on mastery learning revolves around the question of achievement measures used as dependent variables. Most studies of mastery learning use experimenter-made summative achievement tests as the criterion of learning effects. The danger inherent in the use of such tests is that they will correspond more closely to the curriculum taught in the mastery learning classes than to that taught in control classes. Some articles on mastery learning experiments (e.g., Kersh, 1970; Lueckemeyer & Chiappetta, 1981) describe considerable efforts to ensure that experimental and control classes were pursuing the same objectives. Many studies administer the formative tests used in the mastery learning classes as quizzes in the control classes; in theory this should help focus the control classes on the same objectives. On the other hand, many other studies specified that students used the same texts and other materials but did not use formative tests in the control group or otherwise focus the control groups on the same objectives as those pursued in the mastery learning classes (e.g., Cabezon, 1984; Crotty, 1975).

The possibility that experimenter-made tests will be biased toward the objectives taught in experimental groups exists in all educational research that uses such tests, but it is particularly problematic in research on mastery learning, which by its nature focuses teachers and students on a narrow and explicitly defined set of objectives. When careful control of instruction methods, materials, and tests is not exercised, there is always a possibility that the control group is learning valuable information or skills not learned in the mastery learning group but not assessed on the experimenter-made measure.

Even when instructional objectives are carefully matched in experimental and control classes, use of experimenter-made tests keyed to what is taught in both classes can introduce a bias in favor of the mastery learning treatment. As noted earlier, when time for corrective instruction is provided during regular class time (rather than after class or after school), mastery learning trades *coverage* for *mastery* (see Anderson, 1985). The overall effects of this trade must be assessed using broadly based measures. What traditional whole-class instruction is best at, at least in theory, is covering material. Mastery learning proponents point out that material covered is not necessarily material learned. This is certainly true, but it is just as certainly true that material *not* covered is material *not* learned. Holding mastery learning and control groups to the same objectives in effect finesses the issue of instructional pace by measuring only the objectives covered by the mastery learning classes. If the control classes in fact cover more objectives, or could have done so had they not been held to the same pace as the mastery learning classes, this would not be registered on the experimenter-made test.

Two studies clearly illustrate the problems inherent in the use of experimenter-made tests to evaluate mastery learning. One is a year-long study of mastery learning in grades 1-6 by Anderson, Scott, and Hutlock (1976), which is described in detail later in this review. On experimenter-made math tests, the mastery learning classes significantly exceeded control at every grade level (mean effect size = +.64). On a retention test administered 3 months later, the experimental-control differences were still substantial (ES = +.49). However, the experimenters also used the mathematics scales from the standardized California Achievement Test as a dependent variable. On this test the experimental-control differences were effectively zero (ES = +.04).

A study by Taylor (1973) in ninth-grade algebra classes – although not strictly speaking a study of mastery learning – nevertheless illustrates the dilemma involved in the use of experimenter-made tests in evaluation of mastery learning programs. At the beginning of the semester, students in the experimental classes were each given a copy of a "minimal essential skills" test and were told that to pass the course they would need to obtain a score of at least 80% on a parallel form of the test. About 3 weeks before the end of the semester, another parallel form of the final test was administered to students, and the final 3 weeks was spent on remedial work and retesting for students who needed it (while other students worked on enrichment activities). At the end of the semester, the final test was given. A similar procedure was followed for the second semester.

Experimenter-made as well as standardized measures were used to assess the achievement effects of the program. On the minimum essential skills section of the experimenter-made test, scores averaged 87.3% correct, dramatically higher than they had been on the same test in the same schools the previous year (55.4%). On a section of the experimenter-made test covering skills "beyond, but closely related to, minimum essentials," differences favoring the experimental classes were still substantial, 44.6% correct versus 29.2%. Differences on the minimum essentials subtest of the standardized Cooperative Algebra Test also favored the experimental group (ES = +.47). However, on the section of the standardized test covering skills *beyond* minimum essentials, the control group exceeded the experimental group *(ES = −.25)*.

The Taylor (1973) intervention does not qualify as mastery learning because it involved only one feedback-corrective instruction cycle per semester. However, the study demonstrates a problem characteristic of mastery learning studies that use experimenter-made tests as dependent measures. Had Taylor used only the experimenter-made test, his study would have appeared to provide overwhelming support for the experimental procedures. However, the results for the standardized tests indicated that students in the control group (the previous year) were learning materials that did not appear on the experimenter-made tests. The attention and efforts of teachers as well as students were focused on a narrow set of instructional; objectives that constituted only about 30% of the items on the broader-based standardized measure. These observations concerning problems in the use of experimenter-made measures do not imply that all studies that use them should be ignored. Rather, they are meant to suggest extreme caution and careful reading of details of each such study before conclusions are drawn.

Methods

This review uses a method called "best-evidence synthesis," procedures described by Slavin (1986a) for synthesizing large literatures in social science. Best-evidence synthesis essentially combines the quantification of effect sizes and the systematic literature search and inclusion procedures of meta-analysis (Glass et al., 1981) with the description of individual studies and methodological and substantive issues characteristic of traditional literature reviews. In order to allow for adequate description of a set of studies high in internal and external validity, best-evidence synthesis applies well-justified a priori criteria to select studies to constitute the main body of the review (see Slavin, 1986b, for an earlier example of this procedure).

This section, "Methods," outlines the specific procedures used in preparing the review, including such issues as how studies were located, which were selected for inclusion, how effect sizes were computed, how studies were categorized, and how the question of pooling of effect sizes was handled.

Literature Search Procedures

The first step in conducting the best-evidence synthesis was to locate as complete as possible a set of studies of mastery learning. Several sources of references were used. The ERIC system and Dissertation Abstracts produced hundreds of citations in response to the key words "mastery learning." Additional sources of citations included a bibliography of mastery learning studies compiled by Hymel (1982), earlier reviews and meta-analyses on mastery learning, and references in the primary studies. Papers presented at the American Educational Research Association meetings since 1976 were solicited from their authors. Dissertations were ordered from University Microfilms and from the University of Chicago, which does not cooperate with University Microfilms.

Criteria for Study Inclusion

The studies on which this review is primarily based had to meet a set of a priori criteria with respect to germaneness and methodological adequacy.

Germaneness. To be considered germane to the review, all studies had to evaluate group-based mastery learning programs in regular (i.e., nonspecial) elementary and secondary classrooms. "Group-based mastery learning" was defined as any instructional method that had the following characteristics:

1. Students were tested on their mastery of instructional objectives at least once every 4 weeks. A mastery criterion was set (e.g., 80% correct), and students who did not achieve this criterion on an initial formative test received corrective instruction and a second formative or summative test. This cycle could be repeated one or more times. Studies were included regardless of the form of corrective instruction used and regardless of whether corrective instruction was given during or outside of regular class time.

2. Before each formative test, students were taught as a total group. This requirement excluded studies of individualized or continuous progress forms of mastery learning and studies of the Personalized System of Instruction. However, studies in which mastery learning students worked on individualized materials as corrective (not initial) instruction were included.

3. Mastery learning was the only or principal intervention. This excluded comparisons such as those in two studies by Mevarech (1985a, 1985b) evaluating a combination of mastery learning and cooperative learning, and comparisons involving enhancement of cognitive entry behaviors (e.g., Leyton, 1983).

Studies evaluating programs similar to mastery learning but conducted before Bloom (1968) described it were excluded (e.g., Rankin, Anderson, & Bergman, 1936). Other than this, no restrictions were placed on sources or types of publications. Every attempt was made to locate dissertations, ERIC documents, and conference papers as well as published materials.

Methodological Adequacy. Criteria for methodological adequacy were as follows.

1. Studies had to compare group-based mastery learning programs to traditional group-paced instruction not using the feedback-corrective cycle. A small number of studies (e.g., Katims & Jones, 1985; Strasler & Rochester, 1982) that compared achievement under mastery learning to that during previous years (before mastery learning was introduced) were excluded, on the basis that changes in grade-to-grade promotion policies, curriculum alignment, and other trends in recent years make year-to-year changes difficult to ascribe to any one factor.

2. Evidence had to be given that experimental and control groups were initially

equivalent, or the degree of nonequivalence had to be quantified and capable of being adjusted for in computing effect sizes. This excluded a small number of studies which failed either to give pretests or to randomly assign students to treatments (e.g., Dillashaw & Okey, 1983).

3. Study duration had to be at least 4 weeks (20 hours). This restriction excluded a large number of brief experiments that often used procedures that would be difficult to replicate in practice (such as providing 1 hour of corrective instruction for every hour of initial instruction). The reason for this restriction was to concentrate the review on mastery learning procedures that could in principle be used over extended time periods. One 4-week study by Strasler (1979) was excluded on the basis that it was really two 2-week studies on two completely unrelated topics, ecology and geometry. The 4-week requirement caused by far the largest amount of exclusion of studies included in previous reviews and meta-analyses. For example, of 25 elementary and secondary achievement studies cited by Guskey and Gates (1985), 11 (with a median duration of 1 week) were excluded by this requirement. However, it should be noted that most of these brief studies would also have been excluded by other criteria, principally use of individualized rather than group-based forms of mastery learning and inclusion of only one class per treatment (see below).

4. At least two experimental and two control classes and/or teachers had to be involved in the study. This excluded a few studies (e.g., Collins, 1971; Leyton, 1983; Long, Okey, & Yeany, 1981; Mevarech, 1985a; Tenenbaum, 1982) in which, because only one teacher taught in each treatment condition, treatment effects were completely confounded with teacher/class effects. Also excluded were a few studies in which several teachers were involved but each taught a different subject (Guskey, 1982, 1984; Okey, 1974, 1977; Rubovits, 1975). Because it would be inappropriate to compute effect sizes across the different subjects, these studies were seen as a set of two-class comparisons, each of which confounded teacher and class effects with treatment effects.

5. The achievement measure used had to be an assessment of objectives taught in control as well as experimental classes. This requirement was liberally interpreted and excluded only one study, a dissertation by Froemel (1980) in which the mastery learning classes' summative tests were used as the criterion of treatment effects and no apparent attempt was made to see that the control classes were pursuing the same objectives. In cases in which it was unclear to what degree control classes were held to the same objectives as experimental classes and experimenter-made measures were used, the studies were included. These studies are identified and discussed later in this review, and their results should be interpreted with a great deal of caution.

Also excluded were studies that used grades as the only dependent measures (e.g., Mathews, 1982; Wortham, 1980). In group-based mastery learning, grades are increased as part of the treatment, as students have opportunities to take tests over to try to improve their scores. They are thus not appropriate as measures of the achievement effects of the program. Similarly, studies that used time on-task as the only dependent measure were excluded (e.g., Fitzpatrick, 1985).

Computation of Effect Sizes

The size and direction of effects of mastery learning on student achievement are presented throughout this review in terms of effect size. Effect size, as described by Glass et al. (1981), is the difference between experimental and control posttest means divided by the control group's posttest standard deviation. However, this formula was adapted in the present review to take into account pretest or ability differences between the experimental and control groups. If pretests were available, then the formula used was the difference in experimental and control *gains* divided by the control group's posttest standard deviation. If ability measures rather than pretests were presented, then the experimental-control difference on these measures, divided by the control group's standard deviation, was subtracted from the posttest effect size (this was necessary in only one case, a study by Cabezon, 1984). The reason for these adjustments is that in studies of achievement, posttest scores are so highly correlated with pretest levels that any pretest differences are likely to be reflected in posttests, correspondingly inflating or deflating effect sizes computed on posttests alone. These adjustments are not precisely those recommended by Glass et al. (1981), who present formulas for dealing with gain scores that rely on knowledge of pre-post correlations (which are rarely reported). However, the adjustment procedures used in the present paper follow Glass et al. (1981) in accounting for pretest differences while preserving the control group's standard deviation as the common metric of effect size. Such procedures as ignoring pretest information or using standard deviations of gain scores as the denominator in computing effect sizes are often seen in meta-analyses but are explicitly rejected by Glass and his colleagues (see Glass et al., 1981, pp. 115- 119).

Because individual-level standard deviations are usually of concern in mastery learning research, most studies that met other criteria for inclusion presented data sufficient for direct computation of effect size. In many studies, data analyses used *class* means and standard deviations, but individual-level standard deviations were also presented. In every case (following Glass et al., 1981) the individual-level standard deviations were used to compute effect sizes; class-level standard deviations are usually much smaller than individual-level *SD*s, inflating effect size estimates. Also, note that the control group standard deviation, not a pooled standard deviation, was always used, as mastery learning often has the effect of reducing achievement standard deviations.

In the few cases in which data necessary for computing effect sizes were lacking in studies which otherwise met criteria for inclusion, the studies' results were indicated in terms of their direction and statistical significance.

Research on Achievement Effects of Group-Based Mastery Learning

What are the effects of group-based mastery learning on the achievement of elementary and secondary students? In essence, there are three claims that proponents of mastery learning might make for the effectiveness of mastery learning. These are as follows:

1. Mastery learning is more effective than traditional instruction *even when instructional time is held constant and achievement measures register coverage as well as mastery.* This might be called the "strong claim" for the achievement effects of mastery learning. It is clear, at least in theory, that if mastery learning procedures greatly increase allocated time for instruction by providing enough additional time for corrective instruction to bring all students to a high level of mastery, then mastery learning students will achieve more than traditionally taught control students. But it is less obviously true that the additional time for corrective instruction is more productive in terms of student achievement than it would be simply to increase allocated time for the control students. The "strong claim" asserts that time used for corrective instruction (along with the other elements of mastery learning) is indeed more productive than time used for additional instruction to the class as a whole. It is important to note that this "strong claim" might not be endorsed by all mastery learning proponents. For example, Bloom (1976, p. 5) notes that the "time costs [necessary to enable four fifths of students to reach a level of achievement that less than one fifth attain

in nonmastery conditions] are typically of the order of ten to twenty percent additional time over the classroom scheduled time." However, Block and Anderson (1975) describe a form of mastery learning that can be implemented within usual time constraints, and in practice corrective instruction is rarely given during additional time. Similarly, it is clear (in theory) that if students who experienced mastery learning are tested on the specific objectives they studied, they will score higher on those objectives than will students who were studying similar but not identical objectives. Further, it is likely that even if mastery learning and control classes are held to precisely the same objectives but the control classes are not allowed to move ahead if they finish those objectives before their mastery learning counterparts do, then the traditional model is deprived of its natural advantage, the capacity to cover material rapidly. A "fair" measure of student achievement in a mastery learning experiment would have to register both coverage and mastery, so that if the control group covered more objectives than the mastery learning group its learning of these additional objectives would be registered. The "strong claim" would hold that, even allowing control classes to proceed at their own rate and even using such an achievement measure, mastery learning would produce more achievement than control methods.

The best evidence for the "strong claim" would probably come from studies in which mastery learning and control classes studied precisely the same objectives using the same materials and lessons and the same amount of allocated time, but in which teachers could determine their own pace of instruction and achievement measures covered the objectives reached by the fastest-moving class. Unfortunately, such studies are not known to exist. However, a good approximation of these experimental design features is achieved by studies that hold allocated time constant and use standardized tests as the criterion of achievement. Assuming that curriculum materials are not specifically keyed to the standardized tests in either treatment, these tests offer a means of registering both mastery and coverage. In such basic skills areas as mathematics and reading, the standardized tests are likely to have a high overlap with the objectives pursued by mastery learning teachers as well as by control teachers.

2. Mastery learning is an effective means of ensuring that teachers adhere to a particular curriculum and students learn a specific set of objectives (the "curricular focus" claim). A "weak claim" for the effectiveness of mastery learning would be that these methods focus teachers on a particular set of objectives held to be superior to those that might have been pursued by teachers on their own. This might be called the "curricular focus" claim. For example, consider a survey course on U.S. history. Left to their own devices, some teachers might teach details about individual battles of the Civil War; others might entirely ignore the battles and focus on the economic and political issues; and still others might approach the topic in some third way, combine both approaches, or even teach with no particular plan of action. A panel of curriculum experts might determine that there is a small set of critical understandings about the Civil War that all students should have, and they might devise a criterion-referenced test to assess these understandings. If it can be assumed that the experts' judgments are indeed superior to those of individual teachers, then teaching to this test may not be inappropriate, and mastery learning may be a means of holding students and teachers to the essentials, relegating other concepts they might have learned (that are not on the criterion-referenced test) to a marginal status. It is no accident that mastery learning grew out of the behavioral objectives/criterion-referenced testing movement (see Bloom, Hastings, & Madaus, 1971); one of the central precepts of mastery learning is that once critical objectives are identified for a given course, students should be required to master those and only those objectives. Further, it is interesting to note that in recent years the mastery learning movement has often allied itself with the "curriculum alignment" movement, which seeks to focus teachers on objec-

tives that happen to be contained in district- and/or state-level criterion-referenced minimum competency tests as well as norm-referenced standardized tests (see Levine, 1985).

The "curricular focus" claim, that mastery learning may help focus teachers and students on certain objectives, is characterized here as a "weak claim" because it requires a belief that any objectives other than those pursued by the mastery learning program are of little value. Critics (e.g., Resnick, 1977) point out with some justification that a focus on a well-defined set of minimum objectives may place a restriction on the maximum that students might have achieved. However, in certain circumstances it may well be justifiable to hold certain objectives to be essential to a course of study, and mastery learning may represent an effective means of ensuring that nearly all students have attained these objectives.

The best evidence for the "curricular focus" claim would come from studies in which curriculum experts formulated a common set of objectives to be pursued equally by mastery learning and control teachers within an equal amount of allocated time. If achievement on the criterion-referenced assessments were higher in mastery learning than in control classes, we could at least make the argument that the mastery learning students have learned more of the *essential* objectives, even though the control group may have learned additional, presumably less essential concepts.

3. Mastery learning is an effective use of additional time and instructional resources to bring almost all students to an acceptable level of achievement (the "extra time" claim). A second "weak claim" would be that given the availability of additional teacher and student time for corrective instruction, mastery learning is an effective means of ensuring a minimal level of achievement for all students. As noted earlier, in an extreme form this "extra time" claim is almost axiomatically true. Leaving aside cases of serious learning disabilities, it should certainly be possible to ensure that virtually all students can achieve a minimal set of objectives in a new course if an indefinite amount of one-to-one tutoring is available to students who initially fail to pass formative tests. However, it may be that, even within the context of the practicable, providing students with additional instruction if they need it will bring almost all to a reasonable level of achievement.

The reason that this is characterized here as a "weak claim" is that it begs the question of whether the additional time used for corrective instruction is the *best* use of additional time. What could the control classes do if they also had more instructional time? However, the "extra time" issue is not a trivial one, as it is not impossible to routinely provide corrective instruction to students who need it outside of regular class time. For example, this might be an effective use of compensatory (Chapter I) or special education resource pullouts, a possibility that is discussed later.

The best evidence for this claim would come from studies that provided mastery learning classes with additional time for corrective instruction and used achievement tests that covered all topics that could have been studied by the fastest-paced classes (e.g., standardized tests). However, such studies are not known to exist; the best existing evidence for the "extra time" claim is from studies that used experimenter-made achievement measures and provided corrective instruction outside of class time.

Evidence for the "Strong Claim"

Table 1 summarizes the major characteristics and findings of seven mastery learning studies that met the inclusion criteria discussed earlier, provided equal time for experimental and control classes, and used standardized measures of achievement.

Table 1 clearly indicates that the effects of mastery learning on standardized achievement measures are extremely small, at best. The median effect size across all seven studies is essentially zero (ES = +.04). The only study with a nontrivial. effect size (ES = +.25), a semester-long experiment in inner-city Chicago elementary schools by Katims, Smith, Steele, & Wick (1977), also had a serious design flaw. Teachers were allowed to select themselves into mastery learning or control treatments or were assigned to conditions by their principals. It is entirely possible that the teachers who were most interested in using the new methods and materials, or those who were named by their principals to use the new program, were better teachers than were the control teachers. In any case, the differences were not statistically significant when analyzed at the class level, were only marginally significant ($p = .071$) for individual-level gains, and amounted to an experimental-control difference of only 11 % of a grade equivalent.

The Katims et al. (1977) study used a specially developed set of materials and procedures that became known as the Chicago Mastery Learning Reading program, or CMLR. This program provides teachers with specific instructional guides, worksheets, formative tests, corrective activities, and extension materials. A second study of CMLR by Jones, Monsaas, and Katims (1979) compared matched CMLR and control schools over a full year. This study found a difference between CMLR and control students on the Iowa Test of Basic Skills Reading Comprehension scale that was marginally significant at the individual level but quite small (ES = +.09). In contrast, on experimenter-made "end-of-cycle" tests the mastery learning classes did significantly exceed control (ES = +.18). A third study of CMLR by Katims and Jones (1985) did not qualify for inclusion in Table I because it compared year-to-year gains in grade equivalents rather than comparing experimental to control groups. However, it is interesting to note that the difference in achievement gains between the cohort of students who used the CMLR program and those in the previous year who did not was only 0.16 grade equivalents, which is similar to the results found in the Katims et al. (1977) and Jones et al. experimental-control comparisons.

One of the most important studies of mastery learning is the year-long Anderson, Scott, and Hutlock (1976) experiment briefly described earlier. This study compared students in grades 1-6 in one mastery learning and one control school in Lorain, Ohio. The school populations were similar, but there were significant pretest differences at the first- and fourth-grade levels favoring the control group. To ensure initial equality in this nonrandomized design, students were individually matched on the Metropolitan Readiness Test (grades 1-3) or the Otis-Lennon Intelligence Test (grades 4-6). In the mastery learning school, students experienced the form of mastery learning described by Block and Anderson (1975). The teachers presented a lesson to the class and then assessed student progress on specific objectives. "Errors . . . were remediated through the use of both large-group and small-group re-learning and review sessions. After every student had demonstrated mastery on the formative test for each unit, the class moved on to the next unit" (Anderson et al., 1976, p. 4).

One particularly important aspect of the Anderson et al. (1976) study is that it used both standardized tests and experimenter-made, criterion-referenced tests. The standardized tests were the Computations, Concepts, and Problem Solving scales of the California Achievement Test. The experimenter-made test was constructed by the project director (Nicholas Hutlock) to match the objectives taught in the mastery learning classes. Control teachers were asked to examine the list of objectives and identify any they did not teach, and these were eliminated from the test.

The results of the study were completely different for the two types of achievement tests. On the experimenter-made tests, students in the mastery learning classes achieved significantly more than did their matched counterparts at every grade level (mean ES = +.64). A retention test based on the same objectives was given 3 months

Table 1 Equal-time studies using standardized measures

Article	Grades	Location	Sample size	Duration	Design	Treatments	Subjects	Effect sizes By group/measure	Total
Elementary									
Anderson et al., 1976	14	Lorain, OH	2 sch.	1 yr.	Students in matched ML*, control schools matched on ability	ML–Followed Block (1971). Control–Untreated	Math		+.04
Kersh, 1970	5	Suburban Chicago	11 cl.	1 yr.	Teachers/classes randomly assigned to ML, control within each school	ML–Corr. inst. included re-teaching, alternative mtls., peer tutoring; formative tests given every 3-4 wks. Control–Untreated	Math	middle cl. (–) lower cl. (+)	0
Gutkin, 1985	1	Inner-city New York	41 cl.	1 yr.	Schools randomly assigned to ML, control	ML–Formative tests given every month Control–Untreated	Reading		+.12
Katims et al., 1977	Upper elem.	Inner-city Chicago	19 cl.	15 wks.	1 ML, 1 cont. class from each of 10 schools. Trts. self-selected or principal-imposed	ML–specific mtls. provided Control–Untreated	Reading		+.25
Jones et al., 1979	Upper elem.	Inner-city Chicago	4 sch.	1 yr.	2 ML schools matched with 2 control schools	ML–specific mtls. provided Control–Untreated	Reading		+.09
Secondary									
Slavin & Karweit, 1984	9	Inner-city Philadelphia	25 cl.	26 wks.	Teachers/classes randomly assigned to ML, control	ML–Formative tests given every 2-3 wks.; Corr. inst. given by teachers. Control–Used same mtls., tests, procedures as ML except for corr. inst. & summative tests	General math	Hi 0 Lo 0	+.02
Chance, 1980	8	Inner-city New Orleans	6 cl.	5 wks.	Students within each of 3 classes randomly assigned to ML or control	ML–Formative tests given every wk. Mast. crit. = 80-90% Control—Used same mtls., tests, procedures as ML	Reading	Hi 0 Av 0 Lo 0	0

* Mastery learning   (+) Nonsignificant difference favoring ML   0 No difference   (–) Nonsignificant difference favoring control

after the end of the intervention period, and mastery learning classes still significantly exceeded control (ES = +.49). However, on the standardized tests, these differences were not registered. Mastery learning students scored somewhat higher than control on Computations (ES = +.17) and Problem Solving (ES = +.07), but the control group scored higher on Concepts (ES = −.12).

The Anderson et al. (1976) finding of marked differences in effects on standardized and experimenter-made measures counsels great caution in interpreting results of other studies that used experimenter-made measures only. In a year-long study of mathematics, it is highly unlikely that a standardized mathematics test would fail to register any meaningful treatment effect. Therefore, it must be assumed that the strong positive effects found by Anderson et al. on the experimenter-made tests are mostly or entirely due to the fact that these tests were keyed to the mastery learning classes' objectives. It may be that the control classes covered more objectives than the mastery learning classes, and that learning of these additional objectives was registered on the standardized but not the experimenter-made measures.

Another important study of mastery learning at the elementary level is a dissertation by Kersh (1970), in which 11 fifth-grade classes were randomly assigned to mastery learning or control conditions for an entire school year. Two schools were involved, one middle-class and one lower class. Students' math achievement was assessed about once each month in the mastery learning classes, and peer tutoring, games, and other alternative activities were provided to students who did not show evidence of mastery. Control classes were untreated. The study results did not favor either treatment overall on the Stanford Achievement Test's Concepts and Applications scales. Individual-level effect sizes could not be computed, as only class-level means and standard deviations were reported. However, class-level effect sizes were essentially zero in any case (ES = −.06). On an experimenter-made criterion-referenced test not specifically keyed to the mastery objectives, the results were no more conclusive; class-level effects slightly favored the control group (ES = −.20). Effects somewhat favored mastery learning in the lower class school and favored the control group in the middle-class school, but since none of the differences approached statistical significance these trends may just reflect teacher effects or random variation.

In a recent study by Gutkin (1985), 41 first-grade classes in New York City were randomly assigned to mastery learning or control treatments. The article does not describe the mastery learning treatment in detail, except to note that monthly formative tests were given to assess student progress through prescribed instructional units. The mastery learning training also included information on classroom management skills, process-product research, and performance-based teacher education, and teachers received extensive coaching, routine feedback from teacher trainers, and scoring services for formative and summative tests. After one year, mastery learning-control differences did not approach statistical significance in Total Reading on the California Achievement Test (ES = +.12). However, effects were more positive on a Phonics subscale (ES = +.36) than on Reading Vocabulary (ES = +.04) or Reading Comprehension (ES = +.15). Phonics, with its easily measurable objectives, may lend itself better to the mastery learning approach than do reading comprehension or vocabulary.

Studies using standardized measures at the secondary level are no more supportive of the "strong claim" than are the elementary studies. A 26-week experiment in inner-city, mostly black, Philadelphia junior and senior high schools assessed mastery learning in ninth grade "consumer mathematics," a course provided for students who do not qualify for Algebra I (Slavin & Karweit, 1984). Twenty-five teachers were randomly assigned to mastery learning or control treatments, both of which used the same books, worksheets, and quizzes in the same cycle of activities. However, instructional pace was not held constant. After each 1-week unit (approximately), mastery learning classes took a formative test, and then any students who did not achieve a score of at least 80% received corrective instruction from the teacher while those who did achieve at that level did enrichment activities. The formative tests were used as quizzes in the control group, and after taking the quizzes the class went on to the next unit.

Results on a shortened version of the Comprehensive Test of Basic Skills Computations and Concepts and Applications scales indicated no differences between mastery learning and control treatments (ES = +.02), and no interaction with pretest level; neither low nor high achievers benefited from the mastery learning model. It is interesting to note that there were two other treatment conditions evaluated in this study, a cooperative learning method called Student Teams-Achievement Divisions (STAD) (Slavin, 1983), and a combination of STAD and mastery learning. STAD classes did achieve significantly more than control (ES = +.19), but adding the mastery learning component to STAD had little additional achievement effect (ES = +.03).

A 5-week study by Chance (1980) compared randomly assigned mastery learning and control methods in teaching reading to students in an all-black, inner-city New Orleans school. Approximately once each week, students in the mastery learning groups took formative tests on unit objectives. If they did not achieve at 80% on three quizzes or 90% on one, they received tutoring, games, and/or manipulatives to correct their errors and had three opportunities to pass. No effects for students at any level of prior performance were found on the Gates-McGinitie Comprehension Test. However, it may be unrealistic to expect effects on a standardized measure after only 5 weeks.

Overall, research on the effects of mastery learning on standardized achievement test scores provides little support for the "strong claim" that, holding time and objectives constant, mastery learning will accelerate student achievement. The studies assessing these effects are not perfect; particularly when mastery learning is applied on a fairly wide scale in depressed inner-city schools, there is reason to question the degree to which the model was faithfully implemented. However, most of the studies used random assignment of classes or students to treatments, study durations approaching a full school year, and measures that registered coverage as well as mastery. Not one of the seven studies found effects of mastery learning that reached even conventional levels of statistical significance (even in individual-level analyses), much less educational significance. If group-based mastery learning had strong effects on achievement in such basic skills as reading and math, these studies would surely have detected them.

Evidence for the "Curricular Focus" Claim

Table 2 summarizes the principal evidence for the "curricular focus" claim, that mastery learning is an effective means of increasing student achievement of *specific* skills or concepts held to be the critical objectives of a course of study. The studies listed in the table are those that (in addition to meeting general inclusion criteria) used experimenter-made, criterion-referenced measures and apparently provided experimental and control classes with equal amounts of instructional time. It is important to note that the distinction between the equal-time studies listed in Table 2 and the unequal-time studies in Table 3 is often subtle and difficult to discriminate, as many authors did not clarify when or how corrective instruction was delivered or what the control groups were doing during the time when mastery learning classes received corrective instruction.

A total of nine studies met the requirements for inclusion in Table 2. Three of these (Anderson et al., 1976; Jones et al., 1979; Kersh, 1970) were studies that used both standardized and experimenter-made measures and were therefore also included in Table 1 and discussed earlier.

All but one (Kersh, 1970) of the studies listed in Table 2 found positive effects of mastery learning on achievement of specified objectives, with five studies falling in an effect size range from +.18 to +.27. The overall median effect size for the eight studies that used immediate posttests is +.255. However, the studies vary widely in duration, experimental and control treatments, and other features, so this median value should be cautiously interpreted.

Fuchs, Tindal, and Fuchs (1985) conducted a small and somewhat unusual study of mastery learning in rural first-grade reading classes. Students in four classes were randomly assigned to one of two treatments. In the mastery learning classes, students were tested on oral reading passages in their reading groups each week. The whole reading group reviewed each passage until at least 80% of the students could read the passage correctly at 50 words per minute. The control treatment was held to be the form of "mastery learning" recommended by basal publishers. These students were given unit tests every 4-6 weeks, but all students went on to the next unit regardless of score. Surprisingly, the measure on which mastery learning classes exceeded control was "end-of-book" tests provided with the basals (ES = +.35), not passage reading scores that should have been more closely related to the mastery learning procedures (ES = +.05). On both measures it was found that while low achievers benefited from the mastery learning approach, high achievers generally achieved more in the control classes. Since the control teachers were presumably directing their efforts toward the objectives assessed in the end-of-book tests to the same degree as the mastery learning teachers, the results on this measure are probably fair measures of achievement. However, the Fuchs et al. study may be more a study of the effects of repeated reading than of mastery learning per se. Research on repeated reading (e.g., Dahl, 1979) has found this practice to increase comprehension of text.

Another small and unusual study at the elementary level was reported by Wyckoff (1974), who randomly assigned four sixth-grade classes to experimental or control conditions for a 9-week anthropology unit. Following teaching of each major objective, students were quizzed. If the class median was at least 70% correct, the class moved on to the next objective; otherwise, those who scored less than 70% received peer tutoring or were given additional reading or exercises. The control groups used precisely the same materials, tests, and schedule. The achievement results were not statistically significant, but they favored the mastery learning classes (ES = +.24). However, this trend was entirely due to effects on low performance readers (ES = +.58), not high-ability readers (ES = +.03).

One remarkable study spanning grades 3, 6, and 8 was reported in a dissertation by Cabezon (1984). The author, the director of the National Center for Curriculum Development in Chile, was charged with implementation of mastery learning throughout that country. Forty-one elementary schools throughout Chile were selected to serve as pilots, and an additional 2,143 *schools* began using mastery learning 2 years later. Three years after the pilots had begun, Cabezon randomly selected a sample of schools that had been using mastery learning for 3 years, for I year, or not at all. Within each school two classes at the third-, sixth-, and eighth-grade levels were selected.

The form of mastery learning used was not clearly specified, but teachers were expected to assess student progress every 2-3 weeks and to provide corrective instruction to those who needed it. Two subjects were involved, Spanish and mathematics. Unfortunately, the classes that had used mastery learning for 3 years were found to be much higher in socioeconomic status (SES) and mean IQ level than were control classes. Because of this problem these comparisons did not meet the inclusion criteria. However, the classes that had used mastery learning for 1 year were comparable to the control classes in SES and only slightly higher in IQ.

The study results, summarized in Table 2, indicated stronger effects of mastery learning in Spanish than in math, and stronger effects in the early grades than in later ones, with an overall mean of +0.27. However, while all teachers used the same books, it is unclear to what degree control teachers were held to or even aware of the objectives being pursued by the mastery learning schools.

Two studies at the secondary level assessed both immediate and long-term impacts of mastery learning. One was a study by Lueckemeyer and Chiappetta (1981), who randomly assigned 10th graders to six mastery learning or six control classes for a 6-week human physiology unit. In the mastery learning classes, students were given a formative test every 2 weeks. They were then given 2 days to complete corrective activities for any objectives on which they did not achieve an 80% score, following which they took a second form of the test, which was used for grading purposes. Students who achieved the 80% criterion on the first test were given material to read or games to play while their classmates received corrective instruction. The control group studied the same material and took the same tests, but did not receive the 2-day corrective sessions. The control teachers were asked to complete the three 2-week units in 6 weeks, but were not held to the same schedule as the mastery classes. In order to have time to fit in the 2 days for corrective instruction every 2 weeks, the mastery learning classes "had to condense instruction... and to guard carefully against any wasted time" (C. L. Lueckemeyer, personal communication, November 4, 1986).

On an immediate posttest the mastery learning classes achieved significantly more than the control group (ES = +.39), but on a retention test given 4 weeks later the difference had disappeared. The study's authors reported the statistically significant effects on posttest achievement but noted that "it is questionable whether such a limited effect on achievement is worth the considerable time required for the development and management of such an instructional program" (Lueckemeyer & Chiappetta, 1981, p. 273). Further, it is unclear whether the control groups required the full 6 weeks to cover the material. Any additional information students in the control group learned (or could have learned) would of course not have been registered on the experimenter-made test.

In a 15-week experiment in ninth-grade chemistry and physics classes by Dunkelberger and Heikkinen (1984), students were randomly assigned to mastery learning or control classes. In the mastery learning classes, students had several chances to meet an 80% criterion on parallel formative tests. Control students took the tests once and received feedback on their areas of strength and weakness. All students, control as well as experimental, had the same corrective activities available during a regularly scheduled free time. However, mastery learning students took much greater advantage of these activities. The total time *used* by the experimental group was thus greater than that used by control students, but since the total time *available* was held constant, this was categorized as an equal-time study.

For reasons that were not stated, the implementation of the 15-week chemistry and physics unit was concluded in late January, but the posttests were not given until early June, more than 4 months later. For this reason, the program's effects are listed as retention measures only. Effects favored the mastery learning classes (ES = +.26). Two studies by an Israeli student of Bloom, Zemira Mevarech, produced by far the largest effect sizes of all the mastery learning studies that met the inclusion criteria. One of these (Mevarech, 1980) provided additional time for corrective instruction to the mastery learning classes and is therefore included among the "extra time" studies listed in Table 3. The second (Mevarech, 1986) took place in a "desegregated" Israeli junior high school (i.e., Jews of Middle Eastern and European backgrounds attended

Table 2 Equal-time studies using experimenter-made measures

Article	Grades	Location	Sample size	Duration	Design	Treatments	Subjects	By group/measure	Effect sizes Total	Retention
Elementary										
Anderson et al., 1976	1-6	Lorain, OH	2 sch.	1 yr.	(See Table 1)	(See Table 1)	Math	Posttest Retention (3 mo.)	+.64	+.49
Kersh, 1970	5	Suburban Chicago	11 cl.	1 yr.	(See Table 1)	(See Table 1)	Math		(−)	
Jones et al., 1979	Upper elem.	Inner-city Chicago	4 sch.	1 yr.	(See Table 1)	(See Table 1)	Reading		+.18	
Wyckoff, 1974	6	Suburban Atlanta	4 cl.	9 wks.	Teachers/classes randomly assigned to ML*, control	ML–Mastery criterion 70%. Corr. inst. was either reteaching to whole class or peer tutoring Control–Used same mtls., tests, procedures as ML except for corr. inst. and summanve tests	Anthropology	Hi +.03 Lo +.58	+.24	
Fuchs et al., 1985	1	Rural Minnesota	4 cl.	1 yr.	Students randomly assigned to ML, control	ML–Students tested on oral rdg. passages each wk. Whole rdg. grp. reviewed until 80% of students got at least 50 wpm correct Control–students tested every 4-6 wks., all were promoted w/o corr. inst	Reading	Hi (−) Lo +	+.20	
Elementary and secondary										
Cabezon, 1984	3, 6, 8	Chile	46 cl.	1 yr.	Compared classes using ML to classes similar in SES, IQ	ML–Not clearly specified Control–Untreated	Spanish +.40 Math +.14	Gr3 +.47 Gr6 +.22 Gr8 +.12	+.27	

Secondary

Study	Location	Classes	Duration	Design	Subject	Treatment	Outcome	Effect
Lueckemeyer & Chiappetta, 1981 10	Suburban Houston	12 cl.	6 wks.	Students randomly assigned to ML, control Pretest differences favored control	Human Physiology	ML–Formative test given every 2 wks., followed 2 days of corr. inst. (criterion = 80%) Control–Used same mtls., tests, procedures as ML except for corr. inst. and summative tests	Posttest Retention (4 wks.)	+.39 0
Dunkekberger & Heikkinen, 1984 9	Suburban Delaware	10 cl.	15 wks.	Students randomly assigned to ML, control classes Teachers taught ML & control classes. Posnest given > 4 mos. after end of implementation period	Chem., Physics	ML–Students had to meet 80% criterion on repeatable tests to go on. Corrective activities available during free time Control–Used same mtls., procedures, tests. Received detailed feedback and had same corrective mtls. available during free time	Retention (4 mo.)	+.26 +.26
Mevarech, 1986 7	Israel	4 cl.	3 mo.	Students randomly assigned to ML., control classes	Algebra	ML–Students who did not reach 70% crit. on formative tests rec'd corr. inst. from tch. or peers Control–Used same mtls., tests as ML	Lo SES +1.78 Mid SES +. 91 Hi SES +.66	

• Mastery learning
+ Significant difference favoring ML
0 No difference
(–) Nonsignificant difference favoring control

Table 3 Unequal-time studies using experimenter-made measures (secondary)

Article	Grades	Location	Sample size	Duration	Design	Treatments	Extra time	Subjects	Effect sizes By group/measure	Total	Retention
Long et al., 1978	8	Georgia	6 cl.	5 wks.	Students randomly assigned to 3 trts. Teachers rotated across trts.	Teacher-directed ML–Formative tests given every 2 days. Remedial work given as corr. inst. "If problem persists," indiv. tutoring given by teacher. Student-directed ML–Same formative tests used, returned to students for self-correction. Control—Same inst. but no tests, correctives	Not stated	Earth Sci.	Teacher-directed ML vs. Control: Posttest / Retention (12 wks.) / Teacher-directed ML vs. Student-directed ML / Posttest / Retention (12 wks.)	+.43 / +.19	+.08 / −.03
Fagan, 1976	7	Dallas, TX, 1 middle class sch., 1 lower class sch.	17 cl.	5 wks.	4 teachers randomly assigned to ML, control	ML–Formative tests given every wk. Teachers drilled students who failed to achieve 80% criterion, then gave 2nd formative test. Control–Used same mtls. & procedures as ML. Formative tests taken as quizzes	22%	Transp. and Environ.	Posttest / Retention (4 wks.)	−.11	−.15
Hecht, 1980	10	Urban, suburban Midwest	5 cl.	6 wks.	Students randomly assigned to ML, control classes. Two teachers taught ML & control classes.	ML–Formative tests given every 2 wks., followed by "intensive remedial help" Control–Used same mtls. & procedures as ML including both 1st & 2nd formative tests, but no remedial help	Not stated	Geometry		+.31	

| Mevarech, | 9 1980 | Chicago, | 8 cl. middle class sch | 6 wks. | Students randomly assigned in 2 x 2 design to "algorithmic strategy" vs. "heuristic strategy" and to ML vs. control | ML–Formative tests given every 2 wks. Students had 3 chances to obtain 80% criterion. Corr. inst. included grp. inst., peer tutoring, adult tutoring outside of class. Control–Used same mtls. & procedures. took formative tests as quizzes. While ML classes received corr. first, control worked add'l problems. | Not stated | Alg. I | Algorithmic Strategy +.70 Heuristic Strategy +.83 | +77 |

* Mastery learning.

the same school). Students were randomly assigned to heterogeneous mastery learning or control classes. The mastery learning classes received lessons and then took formative tests. Students who achieved a criterion score of 70% on biweekly quizzes received corrective instruction from peers or from the teacher, after which a second form of the test was given. Students who achieved the mastery criterion on the first test either served as tutors or worked on enrichment activities. Control classes received the same lessons and materials and were given the formative tests as quizzes, but did not receive corrective instruction.

The outcome measure was an achievement test constructed by the teachers. At the end of the 3-month experiment, mastery learning students scored much higher than control students on this test (ES = +.90). The results were also broken down by students' socioeconomic level. Students whose fathers did not complete high school (20% of the sample) gained the most from the mastery learning program (ES = +1.78), followed by those whose fathers had a high school degree, 50% of the sample (ES = +.91), and those whose fathers completed college (ES = +.66).

In light of the extraordinary positive effects found in this study, more than three times the median for all the studies in Table 2, it is useful to consider what may be unique about the Mevarech (1986) study. One important factor is that only two teachers were involved in each treatment, and that while *students* were randomly assigned to treatments, *teachers* were not assigned at random. Thus, the possibility of teacher effects cannot be ruled out; it may be that the mastery learning teachers were simply better teachers than those in the control group. Second, as is the case with all mastery learning studies that used experimenter-made measures, there is a possibility that even though all teachers used the same materials, the mastery learning teachers focused on the specific objectives to be tested more than the control classes did. The posttest was described as covering "various aspects of rational numbers." Because there is much more to a typical algebra course, it may be that the mastery learning classes were spending time mastering a limited set of objectives while the control group may have learned a larger set of objectives (though perhaps at a lower level of mastery).

However, bearing in mind these cautions, the 1986 Mevarech study and the 1980 Mevarech study described later in this article provide some grounds for optimism that certain forms of group-based mastery learning may have strong effects on student achievement.

Overall, the effects summarized in Table 2 could be interpreted as supporting the "curricular focus" claim. The effects of mastery learning on experimenter-made, criterion-referenced measures are generally moderate but consistently positive. Two studies found that the effects of mastery learning were greatest for low achievers, as would be expected from mastery learning theory, and one found effects to be greatest for low-SES students.

However, the meaning of the results summarized in Table 2 is far from clear. The near-zero effects of mastery learning on standardized measures (Table 1) and in particular the dramatically different results for standardized and experimenter-made measures reported by Anderson et al. (1976) suggest that the effects of mastery learning on experimenter-made measures result from a shifting of instructional focus to a particular set of objectives not necessarily more valuable than those pursued by the control group. Unfortunately, it is impossible to determine from reports of mastery learning studies the degree to which control teachers were focusing on the objectives assessed on the experimenter-made measures, yet understanding this is crucial to understanding the effects reported in these studies.

Evidence for the Extra-Time Claim

The problem of unequal time for experimental and control groups is a serious one in mastery learning research in general, but the inclusion criteria used in the present review have the effect of eliminating the studies in which time differences are extreme. Mastery learning studies in which experimental classes receive considerably more instructional time than control classes are always either very brief, rarely more than a week (e.g., Anderson, 1975, 1976; Arlin & Webster, 1983), or they involve individualized or self-paced rather than group-paced instruction (e.g., Jones, 1975; Wentling, 1973). In studies of group-paced instruction conducted over periods of at least 4 weeks, extra time for corrective instruction rarely amounts to more than 20-25% of original time. It might be argued that additional instructional time of this magnitude might be a practicable means of ensuring all students a reasonable level of achievement, and the costs of such an approach might not be far out of line with the costs of current compensatory or special education.

Table 3 summarizes the characteristics and outcomes of group-based mastery learning studies in which the mastery learning classes received extra time for corrective instruction. All four of the studies in this category took place at the secondary level, grades 7-10. Also, these studies are distinctly shorter (5-6 weeks) than were most of the studies listed in Tables 1 and 2.

The median effect size for immediate posttests from the five comparisons in four studies is +.31, but none of three retention measures found significant differences (median ES = -.03). However, the four studies differ markedly in experimental procedures, so these medians have little meaning.

The importance of the different approaches taken in different studies is clearly illustrated in a study by Long, Okey, and Yeany (1978). In this study, eighth graders were randomly assigned to six classes, all of which studied the same earth science units on the same schedule. Two classes experienced a mastery learning treatment with teacher-directed remediation. After every two class periods, students in this treatment took a diagnostic progress test. The teacher assigned students specific remedial work, then gave a second progress test. If students still did not achieve at a designated level (the mastery criterion was not described in the article), the teacher tutored them individually. In a second treatment condition, student-directed remediation, students received the same instruction and tests and had the same corrective materials available, but they were asked to use their test results to guide their own learning, rather than having specific activities assigned. These students did not take the second progress test and did not receive tutoring. Students in the third treatment, control, studied the same materials on the same schedule but did not take diagnostic progress tests. Teachers rotated across the three treatments to minimize possible teacher effects. The results of the Long et al. (1978) study indicated that the teacher-directed remediation (mastery learning) group did achieve considerably more than the control group (ES = +.43), but exceeded the student-directed remediation group to a much smaller degree (ES = +.19). What this suggests is that simply receiving frequent and immediate feedback on performance may account for a substantial portion of the mastery learning effect. A replication by the same authors (Long et al., 1981) failed to meet the inclusion criteria because it had only one class per treatment. However, it is interesting to note that the replication found the same pattern of effects as the earlier Long et al. (1978) study; the teacher-directed remediation treatment had only slightly more positive effects on student achievement than the student-directed remediation treatment, but both exceeded the control group.

The Long et al. (1978) study included a retention test, which indicated that whatever effects existed at the end of the implementation period had disappeared 12 weeks later. Retention is especially important in studies in which corrective instruction is given outside of class time, as any determination of the cost-effectiveness of additional time should take into account the lasting impact of the expenditure.

Another extra-time study which assessed retention outcomes was a dissertation by Fagan (1976), who randomly assigned four teachers and their 17 seventh-grade classes to mastery learning or control treatments. The mastery learning treatment essentially followed the sequence suggested by Block and Anderson (1975). Students were quizzed at the end of each week, and teachers worked with students who failed to reach an 80% criterion, after which students took a second formative test. The control classes used the same materials and procedures except that they took the formative tests as quizzes. Teachers scored the quizzes, returned them to students, and then went on to the next unit. The teachers followed the same sequence of activities, but were allowed to proceed at their own pace. As a result, the mastery learning classes took 25 days to complete the five units on "transportation and the environment," whereas control classes took only 20-21 days.

Unfortunately, there were pretest differences favoring the control classes of approximately 40% of a grade equivalent on Iowa Test of Basic Skills vocabulary scores. Analyses of covariance on the posttests found no experimental-control differences; in fact, adjusted scores slightly favored the control group (ES = −.11). On a 4-week retention measure the control group's advantage was slightly greater (ES = −.15). When experimental treatments vary widely in pretests or covariates, statistical adjustments tend to underadjust (see Reichardt, 1979), so these results must be interpreted with caution. However, even discarding the results for the one control teacher whose classes had high pretest scores, differences still favored the control group on the posttest (ES = −.17) and on the retention test (ES = −.23).

A small study by Hecht (1980) compared mastery learning to control treatments in 10th-grade geometry. Students were randomly assigned to treatments, and each of two teachers taught mastery learning as well as control classes. In the mastery learning classes students were given formative tests every 2 weeks that were followed by "intensive remedial help for those who needed it" (mastery criteria and corrective activities were not stated). Results on an experimenter-made test favored the mastery learning classes (ES = +.31).

The largest effect sizes for any of the studies that met the inclusion criteria were found in two studies by Zemira Mevarech. One (Mevarech, 1986) was described earlier. In the second (Mevarech, 1980), students were randomly assigned to eight Algebra I classes in a 2 x 2 factorial design. One factor was "algorithmic" versus "heuristic" instructional strategies. The "algorithmic" treatments emphasized step-by-step solutions of algebraic problems, focusing on lower cognitive skills. The "heuristic" treatments emphasized problem solving strategies such as Polya's (1957) "understanding-planning-carrying out the plan-evaluating" cycle and focused on higher cognitive skills.

The other factor was mastery learning (feedback-correctives) versus nonmastery. In the mastery learning treatments, students were given formative tests every 2 weeks. They then had three chances to meet the mastery criterion of 80% correct. Corrective instruction included group instruction by the teacher, peer tutoring, and tutoring outside of class time by teachers. The amount of additional time allocated to provide this corrective instruction is not stated, but the author claimed the amount of out-of-class tutoring to be small (Z. Mevarech, personal communication, March 16, 1984). In the nonmastery treatments, students studied the same materials and took the formative tests as quizzes. To hold the different classes to the same schedule, nonmastery classes were given additional problems to work while mastery learning classes were receiving corrective instruction.

The relevant comparisions for the present review involve the mastery learning versus nonmastery factor. Within the algorithmic classes, the mastery learning classes exceeded nonmastery on both "lower mental process" items (i.e., algorithms) (ES = +.30) and on "higher mental process" items (ES = +.77). Within the heuristic classes, the effects were even greater for both "lower mental process" (ES = +.66) and "higher mental process" items (ES = +.90).

Overall, the evidence for the "extra time" claim is unclear. Effect sizes for the small number of unequal time studies summarized in Table 3 are no more positive than were those reported for other studies using experimenter-made measures (Table 2), in which mastery learning classes did not receive additional time. In fact, both of the unequal time studies that assessed retention found that any effects observed at posttest disappeared as soon as 4 weeks later. Substantial achievement effects of extra time for corrective instruction appear to depend on provisions of substantial amounts of extra time, well in excess of 20-25%. However, studies in which large amounts of additional time are provided to the mastery learning classes either involved continuous-progress forms of mastery learning or are extremely brief and artificial. What is needed are long-term evaluations of mastery learning models in which corrective instruction is given outside of class time, preferably using standardized measures and/or criterion-referenced measures that register all objectives covered by all classes.

Retention

A total of six comparisons in five studies assessed retention of achievement effects over periods of 4-12 weeks. All six used experimenter-made measures. The median effect size overall is essentially zero, with the largest retention effect (ES = +.49) appearing in the Anderson et al. (1976) study, which found no differences on standardized measures.

Discussion

The best evidence from evaluations of practical applications of group-based mastery learning indicates that effects of these methods are moderately positive on experimenter-made achievement measures closely tied to the objectives taught in the mastery learning classes and are essentially nil on standardized achievement measures. These findings may be interpreted as supporting the "weak claim" that mastery learning can be an effective means of holding teachers and students to a specified set of instructional objectives, but do not support the "strong claim" that mastery learning is more effective than traditional instruction given equal time and achievement measures that assess coverage as well as mastery. Further, even this "curricular focus" claim is undermined by uncertainties about the degree to which control teachers were trying to achieve the same objectives as the mastery learning teachers and by a failure to show effects of mastery learning on retention measures.

These conclusions are radically different from those drawn by earlier reviewers and meta-analysts. Not only would a mean effect size across the 17 studies emphasized in this review come nowhere near the mean of around 1.0 claimed by Bloom (1984a, 1984b), Guskey and Gates (1985), Lysakowski and Walberg (1982), or Walberg (1984), but *no single study* reached this level. Only 2 of the 17 studies both by the same author, had mean effect sizes in excess of the 0.52 mean estimated by Kulik et al. (1986) for precollege studies of mastery testing. How can this gross discrepancy be reconciled? First, these different reviews focus on very different sets of studies. Almost all of the studies cited in this review would have qualified for inclusion in any of the meta-analyses, but the reverse is not true. For example, of 25 elementary and secondary studies cited by Guskey and Gates (1985), only 6 qualified for inclusion in the present review. Of 19 such studies cited by Kulik et al. (1986), only 4 qualified for inclusion in the present review. Only 2 studies, Lueckemeyer and Chiappetta (1981) and Slavin and Karweit (1984), appeared in all three syntheses. The list of mastery learning studies synthesized by Lysakowski and Walberg (1982) is short and idiosyncratic, hardly overlapping at all with any of the other reviews, and Bloom's (1984a) article discusses only a few University of Chicago dissertations.

As noted earlier, the principal reason that studies cited elsewhere were excluded in the present paper is that they did not meet the 4-week duration requirement. The rationale for this restriction is that this review focuses on the effects of mastery learning *in practice,* not in theory. It would be difficult to maintain that a 2- or 3-week study could produce information more relevant to classroom practice than a semester- or year-long study, partly because artificial arrangements possible in a brief study could not be maintained over a longer period. Actually, even 4 weeks could be seen as too short a period for external validity. However, it is useful to examine the results of shorter implementations of mastery learning to be sure that arbitrarily drawing a line at 4 weeks' duration does not misrepresent the evidence. A total of 19 elementary and secondary studies with treatment durations of 1 to 3 weeks were cited by Bloom (1984a, 1984b), Kulik et al. (1986), and/or Guskey and Gates (1985). Most of these would not have been excluded from this review on grounds other than their brevity. For example, nine of the studies used self-instructional or programmed materials rather than group-based mastery learning (e.g., Anderson, 1975; Block, 1972), four more used only one class per treatment (e.g. Swanson & Denton, 1977; Tenenbaum, 1982), one used a procedure only tangentially related to mastery learning (Bryant, Fayne, & Gettinger, 1982), and two failed to provide satisfactory evidence that experimental and control classes were initially equivalent (Hymel & Mathews, 1980; Strasler, 1979).

The remaining three studies are all dissertations by Bloom's students. Two of these used very similar procedures. Anania (1983) randomly assigned students in grades 4, 5, and 8 to three treatments: tutoring, mastery learning, and "conventional" instruction. Only the latter two are relevant to the present review. Students in the mastery learning treatment received two 4-day units and one 3-day unit on probability (grades 4-5) or cartography (grade 8). At the end of this time, students took formative tests and then had an extra period in which they received corrective instruction if necessary; if they did not achieve at an 80% criterion level by this time, students might receive additional tutoring after school. Burke (1983) used nearly identical procedures to teach probability to students in grades 4-5. Nordin (1979) compared mastery learning and control methods to teach a 2-week unit on sets to Malaysian sixth graders. In all three studies, mastery learning students far outperformed control; effect sizes were around 1.0 for the Anania and Burke studies, and exceeded 2.0 for the Nordin study. While all three of these studies are exemplary as basic research, they all have features that severely limit their external validity. First, they all provided significant amounts of extra time for the mastery learning groups, from 125% (Burke, 1983) to 133% (Nordin, 1979) of the time allocated to the control groups. Second, all three selected subject mater that was completely new to students (finessing the issue of student heterogeneity by starting all students at zero) and all three created units that were completely hierarchical, which is to say that learning of the later units depended heavily on mastery of the earlier ones. These procedures are entirely appropriate for theory-building, which was the authors' purpose, but they are hardly representative of conditions in usual classroom teaching, where (for example) students either enter class with different levels of prerequisite skills and/or diverge in their skills over many months or years (see Arlin & Webster, 1983).

In addition to excluding many studies cited elsewhere, the present review included many studies missed in the meta-analyses. These are primarily dissertations and unpublished papers (mostly AERA papers), which comprise 12 of the 17 studies emphasized in this review. Including unpublished studies is critical in any literature review, as they are less likely to suffer from "publication bias," the tendency for studies reporting nonsignificant or negative results not to be submitted to or accepted by journals (see Bangert-Drowns, 1986; Rosenthal, 1979). Other differences in study selection and computation of effect size between the present paper and earlier reviews are important in specific cases. For example, Guskey and Gates (1985) report effect sizes for the Jones, Monsaas, and Katims (1979) study of +.41 for an experimenter-made measure and +.33 for a standardized test, while the present review estimated effect sizes of +.18 and +.09, respectively. The difference is that in the present review pretest differences (in this case favoring the experimental group) were subtracted from the posttest differences. Similarly, Guskey and Gates (1985) report a single effect size of +.58 for the Anderson et al. (1976) study, ignoring the striking difference in effects on standardized as opposed to experimenter-made measures emphasized here.

There are several important theoretical and practical issues raised by the studies of group-based mastery learning reviewed here. These are discussed in the following sections.

Why Are Achievement Effects of Group-Based Mastery Learning So Modest?

The most striking conclusion of the present review is that, other than perhaps focusing teachers and students on a narrow set of objectives, group-based mastery learning has modest to nonexistent effects on student achievement in studies of at least 4 weeks' duration. Given the compelling nature of the theory on which mastery learning is based, it is interesting to speculate on reasons for this.

One possible explanation is that in long-term, practical applications of mastery learning, the quality of training, followup, and/or materials used to support the mastery learning approach are inadequate. One important piece of evidence in support of this possibility comes from a recent study by Dolan and Kellam (1987), who compared an enhanced mastery learning program to the standard mastery learning program used in Baltimore City first grades. The enhanced model provided teachers with 32 hours of instruction in mastery learning principles and practices, monthly progress meetings, classroom visits, files of formative tests, corrective activities, and enrichment activities keyed to school district objectives, and special curriculum materials and other resources to help teachers achieve reading objectives. Teachers using the standard mastery learning procedures also used the teach-test-corrective instruction-test cycle to achieve essentially the same reading objectives, but did not have the additional training, resources, or assistance. A year-long experiment assigned schools and teachers within schools to the two models, and found that the enhanced mastery learning classes gained significantly more on standardized reading tests (ES = +.39).

A particular emphasis of the Dolan and Kellam (1987) enhanced mastery learning model was on the quality of the materials used for corrective instruction. In a letter explaining the extraordinary effects obtained in her studies, Zemira Mevarech (personal communication, January 25, 1987) also emphasized that the quality of the corrective procedures was a key factor in the success of her programs, noting that "corrective activities should be creative, attractive, and designed explicitly to remediate the skills that have not been mastered."

Another possible explanation for the disappointing findings of studies of group-based mastery learning is that it is not only that the *quality* of corrective instruction is lacking, but also that the *amount* of corrective instruction is simply not enough to remediate the learning deficits of low achievers. In none of the studies emphasized in this review did corrective instruction occupy more than one period per week, or 20% of all instructional time. This may be enough to get students up to criterion on very narrowly defined skills, but not enough to identify and remediate serious deficits, particularly when corrective instruction is given in group settings or by peer tutors (as opposed to adult tutors). Studies of students' pace through individualized materials routinely find that the slowest students require 200-600% more time than the fastest students to complete the same amount of material (Arlin & Westbury, 1976; Carroll, 1963; Suppes, 1964), far more than what schools using mastery learning are likely to be able to provide for corrective instruction (Arlin, 1982).

The amount of corrective instruction given in practical applications of group-based mastery learning may be not only too little but also too late. It may be that even 1 or 2 weeks is too long to wait to correct students' learning errors, and many studies provided corrective instruction less frequently, every 3 to 4 weeks. If each day's learning is a prerequisite for the next day's lesson, then perhaps detection and remediation of failures to master individual skills needs to be done daily to be effective. Further, in most applications of mastery learning, students may have years of accumulated learning deficits that 1 day per week of corrective instruction is unlikely to remediate.

Time for corrective instruction in group-based mastery learning is purchased at a cost in terms of slowing instructional pace. If this time does not produce a substantial impact on the achievement of large numbers of students, then a widespread though small negative impact on the learning of the majority may balance a narrow positive impact on the learning of the few students whose learning problems are large enough to need corrective instruction but small enough to be correctable in one class period per week or less.

However, it may be that the feedback-corrective cycle evaluated in the studies reported here is simply insufficient in itself to produce a substantial improvement in student achievement. As Bloom (1980, 1984b) has noted, there are many variables other than feedback-correction that should go into an effective instructional program. Both the process of learning and the process of instruction are so complex that it may be unrealistic to expect large effects on broadly based achievement measures from any one factor; instructional quality, adaptation to individual needs, motivation, and instructional time may all have to be impacted at the same time to produce such effects (see Slavin, 1987).

Is Mastery Learning a Robin Hood Approach to Instruction?

Several critics of mastery learning (e.g., Arlin, 1984a; Resnick, 1977) have wondered whether mastery learning simply shifts a constant amount of learning from high to low achievers. The evidence from the present review is not inconsistent with that view; in several studies positive effects were found for low achievers only. In fact, given that overall achievement means are not greatly improved by group-based mastery learning, the reductions in standard deviations routinely seen in studies of these methods and corresponding decreases in correlations between pretests and posttests are simply statistical indicators of a shift in achievement from high to low achievers. However, it is probably more accurate to say that group-based mastery learning trades coverage for mastery. Because rapid coverage is likely to be of greatest benefit to high achievers, whereas high mastery is of greatest benefit to low achievers, resolving the coverage-mastery dilemma as recommended by mastery learning theorists is likely to produce a "Robin Hood" effect as a by product.

However, it is important to note that few mastery learning studies have found the method to be *detrimental* to the achievement of high achievers. This may be the case because the coverage versus mastery dilemma exists in *all* whole-class, group-paced instruction, including traditional instruction. For example, Arlin and Westbury (1976) compared individualized instruction to whole-class instruction and found that the instructional pace set by the teachers using the whole-class approach was equal to that of students in the 23rd percentile in the individualized classes, supporting Dahlloff's (1971) contention that teachers set their instructional pace according to the needs of a "steering group" of students in the 10th to 25th percentiles of the class ability distribution. Assuming that an instructional pace appropriate for students at the 23rd percentile is too slow for higher achievers (Barr, 1974, 1975), then whole-class instruction in effect holds back high achievers for the benefit of low achievers. Group-based mastery learning may thus be accentuating a "Robin Hood" tendency already present in the class-paced traditional models to which it has been compared. The coverage versus mastery dilemma and the corresponding "Robin Hood" effect are problematic only within the context of group-based mastery learning and (at least in theory) only when instruction time is held constant. In continuous-progress or individualized forms of mastery learning in which students can move through material more or less at their own rates, the coverage-mastery dilemma is much less of a concern (Arlin & Westbury, 1976). This does not imply that continuous-progress forms of mastery learning are necessarily more effective than group-based forms; individualization solves the instructional pace problem but creates new problems, such as the difficulty of providing adequate direct instruction to students performing at many levels (Slavin, 1984b). However, there are examples of continuous-progress mastery learning programs that have positive effects on standardized achievement tests (see, e.g., Cohen, 1977; Cohen & Rodriquez, 1980; Slavin & Karweit, 1985; Slavin & Madden, 1987; Slavin, Madden, & Leavey, 1984).

Importance of Frequent, Criterion-Referenced Feedback

Even if we accept the "weak claim" that mastery learning is an effective means of holding teachers and students to a valuable set of instruction objectives, there is still some question as to which elements of mastery learning account for its effects on experimenter-made, criterion-referenced measures. There is some evidence that much of this effect may be accounted for by frequent testing and feedback to students rather than the entire feedback-corrective cycle. Kulik et al. (1986) report that mastery learning studies that failed to control for frequency of testing produced mean effect sizes almost twice those associated with studies in which mastery learning and control classes were tested with equal frequency. Long et al. (1978) compared mastery learning to a condition with the same frequency of testing and found a much smaller effect than in a comparison with a control group that did not receive tests. Looking across other studies, the pattern is complicated by the fact that most that held testing frequency constant also held the control groups to a slower pace than they might otherwise have attained.

Practical Implications

The findings of the present review should not necessarily be interpreted as justifying an abandonment of mastery learning, either as an instructional practice or as a focus of research. Several widely publicized school improvement programs based on mastery learning principles have apparently been successful (e.g., Abrams, 1983; Levine & Stark, 1982; Menahem & Weisman, 1985; Robb, 1985), and many effective non-mastery learning instructional strategies incorporate certain elements of mastery learning – in particular, frequent assessment of student learning of well-specified objectives and basing teaching decisions on the results of these assessments. Further, the idea that students' specific learning deficits should be remediated immediately instead of being allowed to accumulate into large and general deficiencies makes a great deal of sense. It may be that more positive results are obtained in continuous-progress forms of mastery learning, in which students work at their own levels and rates. Use of remedial (e.g., Chapter I), special education, or other resources to provide substantial amounts of instructional time to help lower achieving students keep up with their classmates in critical basic skills may also increase student achievement (Slavin & Madden, 1987). This review concerns only the achievement effects of the group-based form of mastery learning (Block & Anderson, 1975) most commonly used in elementary and secondary schools.

The "2-Sigma Problem" Revisited

One major implication of the present review is that the "2-sigma" challenge proposed by Bloom (1984a, 1984b) is probably unrealistic, certainly within the context of group-based mastery learning. Bloom's claim that mastery learning can improve achievement by more than 1 sigma (ES = + 1.00) is based on brief, small, artificial studies that provided additional instructional time to the experimental classes. In longer term and larger studies with experimenter-made measures, effects of group-based mastery learning are much closer to 1/4 sigma, and in studies with standardized measures there is no indication of any positive effect at all. The 2-sigma challenge (or l-sigma claim) is misleading out of context and potentially damaging to educational research both within and outside of the mastery learning tradition, as it may lead researchers to belittle true, replicable, and generalizable achievement effects in the more realistic range of 20-50% of an individual-level standard deviation. For example, an educational intervention that produced a reliable gain of .33 each year could, if applied to lower class schools, wipe out the typical achievement gap between lower and middle-class children in 3 years – no small accomplishment. Yet the claims for huge effects made by Bloom and others could lead researchers who find effect sizes of "only" .33 to question the value of their methods.

Clearly, much more research is needed to explore the issues raised in this review. More studies of practical, long-term applications of mastery learning assessing the effects of these programs on broadly based measures of achievement that register coverage as well as mastery are especially needed; idiosyncratic features of the seven studies that used standardized tests preclude any interpretation of those studies as evidence that group-based mastery learning is *not* effective. There is very little known about what would be required to make group-based mastery learning instructionally effective; the Mevarcch (1980, 1986) and Dolan and Kellam (1987) studies provide some clues along these lines, but much more needs to be known. In addition, studies carefully examining instructional pace in mastery and nonmastery models are needed to shed light on the coverage-mastery dilemma discussed here. Mastery learning models in which Chapter I or other remedial teachers provide significant amounts of corrective instruction outside of regular class time might be developed and evaluated, as well as models providing daily, brief corrective instruction rather than waiting for learning deficits to accumulate over 1 or more weeks. The disappointing findings of the studies discussed in this review counsel not a retreat from this area of research but rather a redoubling and redirection of efforts to understand how the compelling theories underlying mastery learning can achieve their potential in practical application.

Mastery learning theory and research has made an important contribution to the study of instructional methods. However, to understand this contribution it is critical to fully understand the conditions under which mastery learning has been studied, the measures that have been used, and other study features that bear on the internal and external validity of the findings. This best-evidence synthesis has attempted to clarify what we have learned from research on mastery learning in the hope that this knowledge will enrich further research and development in this important area.

CHAPTER 11

Policy, Practice, and New Directions

If reform in educational policies and practices is to result in markedly higher achievement and school success for all children, it must begin with a clear conception of how classrooms should be organized and taught to maximize success for all. Only then does it make sense to design schools capable of supporting effective classrooms, school districts capable of supporting effective schools, and state and national systems to support effective schools and districts. To begin with macro-level reforms without compelling evidence about effective practices at the point of contact between teachers and students would be like designing airports before practical airplanes existed.

Our own research, and that my colleagues and I have reviewed (see Slavin, Karweit, & Madden, 1989; Slavin, Karweit, & Wasik, 1994), has focused first on developing or identifying effective instructional strategies, such as various cooperative learning and tutoring programs. Then we began to develop and research schoolwide change models, such as the cooperative school, Success for All, and Roots and Wings, which provide school-level supports for effective classroom practices.

While it is necessary to have validated and replicable classroom and school interventions, this is not sufficient to bring about change in educational practices and student learning on a large scale. Dissemination mechanisms, including well-developed plans for creating awareness and providing high-quality professional development and follow-up, are also necessary (see Slavin, Madden, Dolan, & Wasik, 1996). However, there is also a need for policies and investments at the district, state, and national levels, both to create an infrastructure for development and evaluation of promising programs and practices and to sponsor and support the dissemination process. The following sections present a proposal for such policies.

Vision 2020: How Can We Achieve Success for All?

Fundamental change in the practice of education and the belief systems around it will only come about when there are alternatives to current practices that are known to be very effective, are readily available to every school, and are supported by effective networks of professional development, financial resources, and school district infrastructure. We are very far from such a situation today. What would an education system designed to support fundamental change be like and how could it ensure the success of all children? To create a school system, say by the year 2020, in which schools are routinely able to place a high floor under the achievement of all students while ensuring that all meet their full potential, something like the following set of conditions would have to prevail.

1. Proven, replicable programs would be widely available.

Perhaps the most important prerequisite for an effective system of education is the existence of a broad range of educational programs known (based on rigorous research and evaluation) to be highly effective. This would mean that programs would exist for every subject and grade level capable of ensuring success for all children. These programs would encompass student materials, teachers' manuals, instructional methods, assessment procedures, means of helping students who are struggling and allowing others to go beyond standard expectations, and so on. Their focus might be as broad as teaching reading, writing, and language arts throughout elementary school, or as narrow as a unit on photosynthesis.

Each of these programs would meet two key standards: effectiveness and replicability.

In this scheme, effectiveness would be demonstrated in third party evaluations in which students taught using the program are shown to learn significantly better than students in programs that represent widespread current practice. In such evaluations, students in program and matched control schools would be compared on assessments that reflect current, state-of-the-art standards of educational excellence. When practical, students, teachers, or schools would be randomly assigned to program and control classes, but if this is not practical program schools could be compared to control schools matched on student characteristics, school resources and personnel, and other factors. The third party evaluators would be government contractors using quantitative and descriptive methods to determine the overall effects of the program as well as the conditions of implementation and application in which the program was most and least effective. The evaluations would be large enough and involve enough diversity of schools and students to enable researchers to rule out most sources of potential bias and to determine the generalizability of program impacts for various subgroups and settings. In general, third party evaluations would take place only after program developers had already done their own rigorous evaluations.

For most subjects and grade levels many validated programs would exist at any given time. Educators would be able to choose among them, with confidence that whichever they chose, each program would be effective with their children if they implemented it with at least the quality and integrity involved in the program's evaluations.

It is not enough, of course, that effective programs are known to exist. They must also be replicable. Replicability would be demonstrated by having programs evaluated in typical schools not closely connected with program developers, and without conditions unlikely to exist on a broad scale. Replicability would also depend on the creation of national or regional networks of trainers, support networks among existing program implementors, and in most cases on the development of student materials, teachers' manuals, training videos, and other supports necessary to enable new schools adopting a given program to faithfully recreate or adapt the methods used in the successful evaluations. It is difficult to imagine how programs redeveloped or invented from scratch in every school could be replicated with fidelity on a large scale, although it is certainly likely that school staffs would adapt materials and procedures from well-developed programs according to local needs, resources, or interests. Hopefully, program evaluations would, over time, identify program elements that lend themselves to local adaptations and those that are essential to program outcomes in their original forms.

2. School staffs could choose among effective programs and would be supported in doing so.

Once a variety of well-validated programs existed for each subject and grade level, it would be essential to have systems for making school staffs aware of these programs and enabling them to make informed selections among them. Information on proven programs could be make widely available in print, video, and electronic forms. Teachers and administrators could visit schools already implementing programs either in person or by electronic "tours" that would allow groups of educators to "visit" classrooms and schools through two-way voice and picture links. Educators might attend effective methods fairs, in which they could learn about a range of alternative programs, talk to program representatives and current implementors, and so on. These activities might be followed up by visits to the school by program representatives. In general, schools would adopt programs for the entire school, voting by secret ballot to do so (for example, we require a vote of at least 80% of school staff to implement Success for All and Roots and Wings programs). However, some programs might lend themselves to adoption by departments, by schools-within-schools, or by individual teachers. Regardless of the staff unit making the decision, the principle here is that educators implementing a given program have made an informed, uncoerced and proactive choice to do so, and are supported in this by their peers and administrators.

Educators would in no case be *required* to adopt an externally developed, proven program. School staffs would have an opportunity to propose their own home-grown alternative, or to propose to adopt a program that is not yet validated (but might be at some point in the validation process). However, when many well-validated programs do exist in a give area, school staffs might be asked to justify selecting a less well-validated alternative, and to participate in a local evaluation to compare their model to widely used validated programs.

3. School and district policies would support program adoption

Teachers and schools do not operate in a vacuum. School and district policies, practices, and investments must be aligned with an orientation toward adoption of well-validated programs.

One critical policy requirement for system support for program adoption is the provision of assessments keyed to widely accepted, high-expectations standards. School staffs need to be held accountable for the achievement of their children, and these standards must reflect a full conception of what we believe all children should know and be able to do at a given age. The current movement toward broad, intellectually demanding performance assessments in several U.S. states and other countries is a promising trend in this direction. Consequences for increasing or decreasing scores on these assessments need not be draconian; in general, publishing school means in newspapers and using scores in evaluations of principals are enough to focus school staffs on an outcome orientation.

Beyond assessing student progress, school and district policies and practices can play a major role in providing a supportive infrastructure for responsible reform. For example, district policies can tie resources to adoption of proven programs. In the U.S., funds for Title I, state compensatory education funds, and professional development resources can be linked to adoption of proven programs by requiring that proposals for these funds show how they will be used for program adoption (or for home-grown or newer yet-to-be-evaluated alternatives of equal promise). Of course, special district, state, and national funds could and should be allocated to pay the costs of adopting proven programs, especially the one-time costs for materials, professional development, and so on.

Districts and schools would need to build their own capacities to support proven programs. For example, districts, intermediate units, or states might employ trainers for given programs trained and certified by program developers. Supervisory staffs might be realigned to support effective program implementations; instead of taking responsibility for schools in the eastern part of a district, for example, a supervisor might become an expert on program X, and work with all schools in the district implementing that program. Districts and schools could help maintain program quality over time by creating local support networks of schools implementing a given program, in which teachers and principals could meet on a regular basis to exchange ideas and innovations, share and resolve concerns, and mentor staffs of schools new to the network.

4. National and state governments would support a program adoption infrastructure.

The system of educational reform described here would require major shifts in national and state policies and investments. Clearly, a continued move toward rational achievement standards, performance-based assessments, and school accountability programs require federal and state leadership. National and state governments would need to provide funding for program development, evaluation, and dissemination of proven programs, including helping program designers set up national networks of trainers and implementer support.

The vision outlined here is far from being realized today, although there are elements of it on which progress is being made. For example, there is a national movement toward the use of performance assessments and school accountability systems; current Title I regulations mandate such systems for all states by the year 2000. The National Diffusion Network (NDN) has long promoted a list of "effective programs," both by funding "developer/disseminators" and by providing state facilitators in each state to promote program adoption. However, the evaluation standards for NDN programs are very low, and the system has been chronically underfunded. The New American Schools Development Corporation (NASDC) is currently funding the development and dissemination of seven promising schoolwide reform models.

One way the federal government might support the development of effective programs for each subject and grade level is to hold a series of design competitions, in which developers would be asked to submit plans for development and formative evaluation of models capable of meeting widely accepted performance standards. For example, the U.S. Department of Education might announce a design competition for development of approaches to beginning reading, specifying that all proposers must be prepared to implement their programs in grades K-3 and to evaluate their programs in comparison to control groups on standard measures to be provided by the Department (based on consultation with reading experts). Allowable costs, conditions of evaluation, and other conditions might be specified. A number of design teams might be funded initially with an expectation that teams not moving adequately toward effective and replicable designs might be winnowed out over time. Ultimately, models successful in their formative evaluations would be subjected to third-party evaluations against similar standards.

This design competition approach, used widely in industry and in the military, would unleash enormous creative energies among researchers, developers, and educators, but it could also have several important side benefits. One is that, because programs that succeed in rigorous third-party evaluations would find a ready market, publishers and other private-sector companies might participate in design competitions (and help fund them), just as pharmaceutical companies lavishly fund R & D that may lead to Food and Drug Adminstration approval. Second, I would expect that the public and politicians would be far more willing to pay for design competitions and third-party evaluations likely to produce conclusive solutions to problems of great importance than to fund more basic, less obviously crucial research by university-based investigators. Further, the development, successful evaluation, and widespread dissemination of effective programs for high-priority educational problems would be likely to give all of educational research greater visibility and support, just as public support for basic research in biology is assured by occasional breakthroughs in applied research in medicine.

The New American Schools Development Corporation (NASDC) is currently carrying out a design competition plan somewhat like what I am proposing here. NASDC originally funded eleven design teams (from almost 700 applications) to create whole-school reforms linked to national standards. Over time, four of these fell by the way-side, but the seven remaining designs are now beginning to be disseminated on a broad scale. The NASDC plan is different from what I am proposing in two key respects, however. First, it required whole-school change designs. All designs except our own (Roots and Wings) took on grades K-12 and every subject, and therefore generally could not develop or formatively evaluate specific student materials or other replicable components. Second, the only third-party evaluation (by the RAND Corporation) focuses on implementation quality and collection of local, routinely generated outcome data, not standard agreed-upon achievement indicators. Still, the apparent success of the NASDC approach to support the development of seven quite promising and innovative approaches and to disseminate them to a wide variety of schools hints at the potential of design competitions as a strategy. Before NASDC there were perhaps four major national reform networks capable of working with dozens or hundreds of schools.[1] To add seven whole-school designs to that list is an enormous achievement.

New Directions

The research and development work presented in this volume is continuing to move forward. As of this writing (January, 1996), we are completing the development of our Roots and Wings program. Our intention is to create 6300 hours of top-quality instruction supported by well-developed instructional materials, professional development, and school- and district-level supports. This is the number of hours of academic instruction from prekindergarten through grade 6. We are coming close to achieving this goal, with a full recognition that as soon as we have reached it we will need to begin work immediately on revisions. We also plan to begin development and evaluation of a Roots and Wings middle school design.

Our research continues to focus on building and evaluating school and classroom interventions capable of placing a floor under the achievement of all students while accelerating the learning of all. We are continuing the longitudinal evaluation of Success for All in ten school districts, and of Roots and Wings in three. We are experimenting with variations in the programs and are relating the extent and quality of implementation of various program elements to student outcomes. We are studying the process of "scaling up," focusing in particular on the role of local and national

---

1   Three of these reform networks were involved in NASDC designs; Success for All was the basis of Roots and Wings, and Comer's and Sizer's programs formed part of the basis of the NASDC ATLAS design. However, NASDC support greatly advanced development of these existing designs.

support networks in the quality and maintenance of program implementations. We are experimenting with a variety of means of disseminating Success for All and Roots and Wings on a broader scale.

At the same time as my colleagues and I are working on the nuts and bolts of development and research, we are also working to promote two key ideas. One is the idea, discussed often in this volume, that schools must be fundamentally restructured to abandon the sorting paradigm but instead to begin with the assumption that all children can learn to high levels, and then implement the programs and supports necessary to see that this assumption becomes a reality. The second is the idea that rigorous research, problem-focused development, and third-party evaluation of replicable programs, all tied to national standards and performance assessments, provide the most promising path not only to fundamental and lasting reform, but also to institutionalizing the process of reform. Not until educational systems begin to see evidence of effectiveness for students as a basis for decisions about policy and practice will we get off the pendulum of educational faddism. Not until educators select from among a variety of proven, replicable programs to meet ambitious objectives will the practice of education truly meet its potential to provide effective education for all children.

References

ABRAMI, P.C., LEVENTHAL, L., and PERRY, R.P. (1982). Educational seduction. *Review of Educational Research,52*(3), 446-462.

ABRAMS, J.D. (1983). Overcoming racial stereotyping in Red Bank, New Jersey. *School Administrator, 40*, 5, 7.

ADAMSON, D.P. (1971). *Differented multi-track grouping vs. uni-track educational grouping in mathematics.* Unpublished doctoral dissertation, Brigham Young University.

AIGNER, B.W. (1961). *A statistical analysis of achievement differences of children in a nongraded primary program and traditional classrooms.* Unpublished doctoral dissertation, Colorado State University, Fort Collins.

ALEXANDER, K.L., and COOK, M.A. (1982). Curricula and coursework: A surprise ending to a familiar story. *American Sociological Review, 47*, 626-640.

ALEXANDER, KL, COOK, M.A., and McDILL, E.L. (1978). Curriculum tracking and educational stratification. *American Sociological Review, 43*, 47-66.

ALLEN, S.D. (1991). Ability grouping research reviews: What do they say about grouping and the gifted? *Educational Leadership, 48* (6), 60-65.

ALLEN, W.H., and VAN SICKLE, R.L. (1984). Learning teams and low achievers. *Social Education, 48*, 60-64.

ALLINGTON, R.L. (1980). Poor readers don't get to read much in reading groups. *Language Arts, 57*,872-876.

ALLINGTON, R.L., and McGILL-FRANZEN, A. (1992). Does high-stakes testing improve school effectiveness? *ERS Spectrum, 10* (2), 3-12.

ALLPORT, G. (1954). *The nature of prejudice.* Cambridge, MA: Addison-Wesley.

ALPERT, J.L. (1974). Teacher behavior across ability groups: A consideration of the mediation of Pygmalion effects. *Journal of Educational Psychology, 66*, 348-353.

AMES, C., AMES, R., and FELKER, D. (1977). Effects of competitive reward structure and valence of outcome on children's achievement attributions. *Journal of Educational Psychology, 69*, 1-8.

AMES, G.J., and MURRAY, F.B. (1982). When two wrongs make a right: Promoting cognitive change by social conflict. *Developmental Psychology, 18*, 894-897.

ANANIA, J. (1983). The influence of instructional conditions on student learning and achievement. *Evaluation in Education: An International Review Series, 7*, 3-76.

ANASTASIOW, N.J. (1968). A comparison of two approaches in upgrading reading instruction. *Elementary English, 45*, 495-499.

ANDERSON, L.M., BRUBAKER, N.L., ALLEMAN-BROOKS, J., and DUFFY, G.G. (1985). A qualitative study of seatwork in first-grade classrooms. *Elementary School Journal, 86*, 123-140.

ANDERSON, L.M., EVERTSON, C., and BROPHY, J. (1979). An experimental study of effective teaching in first-grade reading-groups. *Elementary School Journal, 79*(3), 193-223.

ANDERSON, L.W. (1975, April). *Time to criterion: An experimental study.* Paper presented at the annual meeting of the American Educational Research Association, Washington, DC.

ANDERSON, L.W. (1976). An empirical investigation of individual differences in time to learn. *Journal of Educational Psychology, 68*, 226-233.

ANDERSON, L.W. (1985). A retrospective and prospective view of Bloom's "learning for mastery." In M.C. Wang & H.J. Walberg (Eds.), *Adapting instruction to individual differences* (pp. 254-268). Berkeley, CA: McCutchan.

ANDERSON, L.W., SCOTT, C., and HUTLOCK, N. (1976, April). *The effects of a mastery learning program on selected cognitive, affective and ecological variables in grades 1 through 6.* Paper presented at the annual meeting of the American Educational Research Association, San Erancisco.

ARCHER, J.A. (1988). *Feedback effects on achievement, attitude, and group dynamics of adolescents in interdependent cooperative groups for beginning second language and culture study.* Unpublished doctoral dissertation, University of Minnesota.

ARLIN, M. (1973). *Learning rate and learning rate variance under mastery learning conditions.* Unpublished doctoral dissertation, University of Chicago.

ARLIN, M. (1979). Teacher transitions can disrupt time flow in classrooms. *American Educational Research Journal, 16,* 42-56.

ARLIN, M. (1982). Teacher responses to student time differences in mastery learning. *American Journal of Education, 90,* 334-352.

ARLIN, M. (1984a). Time variability in mastery learning. *American Educational Research Journal, 21,* 103-120.

ARLIN, M. (1984b). Time, equality, and mastery learning. *Review of Educational Research,54,* 65-86.

ARLIN, M., and WEBSTER, J. (1983). Time costs of mastery learning. *Journal of Educational Psychology, 75,* 187-195.

ARLIN, M., and WESTBURY, L. (1976). The leveling effect of teacher pacing on science content mastery. *Journal of Research on Science Teaching, 13,* 213-219.

ARMSTRONG, B., JOHNSON, D.W., and BALOW, B. (1981). Effects of cooperative vs. individualistic learning experiences on interpersonal attraction between learning-disabled and normal-progress elementary school students. *Contemporary Educational Psychology, 6,* 102-109.

ARONSON, E., BLANEY, N., STEPHAN, C., SIKES, J., and SNAPP, M. (1978). *The Jigsaw classroom.* Beverly Hills, CA: Sage Publications, Inc.

ARTZT, A.F. (1983). *The comparative effects of the student-team method of instruction and the traditional teacher-centered method of instruction upon student achievement, attitude, and social interaction in high school mathematics courses.* Doctoral dissertation, New York University.

ATKINSON, J.W., and BIRCH, D. (1978). *Introduction to Motivation* (2nd Edition). New York: Van Nostrand.

ATKINSON, J.W., and O'CONNOR, P. (1963). *Effects of ability grouping in schools related to individual differences in achievement-related motivation* (Final Report, Cooperative Research Project No. OE-2-10-024). Washington, DC: U.S. Department of Health, Education and Welfare.

ACHMAN, A.M. (1968). *Factors related to the achievement of junior high school students in mathematics.* Unpublished doctoral dissertation, University of Oregon.

BAILEY, H.P. (1968). A study of the effectiveness of ability grouping on success in first-year algebra. *Dissertation Abstracts International, 28,* 3061A. (University Microfilms No. 681249)

BAKER, L., and BROWN, A.L. (1984). Metacognitive skills and reading. In P.D. Pearson (ed.), *Handbook of reading research.* New York: Longman.

BALLARD, M., CORMAN, L., GOTTLIEB, J., and KAUFFMAN, M. (1977). Improving the social status of mainstreamed retarded children. *Journal of Educational Psychology, 69,* 605-611.

BALOW I.H., and RUDDELL, A.K. (1963). The effects of three types of grouping on achievement. *California Journal of Educational Research, 14,* 108-117.

BALOW, B., and CURTIN, J. (1966). Ability grouping of bright pupils. *Elementary School Journal, 66,* 321-326.

BALOW, I. (1964). The effects of "homogeneous" grouping in seventh grade arithmetic. *Arithmetic Teacher, 12,* 186-191.

BALOW, I.H. (1962). Does homogeneous grouping give homogeneous groups? *Elementary School Journal, 63,* 28-32.

BANGERT-DROWNS, R.L. (1986). A review of developments in meta-analytic method. *Psychological Bulletin, 99,* 388-399.

BANGERT, R., KULIK, J., and KULIK, C. (1983). Individualized systems of imstruction in secondary schools. *Review of Educational Research, 53,* 143-158.

BARKER-LUNN, J.C. (1970). *Streaming in the primary school.* London: National Foundation for Education Research in England and Wales.

BARR, R. (1975). How children are taught to read: Grouping and pacing. *School Review, 83,* 479-498.

BARR, R. (1990). The social organization of literacy instruction. In S. McCormick & J. Zutell (Eds.), *Thirty-ninth yearbook of the National Reading Conference* (pp. 19-33). Chicago: National Reading Conference.

BARR, R. (1992). Teachers, materials, and group composition in literacy instruction. In *Elementary school literacy: Critical issues*, ed. M.J. Dreher and W.H. Slater, 19-33. Norwood, MA: Christopher-Gordon.

BARR, R., and DREEBEN, R. (1983). *How schools work*. Chicago: University of Chicago Press.

BARR, R.C. (1974). Instructional pace differences and their effect on reading acquisition. *Reading Research Quarterly, 9*, 526-554.

BARR, R.C. (1975). How children are taught to read: Grouping and pacing. *School Review, 83*, 479-498.

BARRINGER, C., and GHOLSON, B. (1979). Effects of type and combination of feedback upon conceptual learning by children: Implications for research in academic learning. *Review of Educational Research, 49*(3), 459-478.

BARTH, R. (1979). Home-based reinforcement of school behavior: A review and analysis. *Review of Educational Research, 49*(3), 436-458.

BARTHELMESS, H.M., and BOYER, P.A. (1932). An evaluation of ability grouping. *Journal of Educational Research, 26*, 284-294.

BARTON, D.P. (1964). An evaluation of ability grouping in ninth grade English. *Dissertation Abstracts International, 25*, 1731. (University Microfilms No. 64-9939).

BAUWENS, J., HOURCADE, J.J., and FRIEND, M. (1989). Cooperative teaching: A model for general and special education integration. *Remedial and Special Education, 10*(2), 17-22.

BECK, I., McKEOWN, M., McCASLIN, E., and BURKES, A. (1979). *Instructional dimensions that may affect reading comprehension: Examples from two commercial reading programs* (Technical Report No. 1979/20). Pittsburgh: University of Pittsburgh. Learning and Research and Development Center.

BEGLE, E.G. (1975). *Ability grouping for mathematics instruction: A review of the empirical literature*. Stanford Mathematics Education Study Group, Stanford University. (ERIC Document Reproduction Service No. ED 116 938).

BELL, N., GROSSEN, M., and PERRET-CLERMONT, A-N. (1985). Socio-cognitive conflict and intellectual growth. In M. Berkowitz (ed.), *Peer conflict and psychological growth*. San Francisco: Jossey-Bass.

BEREITER, C., and SCARDAMALIA, M. (1982). From conversation to composition: The role of Instruction in a developmental process. In R. Glaser (ed.), *Advances in instructional psychology*. Hillsdale, NJ: Erlbaum.

BERGMAN, P., and McLAUGHLIN, M (1978). *Federal programs supporting educational change: A model of education change, Vol. VIII: Implementing and sustaining innovations*. Santa Monica, CA: Rand.

BERKUN, M.M., SWANSON, L.W., and SAWYER, D.M. (1966). An experiment on homogeneous grouping for reading in elementary classes. *Journal of Educational Research, 59*, 413-414.

BICAK, L.J. (1962). *Achievement in eighth grade science by heterogeneous and homogeneous classes*. Unpublished doctoral dissertation, University of Minnesota.

BIERDEN, J.E. (1969). Provisions for individual differences in seventh grade mathematics based on grouping and behavioral objectives: An exploratory study. *Dissertation Abstracts International, 30*, 196A-197A. (University Microfilms No. 69-12042).

BIERNACKI, G.J. (1976). *An evaluation of an IGE/MUS-E inner-city elementary school based on selected criteria*. Unpublished doctoral dissertation, University of Toledo.

BILLETT, R.O. (1928). A controlled experiment to determine the advantages of homogeneous grouping. *Educational Research Bulletin, 7*, 139.

BILLETT, R.O. (1932). *The administration and supervision of homogeneous grouping*. Columbus: Ohio State University Press.

BLOCK, J.H. (1972). Student learning and the setting of mastery performance standards. *Educational Horizons, 50*, 183-191.

BLOCK, J.H., and ANDERSON, L.W. (1975). *Mastery learning in classroom instruction*. New York: MacMillan.

BLOCK, J.H., and BURNS, R.B. (1976). Mastery learning. In *Review of Research in Education (Vol. 4)*, ed. L.S. Shulman, 3-49. Itasca, IL: F.E. Peacock.

BLOOM, B.S. (1968). Learning for mastery. *Evaluation Comment, 1*, 2.

BLOOM, B.S. (1976). *Human characteristics and school learning* New York: McGraw-Hill.

BLOOM, B.S. (1980). The new direction in educational research: Alterable variables. *Phi Delta Kappan, 62*, 382-385.

BLOOM, B.S. (1981). *All our children learning*. New York: McGraw-Hill.

BLOOM, B.S. (1984a). The 2 sigma problem: The search for methods of instruction as effective as one-to-one tutoring. *Educational Researcher, 13*(6), 4-16.

BLOOM, B.S. (1984b). The search for methods of group instruction as effective as one-to-one tutoring. *Educational Leadership, 41*(8), 4-17.

BLOOM, B.S., HASTINGS, J.T., and MADAUS, G.F. (1971). *Handbook on formative and summative evaluation of student learning*. New York: McGraw-Hill.

BOCKRATH, S.M.B. (1958). *An evaluation of the ungraded primary as an organizational device for improving learning in Saint Louis Archdiocesan Schools*. Unpublished doctoral dissertation, St. Louis University.

BORG, W., (1965). Ability grouping in the public schools. *The Journal of Experimental Education, 34*, 1-97.

BORG, W., and PRPICH, T. (1966). Grouping of slow learning high school pupils. *Journal of Secondary Education, 41*, 231-238.

BORG, W.R. (1965). Ability grouping in the public schools: A field study. *Journal of Expenmental Education, 34*, 1-97.

BOSSERT, S. (1989). Cooperative activities in the classroom. In E. Rothkopf (Ed.), *Review of research in education* (Vol. 15, pp. 225-250). Washington, DC: American Educational Research Association.

BOWMAN, B.L. (1971). *A comparison of pupil achievement and attitude in a graded school with pupil achievement and attitude in a nongraded school 1968-69*. Unpublished doctoral dissertation, University of North Carolina, Chapel Hill.

BOYKIN, A.W. (1996, April). *A talent development approach to school reform*. Paper presented at the annual meetings of the American Educational Research Association, New York.

BRADDOCK, J.H. (1990). *Tracking Implications for student race-ethnic subgroups* (Technical Report No. 1). Baltimore: Center for Research on Effective Schooling for Disadvantaged Students.

BRADFORD, E.F. (1972). *A comparison of two methods of teaching in the elementary school as related to achievement in reading, mathematics, and self-concept of children*. Unpublished doctoral dissertation, Michigan State University, East Lansing.

BREIDENSTINE, A.G. (1936). The educational achievement of pupils in differentiated and undifferentiated groups. *Journal of Experimental Education, 5*, 91- 135.

BREMER, N. (1958). First grade achievement under different plans of grouping. *Elementary English, 35*, 324-326.

BREWER, D.J., REES, D.I., and ARGYS, L.M. (1995). Detracking America's schools: The reform without cost? *Phi Delta Kappan, 77*(3), 210-215.

BRODY, E.B. (1970). Achievement of first-and second-year pupils in graded and nongraded classrooms. *Elementary School Journal, 70*, 391-394.

BROPHY, J. 1987. Synthesis of research on strategies for motivating students to learn. *Educational Leadership, 5* (Oct.), 40-48.

BROPHY, J.E. and GOOD, T.L. (1986). Teacher behavior and student achievement. In *Handbook of Research on Teaching*, Third Edition, ed. M.C. Wittrock, 328-357. New York: MacMillan.

BROPHY, J.E., and EVERSTON, C.M. (1974). *Process-product correlations in the Texas teacher effectiveness study: Final report* (Research Report No. 74-4). Austin: Research & Development Center for Teacher Education, University of Texas.

BROPHY, J.E., and GOOD, T.C. (1986). Teacher behavior and student achievement. In M.C. Wittrock (Ed.), *Handbook of research on teaching* (3rd ed., pp. 328-375). New York: Macmillan.

BROWN, A., and CAMPIONE, J. (1986). Psychological theory and the study of learning disabilities. *American Psychologist, 14*, 1059-1068.

BROWN, A., and PALINCSAR, A. (1989). Guided cooperative learning and individual knowledge acquisition. In L. Resnick (Ed.), *Knowing, learning, and instruction: Essays in honor of Robert Glaser* (pp. 393-452). Hillsdale, NJ: Erlbaum.

BRYAN, T., BAY, M., LOPEZ-REYNA, N., and DONAHUE, M. (1991). Characteristics of students with learning disabilities: A summary of the extant data base and its implications for educational programs. In J. Lloyd, N. Singh, & A. Repp (Eds.), *The regular education initiative: Alternative perspectives on concepts, issues, and models* (pp. 113-131). Sycamore, IL: Sycamore.

BRYANT, N.D., FAYNE, H.R., and GETTINGER, M. (1982). Applying the mastery model to sight word instruction for disabled readers. *Journal of Experimental Education, 50*, 116- 121.

BRYK, A., RAUDENBUSH, S., SELTZER, M., and CONGDON, R. (1988). *An introduction to HLM: Computer program and user's guide*. Chicago: University of Chicago, Department of Education.

BUFFIE, E.G.W. (1962). *A comparison of mental health and academic achievement: The nongraded school vs. the graded school.* Unpublished doctoral dissertation, Indiana University.

BUREHYETT, J.A. (1972). *A comparison of the effects of non-graded, multi-age, team-teaching vs. the modified self-contained classroom at the elementary school level.* Unpublished doctoral dissertation, Michigan State University, East Lansing.

BURKE, A.J. (1983). *Students' potential for learning contrasted under tutorial and group approaches to instruction.* Unpublished doctoral dissertation, University of Chicago.

BURTLEY, N. (1974). *A comparison of teacher characteristics and student achievement in individually guided education (IGE) and traditional inner city elementary schools.* Unpublished doctoral dissertation, Michigan State University, East Lansing.

CABEZON, E. (1984). *The effects of marked changes in student achievement pattern on the students, their teachers, and their parents: The Chilean case.* Unpublished doctoral dissertation, University of Chicago.

CAMPBELL, A.L. (1965). A comparison of the effectiveness of two methods of class organization for the teaching of arithmetic in junior high school. *Dissertation Abstracts International, 26,* 813-814. (University Microfilms No. 65-6726).

CARBONE, R.F. (1961). A comparison of graded and non-graded elementary schools. *Elementary School Journal, 61,* 82-88.

CARLBERG, C., and KAVALE, K. (1980). The efficacy of special versus regular class placement for exceptional children: A meta-analysis. *Journal of Special Education, 14,* 295-309.

CARROLL, J.B. (1963). A model for school learning. *Teachers College Record, 64,* 723-733.

CARROLL, J.B. (1989). The Carroll model: A 25-year retrospective and prospective view. *Educational Researcher, 18*(1), 26-31.

CARSON, R.M., and THOMPSON, J. (1964). The Joplin Plan and traditional reading groups *Elementary School Journal, 65,* 387-392.

CARTWRIGHT, G.P., and McLNTOSH, D.K (1972). Three approaches to grouping procedures for the education of disadvantaged primary school children. *Journal of Educational Research, 65,* 425-429.

CASE, D.A. (1970). *A comparative study of fifth graders in a new middle school with fifth graders in elementary self-contained classrooms.* Unpublished doctoral dissertation, University of Florida, Gainesville.

CHALFANT, L.S. (1972). *A three year comparative study between students in a graded and nongraded secondary school.* Unpublished doctoral dissertation, Utah State University, Logan.

CHANCE, C.E. (1980). *The effects of feedback/corrective procedures on reading achievement and retention.* Unpublished doctoral dissertation, University of Chicago.

CHASTAIN, C.S. (1961). *An experimental study of the gains in achievement in arithmetic and reading made by the pupils in the intermediate grades in the Rangely, Colorado, Elementary School platoons who were instructed in traditional classrooms, in achievement and in nongraded classrooms.* Unpublished doctoral dissertation, University of Northern Colorado, Greeley.

CHIOTTI, J.F. (1961). *A progress comparison of ninth grade students in mathematics from three school districts in the state of Washington with varied methods of grouping.* Unpublished doctoral dissertation, University of Northern Colorado.

CHISMAR, M.H. (1971). *A study of the effectiveness of cross-level grouping of middle school under-achievers for reading instruction.* Unpublished doctoral dissertation, Kent State University.

CLARKE, S.C.T. (1958). The effect of grouping on variability in achievement at the grade 3 level. *Alberta Journal of Educational Research, 4,* 162-171.

COCHRAN, J. (1961). Grouping students in junior high school. *Educational Leadership, 18,* 414-419.

COHEN, S.A. (1977). Instructional systems in reading: A report on the effects of a curriculum design based on a systems model. *Reading World, 16,* 158-171.

COHEN, S.A., and RODRIQUEZ, S. (1980). Experimental results that question the Ramirez Castaneda model for teaching reading to first grade Mexican Americans. *The Reading Teacher, 34,* 12-18.

COLDIRON, J.R., BRADDOCK, J.H., and McPARTLAND, J.M. (1987, April). *A description of school structures and classroom practices in elementary, middle, and secondary schools.* Paper presented at the annual meeting of the American Educational Research Association Washington, DC.

COLEMAN, J. S., CAMPBELL, E., HOBSON, C., McPARTLAND, J., MOOD, A., WEINFELD, F., and YORK, R. (1966). *Equality of educational opportunity.* Washington, DC: U.S. Department of Health, Education, and Welfare.

COLEMAN, J.S., and KARWEIT, N.L. (1972). *Information systems and performance measures in schools.* Englewood Cliffs, NJ: Educational Technology Publications.

COLLINS, A., BROWN, J., and NEWMAN, S. (1989). Cognitive apprenticeship: Teaching the crafts of reading, writing, and mathematics. In L. Resnick (Ed.), *Knowing, learning, and instruction: Essays in honor of Robert Glaser* (pp. 453-494). Hillsdale, NJ: Erlbaum.

COLLINS, K.M. (1971). *An investigation of the variables of Bloom's mastery learning model for teaching junior high school math.* Unpublished doctoral dissertation, Purdue University, Lafayette, IN.

COMER, J. (1988). Educating poor minority children. *Scientific American, 259,* 42-48.

COOK, T., and LEVITON, L. (1980). Reviewing the literature: A comparison of traditional methods with meta-analysis. *Journal of Personality, 48,* 449472.

COOLEY, W.W., and LEINHARDT, G. (1980). The instruction dimensions study. *Educational Evaluation and Policy Analysis, 2,* 7-25.

COOPER, H. (1989). Synthesis of research on homework. *Educational Leadership, 47*(3), 85-91.

COOPER, H.M. (1984). *The integrative research review: A systematic approach.* Beverly Hills, CA: Sage.

COTTLE, T.J. (1974). What tracking did to Ollie Taylor. *Social Policy, 5,* 21-24.

CRESPO, M., and MICHELNA, J. (1981). Streaming, absenteeism, and dropping out. *Canadian Journal of Education, 6,* 40-55.

CROOKS, T.J. (1988). The impact of classroom evaluation practices on students. *Review of Educational Research, 58*(3), 438-481.

CROTTY, E.K. (1975). An experimental comparison of a mastery learning and lecture-discussion approach to the teaching of world history. *Dissertation Abstracts International, 36,* 7150A. (University Microfilms No. 76-10610)

CTB/McGRAW-HILL. (1979). *California Achievement Tests, Forms C and D, Levels 10-19 Technical Bulletin 1.* Monterey, CA: Author.

CTB/McGRAW-HILL. (1986). *California Achievement Tests Forms E and F: Technical Bulletin 2.* Monterey, CA: CTB/McGraw-Hill.

CUSHENBERRY, D. (1966). Two methods of grouping for reading instruction. *Elementary School Journal, 66,* 267-271.

DAHL, P.R. (1979). An experimental program for teaching high speed word recognition and comprehension skills. In J.E. Button, T.C. Lovitt, & T.D. Rowland (Eds.), *Communication research in learning disabilities and mental retardation* (pp. 33-65). Baltimore, MD: University Park Press.

DAHLLOFF, U. (1971). *Ability grouping content validity, and curriculum process analysis.* New York: Teachers College Press.

DANIELS, J.C. (1961). The effects of streaming in the primary school: Comparison of streamed and unstreamed schools. *British Journal of Educational Psychology, 31,* 119-126.

DAR, Y., and RESH, N. (1986). Classroom intellectual composition and academic achievement. *American Educational Research Journal, 23,* 357-374.

DAVIDSON, N. (1985). Small-group learning and teaching in mathematics: A selective review of research. In R. Slavin, S. Sharan, R. Hertz-Lazarowitz, C. Webb, & R. Schmuck (Eds.), *Learning to cooperate, cooperating to learn* (pp. 147-172). New York: Plenum.

DAVIS, O.L., and TRACY, N.H. (1963). Arithmetic achievement and instructional grouping *Arithmetic Teacher, 10,* 12-17.

DEFORD, D.E., PINNELL, G.S., LYONS, C.A., and YOUNG, P. (1987). *Ohio's Reading Recovery program: Vol. VII, Report of the follow-up studies.* Columbus, OH: Ohio State University.

DEUTSCH, M. (1949). An experimental study of the effects of cooperation and competition upon group processes. *Human Relations, 2,* 199-231.

DEWAR, J. (1964). Grouping for arithmetic instruction in sixth grade. *Elementary School Journal, 63,* 266-269.

DIANDA, M.R. and FLAHERTY, J.F. (April 1995). *Effects of Success for All on the reading achievement of first graders in California bilingual programs.* Paper presented at the annual meeting of the American Educational Research Association, San Francisco.

DIANDA, M.R., MADDEN, N.A., and SLAVIN, R.E. (1993, April). *Lee Conmigo: Success for All in schools serving limited English proficient students.* Paper presented at the annual meeting of the American Educational Research Association, Atlanta.

DILLASHAW, F.D., and OKEY, J.R. (1983). Effects of a modified mastery learning strategy on achievement, attitudes, and on-task behavior of high school chemistry students. *Journal of Research in Science Teaching, 20,* 203-211.

DOLAN, L.J., and KELLAM, S.G. (1987). *Preliminary report of mastery learning intervention during first grade with cohort I students*. Baltimore, MD: Johns Hopkins University, Department of Mental Hygiene, Prevention Center.

DOLAN, L.J., KELLAM, S.G., BROWN, C.H., WERTHAMER-LARSSON, L., REBOK, G.W., MAYER, L.S. LAUDOLFF, J., TURKKAN, J.S., FORD, C. and WHEELER, L. (1993). The short-term impact of two classroom-based preventive interventions on aggressive and shy behaviors and poor achievement. *Journal of Applied Developmental Psychology, 14*(2), 317-345.

DOUGLAS, J.F. (1973). A study of streaming at a grammar school. *Educational Research, 15,* 140-143.

DREWS, E.M. (1963). *Student abilities, grouping patterns, and classroom interaction* (Cooperative Research Project No. 608). Washington, DC: U.S. Department of Health, Education, and Welfare.

DUNKELBERGER, G.E., and HEIKKINEN, H. (1984). The influence of repeatable testing on retention in mastery learning. *School Science and Mathematics, 84,* 590-597.

DUNKIN, M. (1978). Student characteristics, classroom processes, and student achievement. *Journal of Educational Psychology, 70*(6), 998-1009.

DUREN, D. (1981). Reading comprehension instruction in 5 basal reader series. *Reading Research Quarterly, 16,* 515-544.

DURKIN, D. (1978-1979). What classroom observations reveal about reading comprehension instruction. *Reading Research Quarterly, 14,* 481-533.

EASH, M.J. (1961). Grouping: What have we learned? *Educational Leadership, 18,* 429-434.

EDDLEMAN, V.K. (1971). A comparison of the effectiveness of two methods of class organization for arithmetic instruction in grade five. *Dissertation Abstracts International, 32,* 1744A. (University Microfilms No. 71-25035).

EDMINSTON, R.W., and BENFER, J.G. (1949). The relationship between group achievement and range of abilities within groups. *Journal of Educational Research, 42,* 547-548.

EDWARDS, K.J., and DEVRIES, D.L. (1972). *Learning games and student teams: Their effects on student attitudes and achievement* (Report No .147). Baltimore: Johns Hopkins University, Center for Social Organization of Schools.

EDWARDS, K.J., and DEVRIES, D.L. (1974). *The effects of Teams-Games-Tournaments and two structural variations on classroom process, student attitudes, and student achievement* (Report No. 172). Baltimore: Johns Hopkins University, Center for Social Organization of Schools.

EDWARDS, K.J., DEVRIES, D.L., and SNYDER, J.P. (1972). Games and teams: A winning combination. *Simulation and Games, 3,* 247-269.

ELLIS, A.K, and FOUTS, J.T. (1993). *Research on educational innovations*. Princeton Junction, NJ: Eye on Education.

EMMER, E.T., and AUSSIKER, A. (1990). School and classroom discipline programs: How well do they work? In *Student discipline strategies*, ed. O.C. Moles, 111-136. Albany: State University of New York Press.

ENEVOLDSEN, C.L. (1961). *An evaluation of the ungraded primary program in selected schools in the Lincoln, Nebraska, Public School system*. Unpublished doctoral dissertation, University of Nebraska, Teachers College, Lincoln.

ENGEL, B.M., and COOPER, M. (1971). Academic achievement and nongradedness. *The Journal of Experimental Education, 40,* 24-26.

ESPOSITO, D. (1973). Homogeneous and heterogeneous ability grouping: Principal findings and implications for evaluating and designing more effective educational environments. *Review of Educational Research 43,* 163-179.

EVERTSON, C. (1982). Differences in instructional activities in average- and low-achieving junior high English and math classes. *Elementary School Journal, 82,* 329-350.

EVERTSON, C., SANFORD, J., and EMMER, E. (1981). Effects of class heterogeneity in junior high school. *American Educational Research Journal, 18,* 219-232.

EVERTSON, C.M. (1982). Differences in instructional activities in higher- and lower-achieving junior high English and math classes. *Elementary School Journal, 82,* 329-350.

EYSENCK, H.J. (1978). An exercise in mega-silliness. *American Psychologist, 33,* 517.

FAGAN, J.S. (1976). Mastery learning: The relationship of mastery procedures and aptitude to the achievement and retention of transportation-environmental concepts by seventh grade students. *Dissertation Abstracts International, 36,* 5981. (University Microfilms No. 766402).

FANTUZZO, J.W., KING, J.A., and HELLER, L.R (1992). Effects of reciprocal peer tutoring on mathematics and school adjustment: A component analysis. *Journal of Educational Psychology, 84*(3), 331-339.

FANTUZZO, J.W., POLITE, K, and GRAYSON, N. (1990). An evaluation of reciprocal peer tutoring across elementary school settings. *Journal of School Psychology, 28*, 309-323.

FANTUZZO, J.W., RIGGIO, R.E., CONNELLY, S., and DIMEFF, L.A. (1989). Effects of reciprocal peer tutoring on academic achievement and psychological adjustment: A component analysis. *Journal of Educational Psychology, 81*, 173-177.

FARIVAR, S. (1992, April). *Middle school math students' reactions to heterogeneous small group work: They like it!* Paper presented at the annual meeting of the American Educational Research Association, San Francisco.

FELDHUSEN, J., and MOON, S. (1992). Grouping gifted students: Issues and concerns. *Gifted Child Quarterly, 36*(2), 63-67.

FELDHUSEN, J.F. (1989). Synthesis of research on gifted youth. *Educational Leadership, 46*(6), 6-11.

FELMLEE, D., and EDER, D. (1983). Contextual effects in the classroom: The impact of ability groups on student attention. *Sociology of Education, 56*, 77-87.

FERRI, E. (1971). *Streaming: Two years later.* London: National Foundation for Educational Research in England and Wales.

FICK, W.W. (1963). The effectiveness of ability grouping in seventh grade core classes. *Dissertation Abstracts International, 23*, 2753.

FINDLEY, W.G., and BRYAN, M. (1971). *Ability grouping: 1970 status, impact, and alternatives.* Athens: Center for Educational Improvement, University of Georgia. (ERIC Document Reproduction Service No. ED 060 595).

FITZGERALD, J., and SPIEGAL, D. (1983). Enhancing children's reading comprehension through instruction in narrative structures. *Journal of Reading Behavior, 14*, 1-181.

FITZPATRICK, K.A. (1985, April). *Group-based mastery learning A Robin Hood approach to instruction?* Paper presented at the annual meeting of the American Educational Research Association, Chicago.

FLAIR, M.D. (1964). The effect of grouping on achievement and attitudes toward learning of first grade pupils. *Dissertation Abstracts, 25*, 6430. (University Microfilms No. 65-03,259).

FLOWER, L., and HAYES, J. (1980). The dynamics of composing: Making plans and juggling constraints. In L. Gregg and E. Steinberg (eds.), *Cognitive processes in writing.* Hillsdale, NJ: Erlbaum.

FLOWERS, J.R. (1977). *A comparative study of students in open space individually guided education (IGE) and traditional schools.* Unpublished doctoral dissertation, University of Northern Colorado, Greeley.

FLOYD, C. (1954). Meeting children's reading needs in the middle grades: A preliminary report. *Elementary School Journal, 55*, 99-103.

FOGELMAN, K., ESSEN, J., and TIBBENHAM, A. (1978). Ability-grouping in secondary schools and attainment. *Educational Studies, 4*, 201-212.

FORD, S. (1974). *Grouping in mathematics: Effects on achievement and learning environment.* Unpublished doctoral dissertation, Yeshiva Univesity.

FOWLKES, J.G. (1931). Homogeneous or heterogeneous group – which? *The Nation's Schools, 8*, 74-78.

FOX, L.H. (1979). Programs for the gifted and talented: An overview. In A.H. Passow (Ed.) *The gifted and talented: Their education and development* (pp.104-126). Chicago: University of Chicago Press.

FOYLE, H.C., LYMAN, L.R, TOMPKINS, L., PERNE, S., and FOYLE, D. (1993). Homework and cooperative learning: A classroom field experiment. *Illinois School Research and Development, 29* (3), 25-27.

FRANTZ, L.J. (1979). The effects of the student teams achievement approach in reading on peer attitudes. Master's thesis, Old Dominion University.

FREDERICK, W., and WALBERG, H. (1980). Learning as a function of time. *Journal of Educational Research, 73*, 183-194.

FROEMEL, J. (1980). *Cognitive entry behaviors, instructional conditions, and achievement: A study of their interrelationships.* Unpublished doctoral dissertation, University of Chicago.

FUCHS, D., and FUCHS, L. (1991). Framing the REI debate: Abolitionists versus conservationists. In J. Lloyd, N. Singh, & A. Repp (Eds.), *The regular education initiative: Alternative perspectives on concepts, issues, and models* (pp. 241-255). Sycamore, IL: Sycamore.

FUCHS, L.S., TINDAL, F., and FUCHS, D. (1985). *A comparison of mastery learning procedures among high and low ability students.* Unpublished manuscript. Vanderbilt University, Nashville, TN. (ERIC Document Reproduction Service No. ED 259 307).

GALLAGHER, J. (1991). Educational reform, values, and gifted students. *Gifted Child Quarterly, 35*(1), 12-19.

GAMORAN, A. (1987). The stratification of high school learning opportunities. *Sociology of Education, 60*, 135-155.

GAMORAN, A. (1989). Measuring curriculum differentiation. *American Journal of Education, 97*, 129-143.

GAMORAN, A., and BERENDS, M. (1987). The effects of stratification in secondary schools: Synthesis of survey and ethnographic research. *Review of Educational Research, 57*, 415-435.

GAMORAN, A., and MARE, R.D. (1989). Secondary school tracking and educational inequality: Compensation, reinforcement, or neutrality? *American Journal of Sociology, 94*, 1146-1183.

GAO, (1994). *Limited English proficiency: A growing and costly educational challenge facing many school districts.* Washington, DC: United States General Accounting Office.

GARCIA, E.E. (1991). Bilingualism, second language acquisition, and the education of Chicano language minority students. In R.R. Valencia (Ed.), *Chicano school failure and success: Research and policy agendas for the 1990's.* New York: Falmer.

GARCIA, E.E. (1994, April). *The impact of linguistic and cultural diversity on America's schools: A need for new policy.* Paper presented at the annual meeting of the American Educational Research Association, New Orleans.

GEFFNER, R. (1978). The effects of interdependent learning on self-esteem, interethnic relations, and intra-ethnic attitudes of elementary school children: A field experiment. Doctoral dissertation, University of California, Santa Cruz.

GERARD, H.B., and MILLER, N. (1975). *School desegregation: A long-range study.* New York: Plenum.

GERSTEN, R., WALKER, D., and DARCH, C. (1988). Relationship between teachers' effectiveness and their tolerance for handicapped students: An exploratory study. *Exceptional Children, 54*, 433-438.

GETTINGER, M. (1985). Time allocated and time spent relative to time needed for learning as determinants of achievement. *Journal of Educational Psychology, 77*, 3-11.

GIAEONIA, R.M., and HEDGES, L.V. (1982). Identifying features of open education. *Review of Educanonal Research, 52*, 579-602.

GIVENS, H., JR. (1972). *A comparative study of achievement and attitudinal characteristics of black and white intermediate pupils in individualized, multigrade and self-contained instructional programs.* Unpublished doctoral dissertation, St. Louis University.

GLASS, G. and STANLEY, J.C. (1970). *Statistical methods in education and psychology.* Englewood Cliffs, NJ: Prentice-Hall.

GLASS, G., COHEN, L., SMITH, M. L. and FILBY, N. (1982). *School class size.* Beverly Hills, CA: Sage.

GLASS, G.V. (1976). Primary, secondary, and meta-analysis of research. *Educational Research, 5*, 3-8.

GLASS, G.V., McGAW, B., and SMITH, M.L. (1981). *Meta-analysis in social research.* Beverly Hills: Sage.

GOLD, R.M., REILLY, A., SILBERMAN, R., and LEHR, R. (1971). Academic achievement declines under pass-fail grading. *Journal of Experimental Education, 39*(1), 17-21.

GOLDBERG, M.L., PASSOW, A.H., and JUSTMAN, J. (1966). *The effects of ability grouping.* New York: Teachers College Press.

GOLDSTEIN, H., MOSS, J., and JORDAN, J. (1966). *The efficacy of special class training on the development of mentally retarded children* (Cooperative Research Project no. 619). Washington, DC: U.S. Office of Education.

GONZALES, A. (1981). An approach to interdependent/cooperative bilingual education and measures related to social motives. Manuscript, California State University at Fresno.

GONZALES, A. (August, 1979). *Classroom cooperation and ethnic balance.* Paper presented at the annual convention of the American Psychological Association, New York.

GOOD, T. (1987). Teacher expectations. In *Talks to Teachers*, ed. D. Berliner and B. Rosenshine, 159-200. New York: Random House.

GOOD, T., and MARSHALL, S. (1984). Do students learn more in heterogeneous or homogeneous groups? In *The social context of instruction: Group organization and group processes*, ed. P. Peterson, L.C. Wilkinson, and M. Hallinan, 15-38. New York: Academic Press.

GOOD, T.L., and MARSHALL, S. (1984). Do students learn more in heterogeneous or homogeneous groups? In P. Peterson, L.C. Wilkinson, & M. Hallinan (Eds.), *The social context of instruction: Group organization and group processes* (pp. 15-38). New York: Academic Press.

GOOD, T.L., GROUWS, D., and EBMEIER, H. (1983). *Active mathematics teaching.* New York: Longman.

GOOD, T.L., and BROPHY, J.E. (1984). *Looking in classrooms* (third ed .). New York: Harper & Row.

GOOD, T.L., and GROUWS, D. (1977). Teaching effects: A process-product study in fourth grade mathematics classes. *Journal of Teacher Education, 28,* 49-54.

GOOD, T.L., and GROUWS, D. (1979). The Missouri Mathematics Effectiveness Project: An experimental study in fourth grade classrooms. *Journal of Educational Psychology, 71,* 355-362.

GOOD, T.L., and MARSHALL, S. (1984). Do students learn more in heterogeneous or homogeneous groups? In P. Peterson, L. Cherry Wilkinson, & M. Hallinan (Eds.), *Student diversity and the organization, process, and use of instructional groups in the classroom* (pp. 15-37). New York: Academic Press.

GOOD, T.L., GROUWS, D., and EBMEIER, H. (1983). *Active mathematics teaching.* New York: Longman.

GOODLAD, J.I. (1983). *A place called school.* New York: McGraw-Hill.

GOODLAD, J.I., and ANDERSON, R.H. (1959). *The nongraded elementary school.* New York: Harcourt, Brace & Co.

GOODLAD, J.I., and ANDERSON, R.H. (1963). *The nongraded elementary school* (rev. ed.). New York: Harcourt, Brace, & World.

GOODLAD, J.I., and ANDERSON, R.H. (1987). *The nongraded elementary school.* New York: Teachers College Press. (Original work published 1963).

GOODMAN, H., GOTTLIEB, J., and HARRISON, R.H. (1972). Social acceptance of EMRs integrated into a non-graded elementary school. *American Journal of Mental Deficiency, 76,* 412-417.

GOTTFREDSON, S. (1978). Evaluating psychological research reports. *American Psychologist, 33,* 920-934.

GOTTLIEB, J., CORMAN, L., and CURCI, R. (1984). Attitudes toward mentally retarded children. In R. Jones (Ed.), *Attitudes and attitude change in special education: Theory and practice* (pp. 143-156). Reston, VA: Council for Exceptional Children.

GRAVES, D. (1978). *Balance the basics: Let them write.* New York: Ford Foundation.

GRAVES, D. (1983). *Writing: Teachers and children at work.* Exeter, NH: Heinemann.

GRAVES, N., and GRAVES, T. (1985). Creating a cooperative learning environment: An ecological approach. In R.E. Slavin, S. Sharan, S. Kagan, R. Hertz-Lazarowitz, C. Webb, and R Schmuck (eds.), *Learning to Cooperate, Cooperating to Learn.* New York: Plenum.

GRAY, J., and MYERS, M. (1978). Bay Area writing project. *Phi Delta Kappan, 59,* 410-413.

GREEN, D.R., and RILEY, H.W. (1963). Interclass grouping for reading instruction in the middle grades. *Journal of Experimental Education, 31,* 273-278.

GREENWOOD, C.R, DELQUADRI, J.C., and HALL, R.V. (1989). Longitudinal effects of classwide peer tutoring. *Journal of Educational Psychology, 81,* 371-383.

GREENWOOD, C.R., CARTA, J.C., and HALL, R.V. (1988). The use of peer tutoring strategies in classroom management and educational instruction. *School Psychology Review, 17,* 258-275.

GREENWOOD, C.R., DINWIDDIE, G., BAILEY, V., CARTA, J.J., DORSEY, D., KOHLER, F., NELSON, C., ROTHOLZ, D., and SCHULTE, D. (1987). Field replication of classwide peer tutoring. *Journal of Applied BehaviorAnalysis, 20,* 151-160.

GREENWOOD, C.R., DINWIDDIE, G., TERRY, B., WADE, L., STANLEY, S., THIBADEAU, S., and DELQUADRI, J. (1984). Teacher versus peer-mediated instruction: An eco-behavioral analysis of achievement outcomes. *Journal of Applied BehaviorAnalysis, 17,* 521-538.

GUARINO, A.R. (1982). *An investigation of achievement, self-concept, and school related anxiety in graded and nongraded elementary schools.* Unpublished doctoral dissertation, State University of New Jersey, Rutgers.

GUMPPER, D.C., MEYER, J.H., and KAUFMAN, J.J. (1971). *Nongraded elementary education: Individualized learning-teacher leadership-student responsibility.* University Park: Pennsylvania State University, Institute for Research on Human Resources. (ERIC Document Reproduction Service No. ED 057 440).

GUSKEY, T.R. (1982). The effects of change in instructional effectiveness on the relationship of teacher expectations and student achievement. *Journal of Educational Research, 75,* 345-349.

GUSKEY, T.R. (1984). The influence of change in instructional effectiveness upon the affective characteristics of teachers. *American Educational Research Journal, 21,* 245-259.

GUSKEY, T.R., and GATES, S.L. (1985). A synthesis of research on group-based mastery learning programs. Paper presented at the annual convention of the American Educational Research Association, March 28, Chicago.

GUSKEY, T.R., and GATES, S.L. (1986). Synthesis of research on the effects of mastery learning in elementary and secondary classrooms. *Educational Leadership, 43* (8), 73-80.

GUSZAK, F.J. (1967). Teacher questioning and reading. *Reading Teacher, 21*, 27-34.

GUTÍERREZ, R., and SLAVIN, R.E. (1992). Achievement effects of the nongraded elementary school: A best-evidence synthesis. *Review of Educational Research, 62*(4), 333-376.

GUTKIN, J. (1985, March). *The effect of diagnostic inservice training on class reading achievement and the number of lessons covered.* Paper presented at the annual meeting of the American Educational Research Association, Chicago.

HALES, L.W., BAIN, P.T., and RAND, L.P. (1971). An investigation of some aspects of the pass-fail grading system. Paper presented at the annual meeting of the American Educational Research Association, February 17, New York.

HALLER, E., and DAVIS, S. (1980). Does socioeconomic status bias the assignment of elementary school students to reading groups? *American Educational Research Journal, 17*, 409-418.

HALLIWELL, J.W. (1963). A comparison of pupil achievement in graded and nongraded primary classrooms. *Journal of Experimental Education, 32*, 59-64.

HAMBLIN, R.L., HATHAWAY, C.,and WODARSKI, J.S. (1971). Group contingencies, peer tutoring, and accelerating academic achievement. In E. Ramp and W. Hopkins (eds.), *A new direction for education: Behavior analysis* (pp. 41-53). Lawrence: University of Kansas, Department of Human Development.

HANSELL, S., and SLAVIN, R.E. (1981). Cooperative learning and the structure of interracial friendships. *Sociology of Education, 54,* 98-106.

HANSEN, J. (1981). The effects of inference training and practice on young children's reading comprehension. *Reading Research Quarterly, 16*, 391-417.

HARRAH, D.D. (1956). A study of the effectiveness of five kinds of grouping in the classroom. *Dissertation Abstracts International, 17*, 715. (University Microfilms No. 56-1114).

HART, R.H. (1959). The effectiveness of an approach to the problem of varying abilities in teaching reading. *Journal of Educational Research, 52*, 228-231.

HART, R.H. (1962). The nongraded primary school and arithmetic. *The Arithmetic Teacher, 9*, 130 133.

HARTILL, R. (1936). *Homogeneous grouping as a policy in the elementary schools in New York City (Teachers College, Columbia: Contributions to Education,* No. 690). New York: Columbia University, Teachers College.

HAWKINS, J.D., DOUECK, H.J., and LISHNER, D.M. (1988). Changing teacher practices in mainstream classrooms to improve bonding and behavior of low achievers. *American Educational Research Journal, 25*(1), 31-50.

HAYES, L. (1976). The use of group contingencies for behavioral control: A review. *Psychological Bulletin, 83*, 628-648.

HEATHERS, G. (1967). *Organizing schools through the dual progress plan.* Danville, IL: Interstate.

HEATHERS, G. (1969). Grouping. In R. Ebel (Ed.), *Encyclopedia of educational research* (fourth ed., pp. 559-570). New York: Macmillan.

HEATHERS, G. (1969). Grouping. In R. Ebel (Ed.), *Encyclopedia of educational research* (4th ed., pp. 559-570). New York: Macmillan.

HECHT, L.W. (1980, April). *Stalking mastery learning in its natural habitat.* Paper presented at the annual meeting of the American Educational Research Association, Boston.

HEDGES, L., and OLKIN, I. (1985). *Statistical methods for meta-analysis.* New York: Academic Press.

HEDGES, L.V. (1981). Distribution theory for Glass's estimator of effect size and related estimators. *Journal of EducationalStatistics, 6,107-*128.

HELLER, L.R, and FANTUZZO, J.W. (in press). Reciprocal peer tutoring and parent partnership: Does parent involvement make a difference? *School Psychology Review.*

HENN, D.C. (1974). *A comparative analysis of language arts and mathematics achievement in selected IGE/MUS-E and non-IGE/MUS-E programs in Ohio.* Unpublished doctoral dissertation, University of Cincinnati.

HERRINGTON, A.F. (1973). *Perceived attitudes and academic achievement of reported graded and nongraded sixth-year students.* Unpublished doctoral dissertation, University of Miami, Coral Gables.

HERTZ-LAZAROWITZ, R, SHARAN, S., and STEINBERG, R. (1980). Classroom learning styles and cooperative behavior of elementaryschool children. *Journal of Educational Psychology, 72,* 99-106.

HERTZ-LAZAROWITZ, R., IVORY, G., and CALDERON, M. (1993). *The Bilingual Cooperative Integrated Reading and Composition (BCIRC) project in the Ysleta Independent School District: Standardized test outcomes.* Baltimore, MD: Johns Hopkins University Center for Research on Effective Schooling for Disadvantaged Students.

HERTZ-LAZAROWITZ, R., LERNER, M., SCHAEDEL, B., WALK, A., and SARID, M. (in press). Story-related writing: An evaluation of CIRC in Israel. *Helkat-Lashon* (Journal of Linguistic Education, in Hebrew).

HERTZ-LAZAROWITZ, R., SAPIR, C., and SHARAN, S. (1981). Academic and social effects of two cooperative learning methods in desegregated classrooms. Manuscript, Haifa University, Haifa, Israel.

HEYNS, B. (1978). *Summer learning and the effects of schooling* New York: Academic Press.

HICKEY, S.M.P. (1962). *An analysis and evaluation of the ungraded primary program in the diocese of Pinsburgh.* Unpublished doctoral dissertation, Fordham University, Bronx.

HIDI, S., and ANDERSON, V. (1986). Producing written summaries: Task demands, cognitive operations, and implications for instruction. *Review of Educational Research, 56,* 473-493.

HIEBERT, E. (1983). An examination of ability groupings for reading instruction. *Reading Research Quarterly, 18,* 231-255.

HIEBERT, J., WEARNE, D., and TABER, S. (1991). Fourth graders' gradual construction of decimal fractions during instruction using different physical representations. *Elementary School Journal, 91,* 321-341.

HIGGINS, J.J. (1980). *A comparative study between the reading achievement levels of students in a combination/ungraded class and students in a graded class.* Unpublished doctoral dissertation, George Peabody College for Teachers of Vanderbilt University.

HILLSON, M., JONES, J.C., MOORE, J.W., and VAN DEVENDER, F. (1964). A controlled experiment evaluating the effects of a nongraded organization on pupil achievement. *Journal of Educational Research, 57,* 548-550.

HOBSON V. HANSEN, 269 F. Supp. 401 (1967).

HOLY, T.C., and SUTTON, D.H. (1930). Ability grouping in the ninth grade. *Educational Research Bulletin, 9,* 419-422.

HOPKINS, K.D., OLDRIDGE, O.A., and WILLIAMSON, M.L. (1965). An empirical comparison of pupil achievement and other variables in graded and ungraded classes. *American Educational Research Journal, 2,* 207-215.

HORAK, V.M. (1981). A meta-analysis of research findings on individualized instruction in mathematics. *Journal of Educational Research, 74,* 249-253.

HORWITZ, R.I. (1987). Complexity and contradiction in clinical trial research. *American Journal of Medicine, 82,* 498-510.

HOSLEY, C.T. (1954). Learning outcomes of sixth grade pupils under alternate grade organization panerns. *Dissertation Abstracts, 15,* 490-491. (University Microfilms No. 7484).

HUBER, G.L., BOGATZKI, W., and WINTER, M. (1982). *Kooperation als Ziel schulischen Lehrens und lehrens.* (Cooperation: Condition and Goal of Teaching and Learning in Classrooms). Tubingen, West Germany: Arbeitsbereich Padagogische Psychologie der Universitat Tubingen.

HULTEN, B.H., and DEVRIES, D.L. (1976). *Team competition and group practice: Effects on student achievement and attitudes* (Report No. 212). Baltimore: Johns Hopkins University, Center for Social Organization of Schools.

HUMPHREYS, B., JOHNSON, R., and JOHNSON, D.W. (1982). Effects of cooperative, competitive, and individualistic learning on studenrs' achievement in science class. *Journal of Research in Science Teaching, 19,* 351-356.

HUNTER, J. E., SCHMIDT, F. L., and JACKSON, G.B. (1982). *Meta-analysis: Cumulating research findings across studies.* Beverly Hills, CA: Sage.

HYMEL, G.M. (1982). *Mastery learning: A comprehensive bibliography.* New Orleans: Loyola University.

HYMEL, G.M., and MATHEWS, G. (1980). Effects of a mastery approach on social studies achievement and unit evaluation. *Southern Journal of Education Research, 14,* 191-204.

IANO, R., AYERS, D., HELLER, H., McGETTIGAN, T., and WALKER, U. (1974). Sociometric status of retarded children in an integrative program. *Exceptional Children, 41,* 267-271.

INGRAM, V. (1960). Flint evaluates its primary cycle. *Elementary School Journal, 61*, 76-80.

JACKSON, G.B. (1980). Methods for integrative reviews. *Review of Educational Research, 50*, 438-460.

JACKSON, J. (1953). The effect of classroom organization and guidance practice upon the personality adjustment and academic growth of students. *Journal of Genetic Psychology 83*, 159-170.

JACKSON, S.E. (1984, August). *Can metaanalysis be used for theory development in organizational psychology?* Paper presented at the annual convention of the American Psychological Association, Toronto.

JACQUETTE, F.C. (1959). *A five year study to determine the effects of the ungraded classroom organizanon on reading achievement in Grand Junction, Colorado.* Unpublished field study, University of Northern Colorado, Greeley.

JANKE, R. (1978, April). The Teams-Games-Tournaments (TGT) method and the behavioral adjustment and academic achievement of emotionally impaired adolescents. Paper presented at the annual meeting of the American Educational Research Association, Toronto.

JEFFREYS, J.S. (1970). *An investigation of the effects of innovative educational practices on pupilcenteredness of observed behaviors and on learner outcome variables.* Unpublished doctoral dissertation, University of Maryland, College Park.

JENKINS, J., JEWELL, M., LEICESTER, N., O'CONNOR, R., JENKINS, L., and TROUTNER, N. (1994). Accommodations for individual differences without classroom ability groups: An experiment in school restructuring. *Exceptional Children, 60*, 344-358.

JENKINS, J.R, JEWELL, M., LEICESTER, N., JENKINS, L., and TROUTER, N.M. (1991). Development of a school building model for educating students with handicaps and at-risk students in general education classrooms. *Journal of Learning Disabilities, 24(5)*, 311-320.

JOHNSON, D.W., and JOHNSON, R.T. (1981). The integration of handicapped into the regular classroom: Effects of cooperative and individualistic instruction. *Contemporary Educational Psychology, 6*, 344-355.

JOHNSON, D.W., and JOHNSON, R.T. (1987). Research shows the benefits of adult cooperation. *Educational Leadership, 45(3)*, 27-30.

JOHNSON, D.W., and JOHNSON, R.T. (1989). *Cooperation and competition: Theory and research.* Edina, MN: Interaction Book Co

JOHNSON, D.W., JOHNSON, R.T., WARING, D., and MARUYAMA, G. (1986). Different cooperative learning procedures and cross-handicapped relationships. *Exceptional Children, 53*, 247-252.

JOHNSON, D.W., and JOHNSON, R.T. (1974). Instructional structure: Cooperative, competitive or individualistic. *Review of Educational Research, 44*, 213-240.

JOHNSON, D.W., and JOHNSON, R.T. (1975). *Learning Together and Alone.* Englewood Cliffs, NJ: Prentice-Hall.

JOHNSON, D.W., and JOHNSON, R.T. (1979). Conflict in the classroom: Controversy and learning. *Review of Educational Research, 49*, 51-70.

JOHNSON, D.W., and JOHNSON, R.T. (1980). Integrating handicapped children into the mainstream. *Exceptional Children, 47*, 90-98.

JOHNSON, D.W., and JOHNSON, R.T. (1981a). Effects of cooperative and individualistic learning experiences on interethnic interaction. *Journal of Educational Psychology, 73*, 444-449.

JOHNSON, D.W., and JOHNSON, R.T. (1981b). The integration of the handicapped into the regular classroom: Effects of cooperative and individualistic instruction. *Contemporary Educational Psychology, 6*, 344-355.

JOHNSON, D.W., and JOHNSON, R.T. (1985). The internal dynamics of cooperative learning groups. In R.E. Slavin, S. Sharan, S. Kagan, R. Hertz-Lazarowitz, C. Webb, and R. Schmuck (eds.), *Learning to cooperate, cooperating to learn.* New York: Plenum.

JOHNSON, D.W., and JOHNSON, R.T. (1987). *Learning Together and Alone* (2nd ed.). Englewood Cliffs, NJ: Prentice-Hall.

JOHNSON, D.W., and JOHNSON, R.T. (1994). *Learning Together and Alone: Cooperative, competitive, and individualistic learning* (4th ed.). Boston: Allyn & Bacon.

JOHNSON, D.W., and JOHNSON, W. (1972). The effects of attitude similarity, expectation of goal facilitation, and actual goal facilitation on interpersonal attraction. *Journal of Experimental Social Psychology, 8*, 197-206.

JOHNSON, D.W., JOHNSON, R.T. and SCOTT, L. (1978). The effects of cooperative and individualized instruction on student attitudes and achievement. *Journal of Social Psychology, 104*, 207-216.

JOHNSON, D.W., JOHNSON, R.T., and SMITH, K (1991). *Active learning: Cooperation in thc class-room.* Edina, MN: Interaction Book Company.

JOHNSON, D.W., JOHNSON, R.T., BUCKMAN, L.A., and RICHARDS, P. (1986). The effect of pro-longed implementation of cooperative learning on social support within the classroom. *Journal of Psychology, 119,* 405-411.

JOHNSON, D.W., JOHNSON, R.T., HOLUBEC, E.J., and ROY, P. (1984). *Circles of learning.* Alexandria, VA: Association for Supervision and Curriculum Development.

JOHNSON, D.W., JOHNSON, R.T., JOHNSON, J., and ANDERSON, D. (1976). The effects of coop-erative vs. individualized instruction on student prosocial behavior, attitudes toward learn-ing, and achievement. *Journal of Educational Psychology, 68,* 446-452.

JOHNSON, D.W., MARUYAMA, G., JOHNSON, R., NELSON, D., and SKON, L. (1981). Effects of cooperative, competitive, and individualistic goal structures on achievement: A meta-analy-sis. *Psychological Bulletin, 89,* 47-62.

JOHNSON, G.O. (1950). A study of the social position of the mentally retarded child in the reg-ular grades. *American Journal of Mental Deficiency, 55,* 60-89.

JOHNSON, L.C. (1985). *The effects of the groups of four cooperative learning models on student prob-lem-solving achievement in mathematics.* Doctoral disserration, University of Houston.

JOHNSON, L.C., and WAXMAN, H.C. (1985, March). *Evaluating the effects of the "groups of four" program.* Paper presented at the annual meeting of the American Educational Research Association, Chicago.

JOHNSON, R.T., and JOHNSON, D.W. (eds.) (1984). *Structuring cooperative learning: Lesson plans for teachers.* New Brighton, MN: Interaction Book Co.

JOHNSON, R.T., and JOHNSON, D.W. (1981). Building friendships between handicapped and non-handicapped students: Effects of cooperative and individualistic instruction. *American Educational Research Journal, 18,* 415-424.

JOHNSON, R.T., and JOHNSON, D.W. (1982). Effects of cooperative and competitive learning experiences on interpersonal attraction between handicapped and nonhandicapped students. *Journal of Social Psychology, 116,* 211-219.

JOHNSON, R.T., and JOHNSON, D.W. (1983). Effects of cooperative, competitive, and individu-alistic learning experiences on social development. *Exceptional Children, 49,* 323-330.

JOHNSON, R.T., JOHNSON, D.W., SCOTT, L.E., and RAMOLAE, B.A. (1985). Effects of single-sex and mixed-sex cooperative interaction on science achievement and attitudes and cross-hand-icap and cross-sex relationships. *Journal of Research in Science Teaching, 22,* 207-220.

JOHNSTON, H.J. (1973). The effect of grouping patterns on first-grade children's academic achieve-ment and personal and social development. *Dissertation Abstracts International,* 2461A. (University Microfilms No. 73-25, 893).

JOHNSTON, R, ALLINGTON, R., and AFFLERBACH, P. (1985). The congruence of classroom and remedial reading instruction. *Elementary School Journal, 85,* 465-478.

JONES, B.F., MONSAAS, J.A., and KATIMS, M. (1979, April). *Improving reading comprehension: Embedding diverse learning strategies within a mastery learning instructional format.* Paper pre-sented at the annual meeting of the American Educational Research Association, San Francisco.

JONES, C.J., MOORE, J.W., and VAN DEVENDER, F. (1967). Comparison of pupil achievement after one and one-half and three years in a nongraded program. *Journal of Educational Research, 61,* 75-77.

JONES, D.M. (1948). An experiment in adaption to individual differences. *Journal of Educational Psychology, 39,* 257-273.

JONES, D.S.P. (1990). *The effects of contingency based and competitive reward systems on achievement and attitudes in cooperative learning situations.* Unpublished doctoral dissertation, Temple University.

JONES, F.G. (1975). The effects of mastery and aptitude on learning, retention, and time. *Dissertation Abstracts International, 35,* 6547. (University Microfilms No. 75-8126).

JONES, J.C., MOORE, J.W., and VAN DEVENDER, F. (1967). A comparison of pupil achievement after one and one-half and three years in a nongraded program. *The Journal of Educational Research, 61,* 75-77.

JONES, J.D., ERICKSON, E.L., and CROWELL, R. (1972). Increasing the gap between Whites and Blacks: Tracking as a contribution source. *Education and Urban Society, 4,* 339-349.

JOYCE B., and SHOWERS, B. (1980). Improving in-service training: The messages from research. *Educational Leadership, 37,* 379-385.

JOYCE, B., and SHOWERS, B. (1981). Transfer of training: The contribution of "coaching." *Journal of Education, 163*, 163-172.

JOYCE, B., and WEIL, M. (1980). *Models of teaching*. Englewood Cliffs, NJ: Prentice-Hall.

JOYCE, B., HERSH, R., and McKIBBIN, M. (1983). *The structure of school improvement*. New York: Longman.

JUSTMAN, J. (1968). Reading and class homogeneity. *Reading Teacher, 21*, 314-316.

KAGAN, S. (1992). *Cooperative learning resources for teachers*. San Juan Capistrano, CA: Resources for Teachers.

KAGAN, S., and MADSEN, M.C. (1972). Rivalry in Anglo American and Mexican American Children. *Journal of Personality and Social Psychology, 24*, 214-220.

KAGAN, S., ZAHN, G.L., WIDAMAN, K.F., SCHWARZWALD, J., and TYRELL, G. (1985). Classroom structural bias: Impact of cooperative and competitive classroom structures on cooperative and competitive individuals and groups. In R.E. Slavin, S. Sharan, S. Kagan, R. Hertz-Lazarowitz, C. Webb, and R. Schmuck (eds.), *Learning to cooperate, cooperating to learn*. New York: Plenum.

KALLISON, J.M. (1986). Effects of lesson organization on achievement. *American Educational Research Journal, 23*(2), 337-347.

KAMBISS, P.A. (1990). *The effects of cooperative learning on student achievement in a fourth grade classroom*. Research project report, Mercer University.

KAMINSKI, L.B. (1991). *The effect of formal group skill instruction and role development on achievement of high school students taught with cooperative learning*. Unpublished doctoral dissertation, Michigan University.

KAPLAN, R.M., and PASCOE, G.C. (1977). Humorous lectures and humorous examples: Some effects upon comprehension and retention. *Journal of Educational Psychology, 69*(1), 61-65.

KARPER, J., and MELNICK, S.A. (1991, April). *A comparison of team accelerated instruction with traditional instruction in mathematics*. Paper presented at the annual meeting of the American Educational Research Association, Chicago.

KARWEIT, N., and SLAVIN, R.E. (1981). Measurement and modeling choices and studies of time and learning. *American Educational Research Journal, 18*, 157-171.

KARWEIT, N.L. (1989). Time and learning: A review. In *School and Classroom Organization*, ed. R.E. Slavin, 69-95. Hillsdale, NJ: Erlbaum.

KARWEIT, N.L., and WASIK, B.A. (1994). Extra-year kindergarten programs and transitional first grades. In R.E. Slavin, N.L. Karweit, & B.A. Wasik (Eds.), *Preventing early school failure: Research on effective strategies*. Boston: Allyn & Bacon.

KATIMS, M., and JONES, B.F. (1985). Chicago Mastery Learning Reading: Mastery learning instruction and assessment of inner-city schools. *Journal of Negro Education, 54*, 369-387.

KATIMS, M., SMITH, J.K., STEELE, C., and WICK, J.W. (1977, April). *The Chicago Mastery Learning Reading Program: An interim evaluation*. Paper presented at the annual meeting of the American Educational Research Association, New York.

KATZ, L.G., EVANGELOU, D., and HARTMAN, J.A. (1991). *The case for mixed-aged grouping in early childhood education*. Washington, DC: National Association for the Education of Young Children.

KELLER, F.S. (1968). "Good-bye, teacher . . ." *Journal of Applied Behavior Analysis, 1*, 78-89.

KEPLER, K., and RANDALL, J.W. (1977). Individualization: Subversion of elementary schooling. *Elementery School Journal, 11*, 348-363

KERCKHOFF, A.C. (1986). Effects of ability grouping in British secondary schools. *American Sociological Review, 51*, 842-858.

KERSH, M.E. (1970). *A strategy for mastery learning in fifth grade arithmetic*. Unpublished doctoral dissertation, University of Chicago.

KIERSTEAD, R. (1963). A comparison and evaluation of two methods of organization for the teaching of reading. *Journal of Educational Research, 56*, 317-321.

KILLOUGH, C.K. (1971). *An analysis of the longitudinal effects that a nongraded elementary program, conducted in an open-space school, had on the cognitive achievement of pupils*. Unpublished doctoral dissertation, University of Houston.

KINNEY, J.H. (1989, May). *A study of the effects of a cooperative learning program on the achievement of ninth grade multi-cultural general biology classes*. Paper submitted to the Alexandria City, Virginia School Board.

KLAUS, W.D. (1981). *A comparison of student achievement in individually guided education programs and non-individually guided education elementary school programs.* Unpublished doctoral dissertation, University of Missouri, Columbia.

KLAUSMEIER, H.J., QUILLING, M.R., SORENSON, J.S., WAY, R.S., and GLASRUD, G.R. (1971). *Individually guided education and the multiunit elementary school: Guidelines for implementation.* Madison: University of Wisconsin, Wisconsin Research & Development Center for Cognitive Learning.

KLAUSMEIER, H.J., ROSSMILLER, R.A., SAILY, M. (Eds.). (1977). *Individually Guided Elementary Education: Concepts and Practices.* New York: Academic.

KLINE, R.E. (1964). A longitudinal study of the effectiveness of the track plan in the secondary schools of a metropolitan community. *Dissertation Abstracts International, 25,* 324. (University Mierofilms No. 64-4257).

KOHN, A. (1986). *No contest: Thc case against competition.* Boston: Houghton-Mifflin.

KOONTZ, W.F. (1961). A study of achievement as a function of homogeneous grouping. *Journal of Experimental Education, 30,* 249-253.

KOSTERS, A.E. (1990). *The effects of cooperative learning in the traditional classroom on student achievement and attitude.* Unpublished doctoral dissertation, University of South Dakota.

KOZMA, R. (1991). Learning with media. *Review of Educational Research, 61*(2), 179-211.

KUHLMAN, C.L. (1985). *A study of the relationship between organizational characteristics of elementary schools, student characteristics, and elementary competency test results.* Unpublished doetoral dissertation, University of Kansas, Lawrence.

KUHN, D. (1972). Mechanism of change in the development of cognitive structures. *Child Development, 43,* 833-844.

KUKLA, A. (1972). Foundations of an attributional theory of performance. *Psychological Review, 77,* 454-470.

KULIK, C.-L., and KULIK, J.A. (1982). Effects of ability grouping on secondary school students: A meta-analysis of evaluation findings. *American Educational Research Journal, 19,* 415-428.

KULIK, C.-L., and KULIK, J.A. (1984, August). *Effects of ability grouping on elementary school pupils: A meta-analysis.* Paper presented at the annual meeting of the American Psychological Association, Toronto.

KULIK, C.L., and KULIK, J.A. (1984). Effects of aeeelerated instruetion on students. *Review of Educational Research, 54,* 409-425.

KULIK, C.L., and KULIK, J.A. (1982). Effects of ability grouping on secondary school students. A meta-analysis of evaluation findings. *American Educational Research Journal, 19,* 415-428.

KULIK, C.L., KULIK, J.A., and BANGERT-DROWNS, R.L. (1986, April). *Effects of testing for mastery on student learning.* Paper presented at the annual meeting of the American Educational Research Association, San Francisco.

KULIK, J.A., and KULIK, C.L. (1984). Effects of accelerated instruction on students. *Review of Educational Research, 54,* 409-425.

KULIK, J.A., and KULIK, C.L. (1987). Effects of ability grouping on student achievement. *Equity and Excellence, 23,* 22-30.

KULIK, J.A., and KULIK, C.L. (1988). Timing of feedback and verbal learning. *Review of Educational Research, 58,* 79-97.

KULIK, J.A., KULIK, C.L., and COHEN, P.A. (1979). A meta-analysis of outcome studies of Keller's Personalized System of Instruction. *American Psychologist, 34,* 307-318.

LABERGE, D., and SAMUELS, S.J. (1974). Toward a theory of automatic information processing in reading. *Cognitive Psychology, 6,* 293-323.

LAIR, D.P. (1975) . *The effects of graded and nongraded schools on student growth.* Unpublished doctoral dissertation, Baylor University, Waco.

LAMBERIGTS, R, and DIEPENBROEK, J.W. (1992, July). *Implementation and effects of an integrated direct and activative instruction in a cooperative classroom setting.* Paper presented at the International Convention on Cooperative Learning, Utrecht, The Netherlands.

LAND, M.L. (1987). Vagueness and clarity. In *International Encyclopedia of Teaching and Teacher Education,* ed. M.J. Dunkin. New York: Pergamon.

LARRIVEE, B., and HORNE, M.D. (1991). Social status: A comparison of mainstreamed students with peers of different ability levels. *Journal of Special Education, 25,* 90-101.

LATANE, B., WILLIAMS, K., and HARKINS, S. (1979). Many hands make light the work: The causes and consequences of social loafing. *Journal of Personality and Social Psychology, 37,* 822-832.

LAWSON, R.E. (1973). *A comparison of the development of self-concept and achievement in reading of students in the first, third, and fifth year of attendance.* Unpublished doctoral dissertation, Ball State University, Muncie, IN.

LAZAROWITZ, R (1991). Learning biology cooperatively: An Israeli junior high school study. *Cooperative Learning, 11(3),* 19-21.

LAZAROWITZ, R, BAIRD, H., BOWLDEN, V., and HERTZ-LAZAROWITZ, R. (1982). Academic achievements, learning environment, and self-esteem of high school students in biology taught in cooperative-investigative small groups. Manuscript, The Technion, Haifa, Israel.

LAZAROWITZ, R, BAIRD, J. H., HERTZ-LAZAROWITZ, R., and JENKINS, J. (1985). The effects of modified Jigsaw on achievement, classroom social climate, and self-esteem in high-school science classes. In R.E. Slavin, S. Sharan, S. Kagan, R. Hertz-Lazarowitz, C. Webb, and R. Schmuck (eds.), *Learning to cooperate, cooperating to learn.* New York: Plenum.

LAZAROWITZ, R, HERTZ-LAZAROWITZ, R, and BAIRD, J.H. (in press). Learning science in a cooperative setting: Academic achievement and affective outcomes. *Journal of Research in Science Teaching.*

LAZAROWITZ, R., and KARSENTY, G. (1990). Cooperative learning and students self-esteem in tenth grade biology classrooms. In Sharan, S. (ed.). *Cooperative learning, theory and research* (pp. 123-149). New York: Praeger Publishers.

LEIGHTON, M.S., and SLAVIN, R.E. (1988). *Achievement effects of instructional pace and systematic instruction in elementary mathematics.* Paper presented at the annual convention of the American Educational Research Association, March, New Orleans.

LEINHARDT, G., and BICKEL, W. (1987). Instruction's the thing wherein to catch the mind that falls behind. *Educational Psychologist, 22,* 177-207.

LEINHARDT, G., and PALLAY, A. (1982). Restrictive educational settings: Exile or haven? *Review of Educational Research, 52,* 557-578.

LEITER, J. (1983). Classroom composition and achievement gains. *Sociology of Education, 56,* 126-132.

LEVIN, H.M. (1987). Accelerated schools for disadvantaged students. *Educational Leadership, 44* (6), 19-21.

LEVINE, D.U. (Ed.). (1985). *Improving student achievement through mastery learning programs.* San Francisco: Jossey-Bass.

LEVINE, D.U., and EUBANKS, E.E. (1986-1987). Achievement improvement and non-improvement at concentrated poverty schools in big cities. *Metropolitan Education, 3,* 92-107.

LEVINE, D.U., and STARK, J. (1982). Instructional and organizational arrangements that improve achievement in inner-city schools. *Educational Leadership, 39,* 41-46.

LEW, M., MESCH, D., JOHNSON, D.W., and JOHNSON, R.T. (1986). Positive interdependence, academic and collaborative-skills group contingencies, and isolated students. *American Educational Research Journal, 23,* 476-488.

LEYTON, F.S. (1983). *The extent to which group instruction supplemented by mastery of initial cognitive prerequisites approximates the learning effectiveness of one-to-one tutorial methods.* Unpublished doctoral dissertation, University of Chicago.

LIGHT, R.J., and PILLEMER, D.B. (1984). *Summing up: The science of reviewing research.* Cambridge, MA: Harvard University Press.

LINNEMEYER, S. (1992). Cooperative learning: A wolf in sheep's clothing. *Illinois Council for the Gifted Journal, 11,* 62-64.

LITTLE, J.W. (1982). Norms of collegiality and experimentation: Workplace conditions of school success. *American Educational Research Journal, 19,* 325-340

LLOYD, J.W., and GAMBATESE, C. (1991). Reforming the relationship between regular and special education: Background and issues. In J. Lloyd, N. Singh, & A. Repp (Eds.), *The regular education initiative: Alternative perspectives on concepts, issues, and models* (pp. 3-15). Sycamore, IL: Sycamore.

LLOYD, J.W., CROWLEY, E.P., KOHLER, E, and STRAIN, R (1988). Redefining the applied research agenda: Cooperative learning, prereferral, teacher consultation, and peer mediated interventions. *Journal of Learning Disabilities, 21,* 43-52.

LONG, J.C., OKEY, J.R., and YEANY, R.H. (1978). The effects of diagnosis with teacher- or student-directed remediation on science achievement and attitudes. *Journal of Research in Science Teaching, 15,* 505-511.

LONG, J.C., OKEY, J.R., and YEANY, R.H. (1981). The effects of a diagnostic-prescriptive teaching strategy on student achievement and attitude in biology. *Journal of Research in Science Teaching, 18,* 515-523.

LOOMER, B.M. (1962). Ability grouping and its effects upon individual achievement. *Dissertation Abstracts, 23,* 1581. (University Microfilms No. 62-4982).

LORD, F.M. (1960). Large-sample covariance analysis when the control variable is fallible. *Journal of the American Statistical Association, 55,* 307-321.

LOTT, A.J. and LOTT, B.E. (1965). Group cohesiveness as interpersonal attraction: A review of relationships with antecedent and consequent variables. *Psychological Bulletin, 64,* 259-309.

LOVELL, J.T. (1960). The Bay High School experiment. *Educational Leadership, 17,* 383-387.

LUECKEMEYER, C.L., and CHIAPPETTA, E.L. (1981). An investigation into the effects of a modified mastery learning strategy on achievement in a high school human physiology unit. *Journal of Research in Science Teaching, 18,* 269-273.

LYMAN, F. (1981). The responsive classroom discussion. In A.S. Anderson (ed.), *Mainstreaming Digest* College Park: University of Maryland, College of Education.

LYNSYNCHUK, L.M., PRESSLEY, M., and VYE, N.J. (1990). Reciprocal teaching improves standardized reading comprehension performance in poor comprehenders. *Elementary School Journal, 90,* 469-484.

LYSAKOWSKI, R., and WALBERG, H. (1982). Instructional effects of cues, participation, and corrective feedback: A quantitative synthesis. *American Educational Research Journal, 19,* 559-578.

MACHIELE, R.B. (1965). A preliminary evaluation of the non-graded primary at Leal School, Urbana. *Illinois School Research,1,* 20-24.

MADDEN, N.A., and SLAVIN, R.E. (1983). Mainstreaming students with mild academic handicaps: Academic and social outcomes. *Review of Educational Research, 53,* 519-569.

MADDEN, N.A., and SLAVIN, R.E. (1983). Mainstreaming students with mild academic handicaps: Academic and social outcomes. *Review of Educational Research, 84,* 131-138.

MADDEN, N.A., and SLAVIN, R.E. (1983). Mainstreaming students with mild academic handicaps: Academic and social outcomes. *Review of Educational Research, 53,* 519-569.

MADDEN, N.A., and SLAVIN, R.E. (1983a). Effects of cooperative learning on the social acceptance of mainstreamed academically handicapped students. *Journal of Special Education, 17,* 171-182.

MADDEN, N.A., SLAVIN, R.E., and STEVENS, R.J. (1986). *Cooperative Integrated Reading and Comparison: Teacher's manual.* Baltimore: Johns Hopkins University, Center for Research on Elementary and Middle Schools.

MADDEN, N.A., SLAVIN, R.E., KARWEIT, N.L., DOLAN, L.J., and WASIK, B.A. (1993). "Success for All: Longitudinal Effects of a Restructuring Program for Inner-City Elementary Schools." *American Educational Research Journal, 30,* 123-148.

MADDOX, H., and HOOLE, E. (1975). Performance decrement in the lecture. *Educational Review, 28,* 17-30.

MAHEADY, L., MALLETTE, B., HARPER, G.F., and SACCA, K. (1991). Heads together: A peer mediated option for improving the academic achievement of heterogeneous learning groups. *Remedial and Special Education, 12* (2), 25-33.

MALONE, T., and LEPPER, M. (1988). Making learning fun: A taxonomy of intrinsic motivation for learning. In *Aptitude Learning and Instruction, Vol. III: Cognitive and Affective Process Analysis,* ed. R. Snow and M. Farr, 70-81. Hillsdale, NJ: Erlbaum.

MANNING, M.L., and LUCKING, R. (1991). The what, why, and how of cooperative learning. *The Social Studies,* May/June, 120-124.

MARASCUILO, L.A., and McSWEENEY, M. (1972). Tracking and minority student attitudes and performance. *Urban Education, 6,* 303-319.

MARESH, R.T. (1971). *An analysis of the effects of vertical grade groups on reading achievement and attitudes in elementary schools.* Unpublished doctoral dissertation, University of North Dakota, Grand Forks.

MARLIAVE, R., FISHER, C. and DISHAW, M. (1978). *Academic learning time and student achievement in the B-C period.* Far West Laboratory for Educational Research and Development, Technical Note V-29.

MARSH, H., and BALL, S. (1981). Interjudgmental reliability of reviews for the *Journal of Educational Psychology. Journal of Educational Psychology, 73,* 872-880.

MARSTON, D. (1987). The effectiveness of special education. *Journal of Special Education, 21,* 13-27.

MARTIN, W.B. (1959). Effects of ability grouping on junior high school achievement. *Dissertation Abstracts International, 20,* 2810. (University Microfilms No. 59-1108)

MARTIN, W.H. (1927). *The results of homogeneous grouping in the junior high school.* Unpublished doctoral dissertation, Yale University, New Haven, CT.

MARTINEZ, L.J. (1990). *The effect of cooperative learning on academic achievement and self-concept with bilingual third-grade students.* Unpublished doctoral dissertation, United States International University.

MATHEWS, G.S. (1982). *Effects of a mastery learning strategy on the cognitive knowledge and unit evaluation of students in high school social studies.* Unpublished doctoral dissertation, University of Southern Mississippi, Hattiesburg.

MATT, G.E., and COOK, T.D. (1994). Threats to the validity of research and syntheses. In H. Cooper & L.V. Hedges (Eds.), *The handbook of research synthesis* (pp. 503-520). New York: Russell Sage.

MATTINGLY, R.M., and VANSICKLE, R.L. (1991). Cooperative learning and achievement in social studies: Jigsaw II. *Social Education, 55* (6), 392-395.

MAYER, R.E., and GALLINI, J.K. (1990). When is an illustration worth ten thousand words? *Journal of Educational Psychology, 82,* 715-726.

McCUTCHEN, D., and PERFETTI, C. (1983). Local coherence: Helping young writers manage a complex task. *Elementary School Journal, 84,* 71-75.

McLAUGHLIN, M.W. (1990). The Rand change agent study revisited: Macro perspectives and microrealities. *Educational Researcher,19(9),11-16.*

McLOUGHLIN, W.P. (1967). *The nongraded school: A critical assessment.* New York: State Education Department.

McLOUGHLIN, W.P. (1970). Continuous pupil progress in the non-graded school: Hope or hoax? *Elementary School Journal, 71* (2), 90-96.

McPARTLAND, J. (1968). *The segregated student in desegregated schools: Sources of influence on Negro secondary students.* Baltimore, MD: Johns Hopkins University.

McPARTLAND, J.M., COLDIRON, J.R., and BRADDOCK, J.H. (1987). *School structures and classroom practices in elementary, middle, and secondary schools* (Tech. Rep. No. 14). Baltimore, MD: Johns Hopkins University, Center for Research on Elementary and Middle Schools.

MEADOWS, N.B.W. (1988). *The effects of individual, teacher-directed and cooperative learning instructional methods on the comprehension of expository text.* Unpublished doctoral dissertation, University of Washington.

MELOTH, M.S., and DEERING, P.D. (1992). The effects of two cooperative conditions on peer group discussions, reading comprehension, and metacognition. *Contemporary Educational Psychology, 17,* 175-193.

MELOTH, M.S., and DEERING, P.D. (1994). Task talk and task awareness under different cooperative learning conditions. *American Educational Research Journal, 31,* 138-166.

MELTON, R.F. (1978). Resolution of conflicting claims concerning the effect of behavioral objectives on student learning. *Review of Educational Research, 48,* 291-302.

MENAHEM, M., and WEISMAN, L. (1985). Improving reading ability through a mastery learning program: A case study. In D. Levine (Ed.), *Improving student achievement through mastery learning programs* (pp. 223-240). San Francisco: Jossey-Bass.

MERGENDOLLER, J., and PACKER, M.J. (1989). *Cooperative learning in the classroom : A knowledge brief on effective teaching.* San Francisco: Far West Laboratory.

MERRITT, H.P.W. (1973). The effects of variations in instruction and final unit evaluation procedures on community college beginning algebra classes. *Dissertation Abstracts International, 33,* 5978A. (University Microfilms No. 73-11207).

MEVARECH, Z.R. (1980). *The role of teaching-learning strategies and feedback-corrective procedures in developing higher cognitive achievement.* Unpublished doctoral dissertation, University of Chicago.

MEVARECH, Z.R. (1985a). The effects of cooperative mastery learning strategies on mathematics achievement. *Journal of Educational Research, 78,* 372-377.

MEVARECH, Z.R. (1985b, April). *Cooperative mastery learning strategies.* Paper presented at the annual meeting of the American Educational Research Association, Chicago.

MEVARECH, Z.R. (1986). The role of a feedback corrective procedure in developing mathematics achievement and self-concept in desegregated classrooms. *Studies in Educational Evaluation, 12,* 197-203.

MEVARECH, Z.R. (1991). Learning mathematics in different mastery environments. *Journal of Educational Research, 84*(4), 225-231.

MIKKELSON, J.E. (1962). *An experimental study of selective grouping and acceleration in junior high school mathematics.* Unpublished doctoral dissertation, University of Minnesota.

MILLER, R.L. (1976). Individualized instruction in mathematies: A review of research. *The Mathematics Teacher, 69,* 345-351.

MILLER, W.S., and OTTO, H.J. (1930). Analysis of experimental studies in homogeneous grouping. *Journal of Educational Research, 21,* 95-102.

MILLMAN, J., and JOHNSON, M. (1964). Relation of section variance to achievement gains in English and mathematics in grades 7 and 8. *American Educational Research Journal, 1,* 47-51.

MILLS, C., and DURDEN, W. (1992). Cooperative learning and ability grouping: An issue of choice. *Gifted Child Quarterly, 36*(1), 11-16.

MINTZ, J. (1983). Integrating research evidence: A commentary on metaanalysis. *Journal of Consulting and Clinical Psychology, 51,* 71-75.

MOODY, J.D., and GIFFORD, V.D. (1990, November). *The effect of grouping by formal reasoning ability levels, group size, and gender on achievement in laboratory chemistry.* Paper presented at the annual meeting of the Mid-South Educational Research Association, New Orleans.

MOODY, W. B., BAUSELL, R. B., and JENKINS, J. R. (1973). The effect of class size on the learning of mathematics: A parametric study with fourth grade students. *Journal for Research in Mathematics Education, 4,* 170-176.

MOORE, D.I. (1963). *Pupil achievement and grouping practices in graded and ungraded primary schools.* Unpublished doctoral dissertation, University of Michigan, East Lansing.

MOORHOUSE, W.F. (1964). Interclass grouping for reading instruction. *Elementary School Journal, 64,* 280-286.

MORENO, J.L. (1953). *Who shall survive?* New York: Beacon.

MORGAN, E.F., and STUCKER, G.R. (1960). The Joplin Plan of reading vs. a traditional method. *Journal of Educational Psychology, 51,* 69-73.

MORGENSTERN, A. (1963). A comparison of the effects of heterogeneous (ability) grouping on the academic achievement and personal-social adjustment of sixth grade children. *Dissertation Abstracts, 23,* 1054. (University Microfilms No. 63-6560).

MORRIS, V.P. (1969). An evaluation of pupil achievement in a nongraded primary plan after three, and also five years of instruction. *Dissertation Abstracts, 29,* 3809-A. (University Microfilms No. 69-7352).

MORTLOCK, R. (1970). Provision for individual differences in eleventh grade mathematics using flexible grouping based on achievement of behavioral objectives: An exploratory study. *Dissertation Abstracts International, 30,* 3643A. (University Microfilms No. 70-4148).

MOSES, P.J. (1966). A study of inter-class ability grouping on achievement in reading. *Dissertation Abstracts. 26,* 4342.

MOSKOWITZ, J.M., MALVIN, J.H., SCHAEFFER, G.A., and SCHAPS, E. (1983). Evaluation of a cooperative learning strategy. *American Educational Research Journal, 20,* 687-696.

MOSKOWITZ, J.M., MALVIN, J.H., SCHAEFFER, G.A., and SCHAPS, E. (1985). Evaluation of Jigsaw, a cooperative learning technique. *Contemporary Educational Psychology, 10,* 104-112.

MUCK, R.E.S. (1966). *The effect of classroom organizanion on academic achievement in graded and nongraded classes.* Unpublished doctoral dissertation, State University of New York, Buffalo.

MUELLER, D.J. (1976). Mastery learning: Partly boon, partly boondoggle. *Teachers College Record, 78,* 41-52.

MUGNY, B., and DOISE, W. (1978). Socio-cognitive conflict and structuration of individual and collective performances. *European Journal of Social Psychology, 8,* 181-192.

MURRAY, F.B. (1982). Teaching through social conflict. *Contemporary Educational Psychology, 7,* 257-271.

MYCOCK, M.A. (1966). *A comparison of vertical grouping and horizontal grouping in the infant school.* Unpublished master's thesis, University of Manehester.

MYERS, M., and PARIS, S. (1978). Children's metacognitive knowledge about reading. *Journal of Educational Psychology, 70,* 680-690.

NATIONAL ASSODATION FOR THE EDUCATION OF YOUNG CHILDREN. (1989). *Appropriate education in the primary grades.* Washington, DC: Author.

NATIONAL EDUCATION ASSOCIATION. (1968). *Ability grouping research summary.* Washington, DC: Author.

NATRIELLO, G. (1987). The impact of evaluation processes on students. *Educational Psychologist, 22,* 155-175.

NEWBOLD, D. (1977). *Ability-grouping the Banbury enquiry.* Slough, England: National Foundation for Educational Research in England and Wales.

NEWMANN, F.M., and THOMPSON, J. (1987). *Effects of cooperative learning on achievement in secondary schools: A summary of research.* Madison: University of Wisconsin, National Center on Effective Secondary Schools.

NEWPORT, J.F. (1967). The Joplin Plan: The score. *Reading Teacher, 21,* 158-162.

NOLAND, T.K. (1985). *The effects of ability grouping A meta-analysis of research findings.* Unpublished doctoral dissertation, University of Colorado.

NORDIN, A.B. (1979). *The effects of different qualities of instruction on selected cognitive, affective, and time variables.* Unpublished doctoral dissertation, University of Chicago.

OAKES, J. (1982). The reproduction of inequity: The content of secondary school tracking. *Urban Review, 14,* 107-120.

OAKES, J. (1985). *Keeping track: How schools structure inequality.* New Haven, CT: Yale University Press.

OAKES, J. (1987). Tracking in secondary schools: A contextual perspective. *Educational Psychologist, 22,* 129-153.

OAKES, J. (1992). Can tracking research inform practice? Technical, normative, and political considerations. *Educational Researcher, 21*(4), 12-21.

OAKES, J., QUARTZ, K.H., GONG, J., GUITON, G., and LIPTON, M. (1993). Creating middle schools: Technical, normative, and political considerations. *The Elementary School Journal, 93*(5), 461-480.

OICKLE, E. (1980). *A comparison of individual and team learning.* Doctoral dissertation, University of Maryland.

OISHI, S. (1983). *Effects of team assisted individualization in mathematics on cross-race interactions of elementary school children.* Doctoral dissertation, University of Maryland.

OISHI, S., SLAVIN, R.E., and MADDEN, N.A. (1983, April). *Effects of student teams and individualized instruction on cross-race and cross-sex friendships.* Paper presented at the annual meeting of the American Educational Research Association, Montreal.

OKEBUKOLA, P.A. (1984). In search of a more effective interaction pattern in biology laboratories. *Journal of Biological Education, 18,* 305-308.

OKEBUKOLA, P.A. (1985). The relative effectiveness of cooperativeness and competitive interaction techniques in strengthening students' performance in science classes. *Science Education, 69,* 501-509.

OKEBUKOLA, P.A. (1986a). Impact of extended cooperative and competitive relationships on the performance of students in science. *Human Relations, 39,* 673-682.

OKEBUKOLA, P.A. (1986b) . The influence of preferred learning styles on cooperative learning in science. *Science Education, 70,* 509-517.

OKEBUKOLA, P.A. (1986c). The problem of large classes in science: An experiment in co-operative learning. *European Journal of Science Education, 8,* 73-77.

OKEY, J.R. (1974). Altering teacher and pupil behavior with mastery teaching. *School Science and Mathematics, 74,* 530-535.

OKEY, J.R. (1977). The consequences of training teachers to use a mastery learning strategy. *Journal of Teacher Education, 28* (5), 57-62.

OTTENBACHER, R.J., and COOPER, H.M. (1983). Drug treatment of hyperactivity in children. *Developmental Medicine and Child Neurology, 25,* 358-366.

OTTO, H.J. (1969). *Nongradedness: An elementary school evaluation.* Austin: University of Texas.

PACE, A.J. (1981, April). Comprehension monitoring by elementary students: When does it occur? Paper presented at the annual meeting of the American Educational Research Association, Los Angeles.

PAGE, E. (1975). Statistically recapturing the richness within the classroom. *Psychology in the Schools, 12,* 339-344.

PALINSCAR, A.S., and BROWN, A.L. (1984). Reciprocal teaching of comprehension monitoring activities. *Cognition and Instruction, 2,* 117-175.

PALINSCAR, A.S., BROWN, A.L., and MARTIN, S.M. (1987). Peer interaction in reading comprehension instruction. *Educational Psychologist, 22,* 231-253.

PASSOW, A.H. (1962). The maze of research on ability grouping. *Educational Forum, 26,* 281 -288.

PASSOW, A.H. (Ed.) (1979). *The gifted and talented: Their education and development.* Chicago: University of Chicago Press.

PAVAN, B.N. (1972). *Moving elementary schools toward nongradedness: Commitment, assessment, and tactics.* Unpublished doctoral dissertation, Harvard University, Graduate School of Education, Cambridge.

PAVAN, B.N. (1973). Good news: Research on the nongraded elementary school. *Elementary School Journal, 73* (6), 333-342.

PAVAN, B.N. (1977). The nongraded elementary school: Research on academie and mental health. *Texas Tech Journal of Education, 4*, 91-107.

PAVAN, B.N. (1992). The benefids of nongraded schools. *Educational Leadership, 512*(2), 22-25.

PECK, G.L. (1991). *The effects of cooperative learning on the spelling achievement of intermediate elementary students.* Unpublished doctoral dissertation, Ball State University.

PERFETTI, C.A. (1985). *Reading ability.* New York: Oxford University Press.

PERRAULT, R. (1982). An experimental comparison of cooperative learning to noncooperative learning and their effects on cognitive achievement in junior high industrial arts laboratories. Doctoral dissertation, University of Maryland.

PERRET-CLERMONT, A-N. (1980). *Social interaction and cognitive development in children.* London: Academic Press.

PERRIN, J.D. (1969*). A statistical and time change analysis of achievement differences of children in a nongraded and a graded program in selected schools in the Little Rock Public Schools.* Unpublished doctoral dissertation, University of Arkansas, Little Rock.

PERSELL, C.H. (1977). *Education and inequality: The roots and results of stratification in America 's schools.* New York: Free Press.

PERSELL, C.H. (1977). *Education and inequality: A theoretical and empirical synthesis.* New York: Free Press.

PETERS, D., and CECI, S. (1982). Peer-review practices of psychological journals: The fate of published articles, submitted again. *The Behavioral and Brain Sciences, 5*, 187-255.

PETERSON, P.L., WILKINSON, L.C., and HALLINAN, M. (Eds.)(1984). *The social context of education.* New York: Academic Press.

PETERSON, R.L. (1966). *An experimental study of the effects of ability grouping in grades 7 and 8.* Unpublished doctoral dissertation, University of Minnesota.

PHELPS, J.D. (1990). *A study of the interrelationships between cooperative team learning, learning preference, friendship patterns, gender, and achievement of middle school students.* Unpublished doctoral dissertation, Indiana University.

PIAGET, J. (1926). *The language and thought of the child.* New York: Harcourt, Brace.

PILAND, J.C., and LEMKE, E.A. (1971). The effect of ability grouping on concept learning. *Journal of Educational Research, 64*, 209-212.

PINNELL, G.S. (1989). Reading Recovery: Helping at-risk children learn to read. *Elementary School Journal, 90*, 161-182.

PINNELL, G.S., LYONS, C.A., DEFORD, D.E., BRYK, A.S., and SELTZER, M. (1994). Comparing instructional models for the literacy education of high risk first graders. *Reading Research Quarterly, 29*, 8-38.

PINNELL, G.S., LYONS, C.A., DEFORD, D.E., BRYK, A.S., and SELTZER, M. (1991). *Studying the effectiveness of early intervention approaches for first grade children having difficulty in reading.* Columbus: Ohio State University, Martha L. King Language and Literacy Center.

PLATZ, E.F. (1965). The effectiveness of ability grouping in general science classes. *Dissertation Abstracts International, 26*, 1459A-1460A. (University Microfilms No. 65-6914).

POLYA, G. (1957). *How to solve it* (2nd ed.). New York: Doubleday.

POSTLETHWAITE, K., and DENTON, C. (1978). *Streams for the future?* Slough, England: National Foundation for Educational Research.

POWELL, W.R. (1964). The Joplin Plan: An evaluation. *Elementary School Journal, 64*, 387-392.

PRESSLEY, M., WOOD, E., WOLOSHYN, V.E., MARTIN, V., KING, A., and MENKE, D. (1992). Encouraging mindful use of prior knowledge: Attempting to construct explanatory answers facilitates learning. *Educational Psychologist, 27*, 91-109.

PRICE, D.A. (1977). *The effects of individually guided education (IGE) processes on achievement and attitudes of elementary school students.* Unpublished doctoral dissertation, University of Missouri, Columbia.

PROVUS, M.M. (1960). Ability grouping in arithmetic. *Elementary School Journal, 60*, 391-398.

PURDOM, T.L. (1929). *The value of homogeneous grouping.* Baltimore: Warwick & York.

PUTBRESE, L. (1972). An investigation into the effects of selected patterns of grouping upon arithmetic achievement. *Dissertation Abstracts International, 32,* 5113A. (University Microfilms No. 72-8388)

RAMAYYA, D.P. (1972). *Achievement skills, personality variables, and classroom climate in graded and nongraded elementary schools.* Nova Scotia: Dartmouth Public Schools.

RAMSEY, W. (1962). An evaluation of a Joplin Plan of grouping for reading instruction. *Journal of Educational Research, 55,* 567-572.

RANKIN, P.T., ANDERSON, C.T., and BERGMAN, W.G. (1936). Ability grouping in the Detroit individualization experiment. In G.M. Whipple (Ed.), *The grading of pupils: 35th yearbook of the National Society for the Study of Education* (Part 1, pp. 277-288). Bloomington, IL: Public School Publishing Co.

RAPHAEL, T.E. (1980). The effects of metacognitive strategy awareness training on students' question answering behavior. Doctoral dissertation, University of Illinois, Urbana.

RAUDENBUSH, S., and BRYK, A. (1988). Methodological advances in analyzing the effects of schools and classrooms on student learning. In E. Rothkopf (Ed.), *Review of research in education* (Vol. 15, pp. 423-475). Washington DC: American Educational Research Association.

REICHARDT, C.S. (1979). The statistical analysis of data from nonequivalent group designs. In T.C. Cook & D.T. Campbell (Eds.), *Quasi-experimentation: Design and analysis issues for field settings* (pp. 147-205). Chicago: Rand McNally.

REICHART, C.S. (1979). The statistical analysis of data from nonequivalent group designs. In T.C. Cook & D.T. Campbell, *Quasi-experimentation: Design and analysis issues for field settings* (147-205). Chicago: Rand McNally.

REID, B.C. (1973). *A comparative analysis of a nongraded and graded primary program.* Unpublished doctoral dissertation, University of Alabama, Tuscaloosa.

REID, R. (1984). Attitudes toward the learning disabled in school and home. In R. Jones (Ed.), *Attitudes and attitude change in special education: Theory and practice* (157-170). Reston, VA: Council for Exceptional Children.

REMACLE, L.F. (1970). *A comparative study of the differences in attitudes, self-concept, and achievement of children in graded and nongraded elementary schools.* Unpublished doctoral dissertation, University of South Dakota, Vermillion.

RESNICK, L.B. (1977). Assuming that everyone can learn everything, will some learn less? *School Review, 85,* 445-452.

REYNOLDS, M. (1989). A historical perspective: The delivery of special education to mildly disabled and at-risk students. *Remedial and Special Education, 10*(6), 7-11.

RICH, Y., AMIR, Y., and SLAVIN, R.E. (1986). *Instructional strategies for improving children's cross-ethnic relations.* Ramat Gan, Israel: Bar Ilan University, Institute for the Advancement of Social Integration in the Schools.

RICHER, S. (1976). Reference-group theory and ability grouping: A convergence of sociological theory and educational research. *Sociology of Education, 49,* 65-71.

RIST, R. (1970). Student social class and teacher expectations. *Harvard Educational Review, 40,* 411-451.

ROBB, D.W. (1985). Strategies for implementing successful mastery learning programs: Case studies. In D. Levine (Ed.), *Improving student achievement through mastery learning programs* (pp. 255-272). San Francisco: Jossey-Bass.

ROBERTSON, L. (1982). Integrated goal structuring in the elementary school: Cognitive growth in mathematics. Doctoral dissertation, Rutgers University.

ROBINSON, A. (1990). Cooperation or exploitation: The argument against cooperative learning for talented students. *Journal for the Education of the Gifted, 14* (1), 9-27.

ROSENBAUM, J.E. (1976). *Making inequality: The hidden curriculum of high school tracking.* New York: Wiley.

ROSENBAUM, J.E. (1978). The structure of opportunity in school. *Social Forces, 57,* 236-256.

ROSENBAUM, J.E. (1980). Social implications of educational grouping. *Review of Research in Education, 8,* 361-401.

ROSENBAUM, J.E. (1980). Social implications of educational grouping. *Review of Research in Education, 8,* 361-401.

ROSENSHINE, B., and MEISTER, C. (1992). The use of scaffolds for teaching higher level cognitive strategies. *Educational Leadership, 49* (7), 26-33.

ROSENSHINE, B.V., and STEVENS, R.J. (1986). Teaching functions. In *Third handbook of research on teaching,* ed. M.C. Wittrock, 376-391. New York: MacMillan.

ROSENTHAL, R. (1979). The "file-drawer problem" and tolerance for null results. *Psychological Pulletin, 92,* 500-504.

ROSENTHAL, R. (1984). *Meta-analytic procedures for social research.* Beverly Hills, CA: Sage.

ROSS, G.A. (1967). *A comparative study of pupil progress in ungraded and graded primary programs.* Unpublished doctoral dissertation, Indiana University.

ROSS, S.M., SMITH, L.J.& CASEY, J., JOHNSON, B., and BOND, C. (1994, April). *Using "Success for All" to restructure elementary schools: A tale of four cities.* Paper presented at the annual meeting of the American Educational Research Association, New Orleans.

ROSS, S.M., SMITH, L.J., & CASEY, J., and SLAVIN, R.E. (1995). Increasing the academic success of disadvantaged children: An examination of alternative early intervention programs. *American Educational Research Journal, 32,* 773-800.

ROTHROCK, D. (1982). The rise and decline of individualized instruction. *Educational Leadership, 39,* 528-531.

ROTHROCK, D.G. (1961). Heterogeneous, homogeneous, or individualized approach to reading? *Elementary English, 38,* 233-235.

ROWAN, B., and MIRACLE, A. (1983). Systems of ability grouping and the stratification of achievement in elementary schools. *Sociology of Education, 56,* 133-144.

ROWE, M.B. (1974). Wait-time and rewards as instructional variables, their influence on language, logic, and fate control. Part one: wait-time. *Journal of Research in Science Teaching, 11,* 81-94.

RUBOVITS, J.J. (1975). A classroom field experiment in mastery learning. *Dissertation Abstracts International, 36,* 2720-A. (University Microfilms No. 75-24395).

RUCKER, C.N., HOWE, C.E., and SNIDER, B. (1969). The participation of retarded children in juniorhigh academie and non-academic regular classes. *Exceptional Children, 26,* 617-623.

RUSSELL, D.H. (1946). Inter-class grouping for reading instruction in the intermediate grades. *Journal of Educational Research, 39,* 462-470.

RYAN, E. (1982). Identifying and remediating failures in reading comprehension: Toward an instructional approach for poor comprehenders. In G. MacKinnon and T. Walker (eds.), *Advances in Reading Research* (Vol. 3). New York: Academic Press.

RYAN, F., and WHEELER, R. (1977). The effects of cooperative and competitive background experiences of students on the play of a simulation game. *Journal of Educational Research, 70,* 295-299.

SAMUELS, S.J. (1979). The method of repeated readings. *Reading Teacher, 32,* 403-408.

SCARDAMALIA, M., and BEREITER, C. (1986). Research on written composition. In M.C. Wittrock (ed.), *Handbook of research on teaching* (pp. 778-803). New York: Macmillan.

SCARR, S., and WEBER, B. (1978). The reliability of reviews for the *American Psychologist. American Psychologist, 33,* 935.

SCHAEDEL, B., HERTZ-LAZAROWITZ, R., WALK, A., LERNER, M., JUBERAN, S., and SARID, M. (in press). The Israeli CIRC (ALASH): First year achievements in reading and composition. *Helkat-Lashon* (Journal of Lingtustic Education, in Hebrew).

SCHAFER, W., and OLEXA, C. (1971). *Tracking and opportunity: The locking-out process and beyond* Scranton, PA: Chandler.

SCHLAEFLI, A., REST, J.R., and THOMA, S.J. (1985). Does moral education improve moral judgment? A meta-analysis of intervention studies using the defining issues test. *Review of Educational Research, 55,* 319-352.

SCHMUCK, R.A., and SCHMUCK, P.A. (1983). *Group Processes in the Classroom.* Dubuque IA: C. Brown.

SCHNEIDERHAN, R.M. (1973). *A comparison of an individually guided education (IGE) program, an individually guided instruction (IGI) program, and a traditional elementary educational program at the intermediate level.* Unpublished doctoral dissertation, University of Minnesota, Duluth.

SCHOEN, H.L. (1976). Self-paced mathematics instruction: How effective has it been? *Arithmetic Teacher, 23,* 90-96.

SCHRANK, W.R. (1969). Academic stimulation of mathematics pupils from their classroom association with brighter pupils. *Mathematics Teacher, 62,* 474.

SCOTT, R, and McPARTLAND, J. (1982). Desegregation as nationalpolicy: Correlates of racial attitudes. *American Educational Research Journal, 19,* 397-414.

SCOTT, T.J. (1989). *The effects of cooperative learning team vs. traditional classroom/resource room instruction on handicapped student self-esteem and academic achievement.* Unpublished doctoral dissertation, Boston College.

SCRANTON, T., and RYCKMAN, D. (1979). Sociometric status of learning disabled children in an integrative program. *Journal of Learning Disabilities, 12*, 402-407.

SHARAN, S. (1980). Cooperative learning in small groups: Recent methods and effects on achievement, attitudes, and ethnic relations. *Review of Educational Research, 50*, 241-249.

SHARAN, S., and SHACHAR, C. (1988). *Language and learning in the cooperative classroom.* New York Springer-Verlag.

SHARAN, S., and SHARAN, Y. (1976). *Small-group teaching.* Englewood Cliffs, NJ: Educational Technology Publications.

SHARAN, S., HERTZ-LAZAROWITZ, R, and ACKERMAN, Z. (1980). Academic achievement of elementary school children in small group vs. whole class instruction. *Journal of Experimental Education, 48*, 125-129.

SHARAN, S., KUSSELL, P., HERTZ-LAZAROWITZ, R., BEJARANO, Y., RAVIV, S., and SHARAN, Y. (1984). *Cooperative learning in the classroom: Research in desegregated schools.* Hillsdale, NJ. Erlbaum.

SHARAN, Y., and SHARAN, S. (1992). *Group Investigation: Expanding cooperative learning.* New York: Teacher's College Press.

SHATTUCK, M. (1946). Segregation vs. non-segregation of exceptional children. *Journal of Exceptional Children, 12*, 235-240.

SHEPARD, L.A., and SMITH, M.L. (Eds.). (1989). *Flunking grades: Research and policies on retention.* New York: Falmer.

SHERMAN, L.W. (1988). A comparative study of cooperative and competitive achievement in two secondary biology classrooms: The group investigation model versus an individually competitive goal structure. *Journal of Research in Science Teaching, 26*(1), 35-64.

SHERMAN, L.W., and THOMAS, M. (1986). Mathematics achievement in cooperative versus individualistic goal-structured high school classrooms. *Journal of Educational Research, 79*, 169-172.

SHERMAN, L.W., and ZIMMERMAN, D. (1986, November). *Cooperative versus competitive reward-structured secondary science classroom achievement.* Paper presented at the annual meeting of the School Science and Mathematics Association, Lexington, KY.

SHORT, E., and RYAN, E. (1982). *Remediating poor readers' comprehension failures with a story grammar strategy.* Paper presented at the annual meeting of the American Educational Research Association, New York.

SHOWERS, B. (1987). The role of coaching in the implementation of innovations. *Teacher Education Quarterly, 14*, 59-70.

SIE, M.S. (1969). *Pupil achievement in an experimental nongraded elementary school.* Unpublished doctoral dissertation, Iowa State University of Science and Technology, Ames.

SIPERSTEIN, G.N., BOPP, M.J., and BAK, J.J. (1978). Social status of learning disabled children. *Journal of Learning Disabilities, 11*, 98-102.

SIZER, T. (1984). *Horace's compromise: The dilemma of the American high school.* Boston: Houghton Mifflin.

SKAPSKI, M.K. (1960). Ungraded primary reading program: An objective evaluation. *Elementary School Journal, 61*, 41-45.

SLAVIN R.E. (1985, March). *Quantitative review.* Paper presented at the annual meeting of the American Educational Research Association, Chicago.

SLAVIN, R E. (1986a) . *Using Student Team Learning* (3rd ed.). Baltimore: Johns Hopkins University, Center for Research on Elementary and Middle Schools.

SLAVIN, R E. (1986b). Best-evidence synthesis: An alternative to meta-analytic and traditional reviews. *Educationa l Researcher, 15*(9), 5-11.

SLAVIN, R.E. (1984). Team assisted individualized instruction: Cooperative learning and individualized instruction in the mainstreamed classroom. *Remedial and Special Education, 5*, 33-42.

SLAVIN, R.E. (1986). *Using student team learning.* Baltimore, MD: Johns Hopkins University, Center for Research on Elementary and Middle Schools.

SLAVIN, R.E. (1987). Cooperative learning and the cooperative school. *Educational Leadership, 45*, 7-13.

SLAVIN, R.E. (1991). Are cooperative learning and "untracking" harmful to the gifted? A response to Allan. *Educational Leadership, 48*(6), 68-71.

SLAVIN, R. (1994). *Cooperative learning: Theory, research, & practice* (2nd ed.). Boston: Allyn & Bacon.

SLAVIN, R.E. (1989a). Cooperative learning and student achievement. In R.E. Slavin (ed.), *School and classroom organization.* Hillsdale, NJ: Erlbaum.

SLAVIN, R.E., and KARWEIT, N. (1984). Mastery learning and student teams: A factorial experiment in urban general education mathematics classes. *American Educational Research Journal, 21*, 725-736.

SLAVIN, R.E., and KARWEIT, N.L. (1983). *Ability Grouped Active Teaching (AGA T): Teacher's manual.* Baltimore, MD: Johns Hopkins University, Center for Social Organization of Schools.

SLAVIN, R.E., and KARWEIT, N. (1985). Effects of whole-class, ability grouped, and individualized instruction on mathematics achievement. *American Educational Research Journal, 22*, 351-367.

SLAVIN, R.E., and STEVENS, R. (1991). Cooperative learning and mainstreaming. In J. Lloyd, N. Singh, & A. Repp (Eds.), *The regular education initiative: Alternative perspectives on concepts, issues, and models* (pp. 177-192). Sycamore, IL: Sycamore.

SLAVIN, R.E., LEAVEY, M., and MADDEN, N. (1984). Combining cooperative learning and individualized instruction: Effects on student mathematics achievement, attitudes, and behaviors. *Elementary School Journal, 84*, 409-422.

SLAVIN, R.E., STEVENS, R., and MADDEN, N. (1988). Accommodating student diversity in reading and writing instruction: A cooperative learning approach. *Remedial and Special Education, 9*, 60-66.

SLAVIN, R.E. (1995a). *Cooperative learning: Theory, research, and practice* (2nd Ed.) Boston: Allyn & Bacon.

SLAVIN, R.E. (1975). Classroom reward structure: Effects on academic performance, social connectedness, and peer norms. Doctoral dissertation, Johns Hopkins University.

SLAVIN, R.E. (1977d). *A new model of classroom motivation.* Paper presented at the annual convention of the American Educational Research Association, April, New York.

SLAVIN, R.E. (1977a). Classroom reward structure: An analytic and practical review. *Review of Educational Research, 47*, 633-650.

SLAVIN, R.E. (1977b). A student tearn approach to teaching adolescents with special emotional and behavioral needs. *Psychology in the Schools, 14*(1), 77-84.

SLAVIN, R.E. (1977c). *Student team learning techniques: Narrowing the achievement gap between the races* (Report No.228). Baltimore: Johns Hopkins University, Center for Social Organization of Schools.

SLAVIN, R.E. (1978a). Student teams and achievement divisions. *Journal of Research and Development in Education, 12*, 39-49.

SLAVIN, R.E. (1978b). Student teams and comparison among equals: Effects on academic performance and student attitudes. *Journal of Educational Psychology, 70*, 532-538.

SLAVIN, R.E. (1979). Effects of biracial learning teams on cross-racial friendships. *Journal of Educational Psychology, 71*, 381-387.

SLAVIN, R.E. (1980a). Effects of individual learning expectations on student achievement. *Journal of Educational Psychology, 72*, 520-524.

SLAVIN, R.E. (1980b). Effects of student teams and peer tutoring on academic achievement and time on-task. *Journal of Experimental Education, 48*, 252-257.

SLAVIN, R.E. (1980c). Cooperative learning. *Review of Educational Research, 50*, 315-342.

SLAVIN, R.E. (1983a). *Cooperative learning.* New York: Longman.

SLAVIN, R.E. (1983b). When does cooperative learning increase student achievement? *Psychological Bulletin, 94*, 429-445.

SLAVIN, R.E. (1984a). Component building: A strategy for research-based instructional improvement. *Elementary School Journal, 84*, 255-269.

SLAVIN, R.E. (1984b). Meta-analysis in education: How has it been used? *Educational Researcher, 13*(8), 6-15, 24-27.

SLAVIN, R.E. (1984c). Team assisted individualization: Cooperative learning and individualized instruction in the mainstreamed classroom. *Remedial and Special Education, 5*(6), 33-42.

SLAVIN, R.E. (1984d). Component building: A strategy for research-based instructional improvement. *Elementary School Journal, 84*, 255-269.

SLAVIN, R.E. (1984e). *Research methods in education: A practical guide.* Englewood Cliffs, NJ: Prentice-Hall.

SLAVIN, R.E. (1985a). Team-Assisted Individualization: Combining cooperative learning and individualized instruction in mathematics. In R.E. Slavin, S. Sharan, S. Kagan, R. Hertz-Lazarowitz, C. Webb, and R. Schmuck (eds.), *Learning to cooperate, cooperating to learn* (pp. 177-209). New York: Plenum.

SLAVIN, R.E. (1985b). Cooperative learning: Applying contact theory in desegregated schools. *Journal of Social Issues, 41*(3), 45-62.

SLAVIN, R.E. (1986a) Best-evidence synthesis: An alternative to meta-analyitic and traditional reviews. *Educational Researcher, 15*(9), 5-11.

SLAVIN, R.E. (1986b). *Ability grouping and student achievement in elementary schools: A best-evidence synthesis* (Tech. Rep. No. 1). Baltimore, MD: Center for Research on Elementary and Middle Schools, Johns Hopkins University.

SLAVIN, R.E. (1986c). The Napa evaluation of Madeline Hunter's ITIP: Lessons learned. *Elementary School Journal, 87*, 165-171.

SLAVIN, R.E. (1987). A theory of school and classroom organization. *Educational Psychologist, 22*, 89-108.

SLAVIN, R.E. (1987). Ability grouping and student achievement in elementary schools: A best-evidence synthesis. *Review of Educational Research, 57*, 347-350.

SLAVIN, R.E. (1987a). Ability grouping and student achievement in elementary schools: A best-evidence synthesis. *Review of Educational Research, 57*, 293-336.

SLAVIN, R.E. (1987b). Mastery learning reconsidered. *Review of Educational Research, 57*, 175-213.

SLAVIN, R.E. (1987c). A theory of school and classroom organization. *Educational Psychologist, 22*, 89-108.

SLAVIN, R.E. (1987d). Mastery learning reconsidered. *Review of Educational Research, 57*, 175-213.

SLAVIN, R.E. (1987e). Cooperative learning: Where behavioral and humanistic approaches to classroom motivation meet. *Elementary School Journal, 88*, 29-37.

SLAVIN, R.E. (1989a). Achievement effects of substantial reductions in class size. In R.E. Slavin (Ed.), *School and Classroom Organization* (pp. 247-257). Hillsdale, NJ: Erlbaum.

SLAVIN, R.E. (1989b). PET and the pendulum: Faddism in education and how to stop it. *Phi Delta Kappan, 70*, 752-758.

SLAVIN, R.E. (1989c). Cooperative learning and achievement: Six theoretical perspectives. In C. Ames and M.L. Maehr (eds.), *Advances in motivation and achievement*. Greenwich, CT: JAI Press.

SLAVIN, R.E. (1990). Ability grouping and student achievement in secondary schools: A best-evidence synthesis. *Review of Educational Research, 60*, 471-499.

SLAVIN, R.E. (1990). *Cooperative learning: Theory, research, and practice*. Englewood Cliffs, NJ: Prentice-Hall.

SLAVIN, R.E. (1990a). Achievement effects of ability grouping in secondary schools: A best evidence synthesis. *Review of Educational Research, 60*(3), 471-499.

SLAVIN, R.E. (1990d). On making a difference. *Educational Researcher, 19*(3), 30-34.

SLAVIN, R.E. (1991). Are cooperative learning and untracking harmful to the gifted? *Educational Leadership, 48*(6), 68-71.

SLAVIN, R.E. (1992). When and why does cooperative learning increase achievement? Theoretical and empirical perspectives. In R. Hertz-Lazarowitz and N. Miller (eds.), *Interaction in cooperative: The theoretical anatomy of group learning* (pp. 145-173). New York: Cambridge University Press.

SLAVIN, R.E. (1993, April). *Cooperative learning and achievement: an empirically-based theory*. Paper presented at the annual meeting of the American Educational Research Association, Atlanta.

SLAVIN, R.E. (1994a). *Educational psychology: Theory and practice* (4th ed.). Boston: Allyn & Bacon.

SLAVIN, R.E. (1994b). School and classroom organization in beginning reading: Class size, aides, and instructional grouping. In R.E. Slavin, N.L. Karweit, B.A. Wasik, & N.A. Madden (Eds.), *Preventing early school failure: Research on effective strategies*. Boston: Allyn & Bacon.

SLAVIN, R.E. (1995b). Detracking and its detractors: Flawed evidence, flawed values. *Phi Delta Kappan, 77*(3), 220-221.

SLAVIN, R.E. (1995a). *Cooperative Learning: Theory, research, and practice* (2nd Ed.). Boston: Allyn & Bacon.

SLAVIN, R.E. (1995c). A model of effective instruction. *The Educational Forum, 59*(2), 166-176.

SLAVIN, R.E. (1995d). Enhancing intergroup relations in schools: Cooperative learning and other strategies. In W.D. Hawley & A.W. Jackson (Eds.), *Toward a common destiny: Improving race and ethnic relations in America*. San Francisco: Jossey-Bass.

SLAVIN, R.E. (in press). Research on cooperative learning and achievement: What we know, what we need to know. *Contemporary Educational Psychology*.

SLAVIN, R.E. (1987f). Combining cooperative learning and individualized instruction. *Arithmetic Teacher, 35*, 14-16.

SLAVIN, R.E. (in press). Cooperative learning and intergroup relations. In J. Banks (ed.), *Handbook of Research on Multicultural Education*. New York: Macmillan.

SLAVIN, R.E., and KARWEIT, N.L. (1984). Mastery learning and student teams: A factorial experiment in urban general madhematics classes. *American Educational Research Journal, 21*, 725-736.

SLAVIN, R.E., and KARWEIT, N.L. (1984, April). *Within-class ability grouping and student achievement: Two field experiments*. Paper presented at the annual meeting of the American Educational Research Association, New Orleans.

SLAVIN, R.E., and KARWEIT, N.L. (1985). Effects of whole-class, ability grouped, and individualized instruction on mathematics achievement. *American Educational Research Journal, 22*, 351-367.

SLAVIN, R.E., and KARWEIT, N.L. (1981). Cognitive and affective outcomes of an intensive student team learning experience. *Journal of Experimental Education, 50*, 29-35.

SLAVIN, R.E., and MADDEN, N.A. (1979). School practices that improve race relations. *American Educational Research Journal, 16*(2), 169-180.

SLAVIN, R.E., and MADDEN, N.A. (1987, April). *Effective classroom programs for students at risk.* Paper presented at the annual meeting of the American Educational Research Association, Washington, DC.

SLAVIN, R.E., and MADDEN, N.A. (1991). Modifying Chapter 1 program improvement guidelines to reward appropriate practices. *Educational Evaluation and Policy Analysis, 13*, 369-379.

SLAVIN, R.E., and MADDEN, N.A. (1993, April). *Multi-site replicated experiments: An application to Success for All.* Paper presented at the annual meeting of the American Educational Research Association, Atlanta.

SLAVIN, R.E., and MADDEN, N.A. (1994). *Implementing Success for All in the Philadelphia Public Schools* (Final report to the Pew Charitable Trusts). Baltimore, MD: Johns Hopkins University, Center for Research on Effective Schooling for Disadvantaged Students.

SLAVIN, R.E., and MADDEN, N.A. (1995, April). *Effects of Success for All on the Achievement of English Language Learners.* Paper presented at the annual meeting of the American Educational Research Association, San Francisco.

SLAVIN, R.E., and OICKLE, E. (1981). Effects of cooperative learning teams on student achievement and race relations: Treatment by race interactions. *Sociology of Education, 54*, 174-180.

SLAVIN, R.E., and STEVENS, R.J. (1991). Cooperative learning and mainstreaming. In J.W. Lloyd, N.N. Singh, and A.C. Repp (eds.), *The regular education initiative: Alternative perspectives on consepts, issues, and models,* (pp.177-191). Sycamore, IL: Sycamore.

SLAVIN, R.E., and YAMPOLSKY, R. (1991). *Effects of Success for All on students with limited English proficiency: A three-year evaluation.* Baltimore, MD: Johns Hopkins University, Center for Research on Effective Schooling for Disadvantaged Students.

SLAVIN, R.E., BRADDOCK, J.H., HALL, C., and PETZA, R.J. (1989). *Alternatives to ability grouping* Baltimore: Johns Hopkins University, Center for Research on Effective Schooling for Disadvantaged Students.

SLAVIN, R.E., DEVRIES, D.L., and HULTEN, B.H. (1975). *Individual vs. team competition:The interpersonal consequences of academic performance* (Report No. 188). Baltimore: Johns Hopkins University, Center for Social Organization of Schools.

SLAVIN, R.E., KARWEIT, N.L., and WASIK, B.A. (1994). *Preventing early school failure: Research on effective strategies.* Boston: Allyn & Bacon.

SLAVIN, R.E., KARWEIT, N.L., and WASIK, B.A., (1992/93). Preventing early school failure: What works? *Educational Leadership, 50*(4), 10-18.

SLAVIN, R.E., LEAVEY, M., and MADDEN, N.A. (1984). Combining cooperative learning and individualized instruction: Effects on student mathematics achievement, attitudes, and behaviors. *Elementary School Journal, 84*, 409422.

SLAVIN, R.E., LEAVEY, M.B., and MADDEN, N.A. (1986). *Team Accelerated Instruction: Mathematics.* Watertown, MA: Charlesbridge.

SLAVIN, R.E., MADDEN, N.A., and LEAVEY, M. (1984). Effects of Team Assisted Individualization on the mathematics achievement of academically handicapped students and nonhandicapped students. *Journal of Educational Psychology, 76*, 813-819.

SLAVIN, R.E., MADDEN, N.A., DOLAN, L.J., and WASIK, B.A. (1992). *Success for All: A Relentless Approach to Prevention and Early Intervention in Elementary Schools.* Arlington, Va.: Educational Research Service.

SLAVIN, R.E., MADDEN, N.A., KARWEIT, N.L., DOLAN, L., & WASIK, B.A., ROSS, S.M., and SMITH, L.J. (1994). "Whenever and wherever we choose...:" The replication of Success for All. *Phi Delta Kappan, 75*(8), 639-647.

SLAVIN, R.E., MADDEN, N.A., KARWEIT, N.L., DOLAN, L., & WASIK, B.A., ROSS, S.M., and SMITH, L.J. (1994, April). *Success for All: Longitudinal effects of systemic school-by-school reform in seven districts.* Paper presented at the annual meeting of the American Educational Research Association, New Orleans.

SLAVIN, R.E., MADDEN, N.A., KARWEIT, N.L., DOLAN, L., and WASIK, B.A., SHAW, A., MAINZER, K.L., and HAXBY, B. (1991). Neverstreaming: Prevention and early intervention as alternatives to special education. *Journal of Learning Disabilities, 24,* 373-378.

SLAVIN, R.E., MADDEN, N.A., KARWEIT, N.L., LIVERMON, B.J., and DOLAN, L. (1990). Success for All: First-year outcomes of a comprehensive plan for reforming urban education. *American Educational Research Journal, 27,* 255-278.

SLAVIN, R.E., STEVENS, R.J., and MADDEN, N.A. (1988). Accommodating student diversity in reading and writing instruction: A cooperative learning approach. *Remedial and Special Education, 9,* 60-66.

SLAVIN. R.E., and KARWEIT, N.L. (1982, August). *School organizational vs. developmental effects on attendance among young adolescents.* Paper presented at the annual meeting of the American Psychological Association, Washington, DC.

SMITH, J.K. (1981, April). *Philosophical considerations of mastery learning theory.* Paper presented at the annual meeting of the American Educational Research Association, Los Angeles.

SMITH, K.A., JOHNSON, D.W., and JOHNSON, R.T. (1981). Can conflict be constructive? Controversy versus concurrence seeking in learning groups. *Journal of Educationel Pychology, 73,* 651-663.

SMITH, L.J., and ROSS, S.M., and CASEY, J.P. (1994). *Special education analyses for Success for All in four cities.* Memphis: University of Memphis, Center for Research in Educational Policy.

SMITH, L.R., and COTTON, M.L. (1980). Effect of lesson vagueness and discontinuity on student achievement and attitudes. *Journal of Educational Psychology, 72,* 670-675.

SMITH, W.M. (1960). The effect of intra-class ability grouping on arithmetic achievement in grades two through five. *Dissertation Abstracts International, 21,* 563-564.

SNAKE RIVER SCHOOL DISTRICT. (1972). *Curriculum change through nongraded individualizanon.* Blackfoot, ID: School District 52.

SOLOMON, D., WATSON, M., SCHAPS, E., BATTISTICH, V., and SOLOMON, J. (1990). Cooperative learning as part of a comprehensive classroom program designed to promote prosocial development. In S. Sharan (ed.), *Recent research on cooperative learning.* New York: Praeger.

SOLOMON, D., WATSON, M.S., DELUCCHI, K.I., SCHAPS, E., and BATTISTICH, V. (1988). Enhancing children's prosocial behavior in the classroom. *American Educational Research Journal, 25,* 527-554.

SORENSEN, A.B. (1970). Organizational differentiation of students and educational opportunity. *Sociology of Education, 43,* 355-376.

SORENSEN, A.B., and HALLINAN, M.T. (1986). Effects of ability grouping on growth in academic achievement. *American Educational Research Journal, 23,* 519-542.

SOUMOKIL, P.O. (1977). *Comparison of cognitive and affective dimensions of individually guided education (IGE) and standard elementary school programs.* Unpublished doctoral dissertation, University of Missouri, Columbia.

SPARKS, G. (1986). The effectiveness of alternative training activities in changing teaching practices. *American Educational Research Journal, 23,* 217-225.

SPARKS, G., and BURDER, S. (1987). Before and after peer coaching. *Educational Leadership, 45*(3), 54-57.

SPENCE, E.S. (1958). Intra-class grouping of pupils for instruction in arithmetic in the intermediate grades of the elementary school. *Dissertation Abstracts International, 19,* 1682. (University Microfilms No. 58-5635).

SPIVAK, M.L. (1956). Effectiveness of departmental and self-contained seventh- and eighth grade classrooms. *School Review, 64,* 391-396.

STAINBACK, W., and STAINBACK, S. (1984). A rationale for the merger of special and regular education. *Exceptional Children, 51,* 102-111.

STAINBACK, W., and STAINBACK, S. (1991). Rationale for integration and restructuring: A synopsis. In J. Lloyd, N. Singh, & A. Repp (Eds.), *The regular education initiative: Alternative perspectives on concepts, issues, and models* (pp.225-240). Sycamore, IL: Sycamore.

STALLINGS, J., and KRASAVAGE, E.M. (1986). Program implementation and student achievement in a four-year Madeline Hunter Follow Through Project. *Elementary School Journal, 87,* 117-138.

STEIN, N.L., and GLENN, C.G. (1979). An analysis of story comprehension in elementary school children. In R. Freedle (ed.), *New directions in discourse processing* (Vol. 2, 53-120). Norwood, NJ: Ablex.

STENDLER, C., DAMRIN, D., and HAINES, A.C. (1951). Studies in cooperation and competition: I. The effects of working for group and individual rewards on the social climate of children's groups. *Journal of Genetic Psychology, 79,* 173-197.

STERN, A.M. (1972). Intraclass grouping of low achievers in mathematics in the third and fourth grades. *Dissertation Abstracts International, 32,* 5539A. (University Microfilms No. 72- 11900).

STEVENS, R. (1994). Using cooperative learning in literacy instruction: Theory, research, and application. In N. Ellsworth, C. Hedley, & A. Baratta (Eds.), *Literacy: A redefinition.* Hillsdale, NJ: Erlbaum.

STEVENS, R., and SLAVIN, R. (1991). When cooperative learning improves the achievement of students with mild disabilities: A response to Tateyama-Sniezak. *Exceptional Children, 57,* 276-280.

STEVENS, R., and SLAVIN, R. (1995). Effects of a cooperative learning approach in reading and writing on academically handicapped and nonhandicapped students' achievement and attitudes. *Elementary School Journal, 95*(3).

STEVENS, R., MADDEN, N., SLAVIN, R., and FARNISH, A. (1987). Cooperative Integrated Reading and Composition: Two field experiments. *Reading Research Quarterly 22,* 433-454. '

STEVENS, R., SLAVIN, R., and FARNISH, A. (1991). The effects of cooperative learning and direct instruction in reading comprehension strategies on main idea identification. *Journal of Educational Psychology, 83,* 8-16.

STEVENS, R.J., and DURKIN, S. (1992). *Using student team reading and student team writing in middle schools; Two evaluations.* Baltimore, MD: Johns Hopkins University, Center for Research on Effective Schooling for Disadvantaged Students. Report No. 36.

STEVENS, R.J., and MADDEN, N.A., SLAVIN, R.E., and FARNISH, A.M. (1987). Cooperative Integrated Reading and Composition: Two field experiments. *Reading Research Quarterly, 22,* 433-454.

STEVENS, R.J., and SLAVIN, R.E. (1993). *The cooperative elementary school: Effects on students' achievement, attitudes, and social relations.* Submitted for publication.

STEVENS, R.J., and SLAVIN, R.E. (1995). The effects of Cooperative Integrated Reading and Composition (CIRC) on academically handicapped and nonhandicapped students' achievement, attitudes, and metacognition in reading and writing. *Elementary School Journal, 95,* 241-262.

STEVENS, R.J., MADDEN, N.A., SLAVIN, R.E., and FARNISH, A.M. (1987). Cooperative integrated reading and composition: Two field experiments. *Reading Research Quarterly, 22,* 433-454.

STEVENS, R.J., SLAVIN, R.E., and FARNISH, A.M. (1991). The effects of cooperative learning and direct instruction in reading comprehension strategies on main idea identification. *Journal of Educational Psychology, 83*(1), 8-16.

STEVENS, R.J., SLAVIN, R.E., FARNISH, A.M., and MADDEN, N.A. (April, 1988). *Effects of cooperative learning and direct instruction in reading comprehension strategies on main idea identification.* Paper presented at the Annual Conference of the American Educational Research Association, New Orleans.

STIPEK, D.J. (1993). *Motivation to learn: From theory to practice* (2nd ed.). Boston: Allyn & Bacon.

STOAKES, D.W. (1964). *An educational experiment with the homogeneous grouping of mentally advanced and slow learning students in the junior high school.* Unpublished doctoral dissertation, University of Colorado.

STOKES, D.B. (1990). *Cooperative vs. traditional approaches to teaching mathematics in the third grade.* Unpublished doctoral dissertation, University of Southern Mississippi.

STRASLER, G.M. (1979, April). *The process of transfer in a learning for mastery setting* Paper presented at the annual meeting of the American Educational Research Association, San Francisco.

STRASLER, G.M., and ROCHESTER, M. (1982, April). *A two-year evaluation of a competency-based secondary school project in a learning for mastery setting.* Paper presented at the annual meeting of the American Educational Research Association, New York.

SUPPES, P. (1964). Modern learning theory and the elementary school curriculum *American Educational Research Journal, 2,* 79-93.

SVENSSON, N.E. (1962). *Ability grouping and scholastic achievement.* Stockholm: Almqvist & Wirsell.

SWANSON, D.H., and DENTON, J.J. (1977). Learning for mastery versus personalized system of instruction: A comparison of remediation strategies with secondary school chemistry students. *Journal of Research in Science Teaching, 14,* 515-524.

TALMAGE, H., PASCARELLA, E., and FORD, S. (1984). The influence of cooperative learning strategies on teacher practices, student perceptions of the learning environment, and academic achievement. *American Educational Research Journal, 21,* 163-179.

TAYLOR, B.R. (1973). *The effect of a mastery learning of minimum essentials requirement on achievement in ninth grade algebra.* Unpublished doctoral dissertation, University of Minnesota, Minneapolis.

TENENBAUM, G. (1982). *A method of group instruction which is as effective as one-to-one tutorial instruction.* Unpublished doctoral dissertation, University of Chicago

TERWEL, J. (1988, April). *Implementation and effects of a program for mixed ability teaching in secondary mathematics education.* Paper presented at the annual meeting of the American Educational Research Association, New Orleans.

TERWEL, J., HERFS, P.G.P., MERTENS, E.H.M., and PERRENT, J.C. (in press). Co-operative learning and adaptive instruction in a mathematics curriculum. *Journal of Curriculum Studies.*

The effects of segregation and consequences of desegregation: A social science statement. Appendix to appellant's briefs: Brown v. Board of Education of Topeka, Kansas (1953). *Minnesota Law Review, 37,* 427-439.

THOMAS, E.J. (1957). Effects of facilitative role interdependence on group functioning. *Human Relations, 10,* 347-366.

THOMPSON, G.W. (1974). The effects of ability grouping upon achievement in eleventh grade Amencan history. *Journal of Experimental Education, 42,* 76-79.

THURLOW, M., GRODEN, J., YSSELDYKE, J., and ALGOZZINE, R. (1984). Student reading during class: The lost activity in reading instruction. *Journal of Educational Research, 77,* 267-272.

TOBIN, J.F. (1966). *An eight year study of classes grouped within grade levels on the basis of reading ability.* Unpublished doctoral dissertation, Boston University. (University Microfilms No. 66-345)

TOMBLIN, E.A., and DAVIS, B.R. (1985). *Technical report of the evaluation of the race/human relations program: A study of cooperative learning environment strategies.* San Diego: San Diego Public Schools.

TORGELSON, J.W. (1963). *A comparison of homogeneous and heterogeneous grouping for belowaverage junior high school students.* Unpublished doctoral dissertation, University of Minnesota.

TRIMBLE, K.D., and SINCLAIR, R.L. (1987). On the wrong track: Ability grouping and the threat to equity. *Equity and Excellence, 23,* 15-21.

TURNEY, A.H. (1931). The status of ability grouping. *Educational Administration and Supervision, 17,* 21-42, 110-127.

U.S. DEPARTMENT OF EDUCATION v. Dillon County School District No. 1. (1986). Initial decision in compliance proceeding under Title VI of the Civil Rights Act of 1964, 42 U.S.C. Sec. 200d et seq.

VAKOS, H.N. (1969). The effect of part-time grouping on achievement in social studies. *Dissertation Abstracts International, 30,* 2271. (University Microfilms No. 69-20,066).

VAN OUDENHOVEN, J.P., VAN BERKUM, G., and SWEN-KOOPMANS. T. (1987) . Effect of cooperation and shared feedback on spelling achievement. *Journal of Educational Psychology, 79,* 92-94.

VAN OUDENHOVEN, J.P., WIERSMA, B., and VAN YPEREN, N. (1987). Effects of cooperation and feedback by fellow pupils on spelling achievement. *European Journal of Psychology of Education, 2,* 83-91.

VANFOSSEN, B.E., JONES, J.D., and SPADE, J.Z. (1987). Curriculum tracking and status maintenance. *Sociology of Education, 60,* 104-122.

VEDDER, P.H. (1985). *Cooperative learning: A study on processes and effects of cooperation between primary school children.* The Hague: Stichting Voor Onderzoek Van Het Onderwijs.

VERDUCCI, F. (1969). Effects of class size on the learning of a motor skill. *Research Quarterly, 40,* 391-395.

VOGEL, F.S., and BOWERS, N.D. (1972). Pupil behavior in a multiage nongraded school. *The Journal of Experimental Education, 41,* 78-86.

VROOM, V.H. (1969). Industrial social psychology. In G. Lindzey and E. Aronson (eds.), *The hand-book of social psychology* (Vo1. 5, 2nd ed.). Reading, MA: Addison-Wesley.

VYGOTSKY, L. (1978). *Mind in society: The development of higher psychological processes.* Cambridge, MA: Harvard University Press.

WADSWORTH, B.J. (1984). *Piaget's theory of cognitive and affective development* (3rd ed.). New York: Longman.

WALBERG, H.J. (1984). Improving the productivity of America's schools. *Educational Leadership, 41* (8), 19-27.

WALKER, W.E. (1973). *Long-term effects of graded and nongraded primary programs.* Unpublished doctoral dissertation, George Peabody College for Teachers of Vanderbilt University.

WALLEN, N.E., and VOWLES, R.O. (1960). The effect of intraclass ability grouping on arithmetic achievement in the sixth grade. *Journal of Educational Psychology, 51,* 159-163.

WARD, D.N. (1969). *An evaluation of a nongraded school program in grades one and two.* Unpublished doctoral dissertation, University of Texas, Austin.

WARD, P.E. (1970). A study of pupil achievement in departmentalized grades four, five, and six. *Dissertation Abstracts, 30,* 4749A. (University Microfilms No. 70-1201).

WASIK, B.A., and SLAVIN, R.E. (1993). Preventing early reading failure with one-to-one tutoring: A review of five programs. *Reading Research Quarterly, 28,* 178-200.

WAXMAN, H., and WALBERG, H. (1982). The relation of teaching and learning: A review of reviews of process-product research. *Contemporary Education Review, 1,* 103-120.

WAXMAN, H.C., WANG, M.C., ANDERSON, K.A., and WALBERG, H.J. (1985). Adaptive education and student outcomes: A quantitative synthesis. *JournalofEducationalResearch, 78,* 228-236.

WEBB, N. (1985). Student interaction and learning in small groups: A research summary. In R.E. Slavin, S. Sharan, S. Kagan, R. Hertz-Lazarowitz, C. Webb, and R. Schmuck (eds.), *Learning to cooperate, cooperating to learn* (pp. 147-172). New York: Plenum.

WEBB, N.M. (1989). Peer interaction and learning in small groups. *International Journal of Educational Research, 13,* 21-39.

WEBB, N.M. (1992). Testing a theoretical model of student interaction and learning in small groups. In R. Hertz-Lazarowitz and N. Miller (eds.), *Interaction in cooperative groups: The theoretical anatomy of group learning* (pp. 102-119). New York: Cambridge University Press.

WEIGEL, R. H., WISER, P. L., and COOK, S. W. (1975). Impact of cooperative learning experiences on cross-ethnic relations and *attitudes. Journal of Social Issues, 31*(1), 219-245.

WEINER, B. (1979). A theory of motivation for some classroom experiences. *Journal of Educational Psychology, 71,* 3-25.

WEINER, B., and KUKLA, A. (1970). An attributional analysis of achievement motivation. *Journal of Personality and Social Psychology, 15,* 1-20.

WEINSTEIN, C.E. (1982). Training students to use elaboration learning strategies. *Contemporary Educational Psychology, 7,* 301-311.

WENTLING, T.L. (1973). Mastery versus nonmastery instruction with varying test item feedback treatments. *Journal of Educational Psychology, 65,* 50-58.

WHEELER, R, and RYAN, F.L. (1973). Effects of cooperative and competitive classroom environments on the attitudes and achievement of elementary school students engaged in social studies inquiry activities. *Journal of Educational Pychology, 65,* 402-407.

WHEELOCK, A. (1992). *Crossing the tracks: How "untracking" can save America's schools.* New York: New Press.

WIATROWSKI, M., HANSELL, S., MASSEY, C.R., and WILSON, D.L. (1982). Curriculum tracking and delinquency. *American Sociological Review, 47,* 151-160.

WILBS, S. (1991). Breaking down grade barriers: Interest in nongraded classrooms on the rise. *ASCD Update, 33* (3), 1, 4.

WILCOX, J. (1963). A search for the multiple effects of grouping upon the growth and behavior of junior high school pupils. *Dissertation Abstracts International, 24,* 205.

WILKINSON, L.C. (1986, June). *Grouping students for instruction.* Paper presented at the Conference on the Effects of Alternative Designs in Compensatory Education, Washington, DC.

WILL, M. (1986). Educating children with learning problems: A shared responsibility. *Exceptional Children, 52,* 411-416.

WILLCUTT, R.E. (1969). Ability grouping by content topics in junior high school mathematics. *Journal of Educational Research, 63,* 152-156.

WILLIAMS, W. (1966). Academic achievement in a graded school and in a non-graded school. *Elementary School Journal, 66*, 135-139.

WILLIAMSON, L., and RUSSELL, D. (1990). Peer coaching as follow-up training. *Journal of Staff Development, 11*(2), 2-4.

WILLIG, A.C. (1985). A meta-analysis of selected studies on the effectiveness of bilingual education. *Review of Educational Research, 55*, 269-317.

WILLIG, A.C. (1985). A meta-analysis of selected studies on the effectiveness of bilingual education. *Review of Educational Research, 55*, 269-317.

WILSON, B., and SCHMITS, D. (1978). What's new in ability grouping? *Phi Delta Kappan, 60*, 535-536.

WILSON, G.T., and RACHMAN, S.J. (1983). Meta-analysis and the evaluation of psychotherapy outcome: Limitations and liabilities. *Journal of Consulting and Clinical Psychology, 51*, 54-64.

WILT, H.J. (1970). *A comparison of student attitudes toward school, academic achievement, internal structures and procedures: The nongraded school vs. the graded school.* Unpublished doctoral dissertation, University of Missouri, Columbia.

WITTROCK, M. (1986). Students' thought processes. In M. Wittrock (Ed.), *Handbook of research on teaching* (pp. 297-314). New York: Macmillan.

WITTROCK, M.C. (1978). The cognitive movement in instruction. *Educational Psychologist, 13*, 15-29.

WOLFE, J.A., FANTUZZO, J.W., and WOLFE, P.K (1986). The effects of reciprocal peer management and group contingencies on the arithmetic proficiency of underachieving students. *Behavior Therapy, 17*, 253-265.

WONG-FILLMORE, L., and VALADEZ, C. (1986). Teaching bilingual learners. In M.C. Wittrock (Ed.), *Handbook of Research on Teaching* (3rd Ed.). New York: Macmillan.

WORTHAM, S.C. (1980). *Mastery learning in secondary schools: A first year report.* Unpublished manuscript, San Antonio (TX) Public Schools. (ERIC Document Reproduction Service No. ED 187 387).

WYCKOFF, D.B. (1974). *A study of mastery learning and its effects on achievement of sixth grade social studies students.* Unpublished doctoral dissertation, Georgia State University, Atlanta.

YAGER, S., JOHNSON, D.W., and JOHNSON, R.T. (1985). Oral discussion, group-to-individual transfer, and achievement in cooperative learning groups. *Journal of Educational Psychology, 77*, 60-66.

YAGER, S., JOHNSON, R.T., JOHNSON, D.W., and SNIDER, B. (1986). The impact of group processing on achievement in cooperative learning. *Journal of Social Psychology, 126*, 389-397.

ZERBY, J.R. (1960). *A comparison of academic achievement and social adjustment of primary school-children in the graded and nongraded school programs.* Unpublished doctoral dissertation, Pennsylvania State University, University Park.

ZIEGLER, S. (1981). The effectiveness of cooperative learning teams for increasing cross-ethnic friendship: Additional evidence. *Human Organization, 40*, 264-268.

ZWEIBELSON, I., BAHNMULLER, M., and LYMAN, L. (1965). Team teaching and flexible grouping in the junior high-school social studies. *Journal of Experimental Education, 34*, 20-32.

Index